BREAKING AWAY

BREAKING AWAY

The Engineer's Guide
to a Successful Consulting
Practice

R. Matthiew Seiden, P.E.

Prentice-Hall, Inc., Englewood Cliffs, New Jersey 07632

Prentice-Hall International, Inc., *London*
Prentice-Hall of Australia, Pty. Ltd., *Sydney*
Prentice-Hall Canada, Inc., *Toronto*
Prentice-Hall of India Private Ltd., *New Delhi*
Prentice-Hall of Japan, Inc., *Tokyo*
Prentice-Hall of Southeast Asia Pte. Ltd., *Singapore*
Editora Prentice-Hall do Brasil Ltda., *Rio de Janeiro*
Prentice-Hall Hispanoamericana, S.A., *Mexico*

© 1987 *by*

PRENTICE-HALL, INC.

Englewood Cliffs, N.J.

All rights reserved. No part of this book may be reproduced in any form or by any means, without permission in writing from the publisher.

Library of Congress Catalog Card Number: 87-061012

ISBN 0-13-081548-9

Printed in the United States of America

DEDICATION AND ACKNOWLEDGMENT

To all those engineers who have what it takes to be successful as consultants, part- or full-time, but who have not had adequate or proper encouragement to practice as independent professionals; to all those clients, wherever they may be, who have needs that force the engineering profession to strive toward the highest levels of independent professionalism, and whose faith and confidence in the potential and promise of engineering technology bring out the best in us; to the public-at-large, whose mandate to the engineering profession to promote and hold paramount the public safety, health, and welfare guides and compels us to develop ever higher levels of expertise and skill to assure a better world for ourselves, our children, and theirs; and last but not least, to my wife, Gella, whose encouragement, indulgence, and sacrifice made this book possible.

INTRODUCTION

This practical, hands-on book will show you, step by step, how to be successful as a consulting engineer or scientist. It starts at the beginning and is written in a no-nonsense, conversational style that will conserve your time while accomplishing its objective. Whether your objective is $10,000 of supplementary income annually, or $100,000 or more in a full-time engineering consulting practice, you have come to the right place.

Perhaps you already have had a few part-time engagements and wish to continue on this basis but want to polish up your act, or maybe you want to do part-time consulting but haven't yet gotten your feet wet. Or perhaps you are seriously considering career options, are dissatisfied with remaining an employee for the rest of your life, or are in full-time consulting and want to "compare notes." Whatever the case may be, this book has been written just for you and, importantly, if you are between jobs or dead-ended in your present job, here is a "real world" option that can even make you wealthy, depending upon your commitment to what you have to do to succeed. In reality, you are already on your way, because you know you have the pure technical skills necessary to resolve problems in your specialty and also learn new subject matter whenever necessary.

Age is no barrier. If you are retired, you have unique opportunities. And you can even consult from that land of sunshine and bliss a few times a year, or whenever you choose. Or, if you are barely thirty but have some unusual engineering or scientific specialty, interest or knack, you can put it to work for you...profitably.

If you are presently a valued "internal" consultant, why not become "your own man" and do *real* consulting for your own account? A consulting career is what *you* make it. Challenge, pay, control over your future, greater utilization of your expertise, talents and capabilities, wider exposure and involvement in your professional world . . . what do you want? It can be yours for the taking.

All you need is motivation, willingness to work hard and long hours (you're probably already doing that right now for someone else!), high technical expertise, good educational and/or experiential credentials, and knowledge of how to go about it.

This book will give you that knowledge. Here are just a few examples of specific ways in which it will become invaluable to you:

Chapters 1 through 3 will convince you that technical consulting is a "real world," potentially lucrative opportunity for engineers and scientists. They will also show you how to avoid pitfalls, adopt winning attitudes and habits, and acquire profit-making skills, services, and resources that can make all the difference between success and failure.

Chapters 4 and 5 will explain how consultants get clients, sell to them and keep them, and have them come back time and time again. These chapters will show you how a consulting engagement works, how to set fees, avoid collection problems, and talk dollars with clients without feeling embarrassed or intimidated.

Chapters 6 through 10 show you how to plan, control, and operate your consulting practice. They emphasize the financial principles and imperatives of good management for survival and success. Knowledge of the competition must be a vital part of your business strategy. Therefore, this subject is also explored in a down-to-earth way that will help you catch up and pull ahead of your competition. You will also learn how to use reports to build business.

Chapters 11 and 12 introduce you to the remarkable world of forensic consulting. The special nature of this field, its unique demands and opportunities, and the best way for you to start are all described in these practical chapters that can mean significant income to you.

Chapters 13 and 14 introduce you to the vitally important concepts and principles of safety engineering so you will be able to avoid potential errors in design that can lead to lawsuits against both you and your clients. Chapter 15 offers you tips and strategies for making your consulting practice grow. Finally, Chapter 16 summarizes the book.

Introduction

This book will open your eyes. You won't learn how to become a consultant in school, or even in industry. Do your apprenticeship. Keep your eyes and ears open. Then, when you spot the right opportunity on a part-time or full-time basis, make your move. Your eyes will be wide open. You'll be ready. In fact, you will have learned how to spot the right opportunity in the first place.

From now on, science and technology are forever. There is no turning back. Progress is cumulative and unrelenting. A technology-driven Society is beating a path to YOUR door. Engineers and scientists, as never before, are at both the cutting edge and dead center of our civilization. Entire industries—and even nations—grow, decay, or become transformed due to technological and technologically-influenced managerial factors.

By becoming your own "boss", and doing so successfully and profitably, you can achieve the ultimate promise of your technical and professional training. You will be able to utilize your knowledge as an independent professional to advance industry, the public interest and human progress, always holding paramount the public safety, health and welfare. And you will be able to earn a professional level income while doing it.

By reading and studying this book, your chances for succeeding will be increased manyfold. You will learn how to reach out for and grasp opportunities, avoid or accommodate setbacks, minimize wasted efforts and costs, and recognize and cope with problems peculiar to the consulting business. And you will learn how to recognize and be ready for opportunities when they do come along.

The bottom line is that this book is the kind I wish someone had written for me when I first entered engineering practice and started looking for my very first job. It will tell you everything you always wanted to know about consulting but were afraid to ask and nobody would tell you.

CONTENTS

INTRODUCTION: A TECHNOLOGY-DRIVEN SOCIETY IS BEATING A PATH TO YOUR DOOR

1 **NOW IS THE BEST OF TIMES TO BE A CONSULTING ENGINEER OR SCIENTIST** 1

The Need 1 The Consulting Profession 3 Income in Private Practice 4 Test the Market and the Possibilities 8 Age Is Neither Barrier nor Criterion 9 Some Sad Cases 10 Engineers and Scientists Are Not Trained to Be Entrepreneurs 10 What Does It Take? 11 Is "Intrapreneuring" for You? 12 Set Your Sights 12 Define Your Problem and Focus on it 13

2 **BUILD UPON THE COMMON BODY OF KNOWLEDGE FOR CONSULTANTS** 15

The Technical Content of "Consulting" 15 More Similarities Than Differences Among Consulting Types 17 The Common Body of Knowledge and Your Library 16 Your Technical Library 19 The "Business" Side of Your Library 20 The First Ten Steps You Should Take 21 Ready to Roll. You Are Wearing All the Hats 26 Characteristics of the Small Business 27

xi

3 **PREPARATION, INVOLVEMENT, AND GOOD HABITS THAT ASSURE SUCCESS** 29

The First Order of Business 29 Successful Is as Successful Does 30 Break Three Bad Habits 31 Details Are Critical 33 Be Busy Even When You're Not Busy 34 Break More Bad Habits 36 Circulate, Listen, and Learn 38 Your Support Systems 39 Your Home Office 39 Pipelines and Windows for Success 40 Office hardware and software 41 Be a "clipper" 42 Keeping Track 42 Project Indexing 44 Taking Stock 46 Revelation 46 Efficient Is Not Effective 47

4 **HOW TO LOCATE, SIZE UP, SELL, AND KEEP CLIENTS** 49

Attracting Clients 49 Prospecting 53 Part-time Prospecting 60 Client Lists 61 Advertising 61 Investments That Mean Business 63 Beware "Dabbling" 66 Credibility Sells 67 Client Retention Factors 67 Keep the Client Up to Date 68 Follow-up Is Important 69 Don't Be Intimidating 69 "Qualifying" Clients and Cases 70 Limited Engagements 71 Beware the Long-term Offer 72

5 **ECONOMICS AND MANAGEMENT OF THE CONSULTING ENGAGEMENT** 75

Types of Engineering and Scientific Consulting 75 Fee Arrangements 76 Fee Components 77 Salary, Fee, and Profit Levels 79 What Will the Traffic Bear? 81 You Cannot Afford to Work "Cheap" 82 Retainers 83 The Proposal, Contract, and Nondisclosure Morass 84 Invoicing for Balances Due 86 Contingency Fees 86 Relationships 87 Scope of Work 88 Milestone Management 88 Engagement Quality Control 89 A Strategy for Built-in Engagement Viability 90

6 **PLANNING, OPERATING AND CONTROLLING YOUR CONSULTING PRACTICE** 93

The Grand Illusion 93 When You Are the Boss 94 The Scientific Method 95 The Thirteen Rules for Business Survival 95 The Function of Management 96 Management

Contents   xiii

by Objectives *96* Managerial Control *96* Conflicts and Styles *98* Mountain Goat Management *98* The Moral *99* "Hassle-free" Management *99* "Voodoo" Management *100* Week-end Projects *101* Think and Reach *101* "Intensity" Factors *102* The Planning Process *102* Performance Measurement *103* Don't Be OBE *104* The Future Is What You Make It *104* Practical Planning *105*

7 **FINANCIAL ENGINEERING FOR EFFECTIVE BUSINESS MANAGEMENT** *107*

Financial and Profit Planning *107* Don't Look at It as Your Money *109* Risk Financing *109* Watch Out for Deficits and Cash Flow *111* Enterprise Structural Profiles: Management Mind-Expanding *114* The Working Profitgraph *122* The Working Capitalgraph *124* Efficiency of Performance: A Still Closer Look *126* Uncertainty *132* The Case for Comparative Analysis *132* Charting Total Return on Capital *134*

8 **FINANCIAL ENGINEERING AND THE PROFITOGRAPHY SYSTEM** *141*

The Need for Outside Capital *141* Cash Flow *147* Surpluses and Deficits: A Closer Look *149* Pickup Characteristics of Industrial Cost and Capital Components *150* The Profitography System of Financial Analysis and Capital Budgeting *151* The Concept of Leverage *154* Examples *165* Summary of Some Important Relationships *173* Basic Ratio Analysis *173* Summary of Financial Engineering Benefits *179* We're Not Quite Finished *180*

9 **BUSINESS-BUILDING REPORTS** *183*

The Craft of Writing *183* The Engineering and Scientific Advantage *184* Some Hints *185* Thirteen Rules for Effective Writing *186* The Technical Report *188* Report Appearance and Visual Aids *189* The Report as Prelude *190* Problems as Opportunities *191* Evaluate Your Report *192* Be Responsive to Client Needs *193*

10 IDENTIFYING, UNDERSTANDING, AND SURPASSING YOUR COMPETITION 195

Getting a Fix on Your Competition 195 You versus Them 197 Part-time versus Full-time Consulting 198 Staying Power and "Firepower" 198 "Contain the Competition" by Increasing Your Market Share 199 Moving into the Territory 200 Service Is the Key 201 Don't Compete on Fees 201 Surpass Your Competition by Surpassing Yourself 202

11 UNCOMMON OPPORTUNITIES IN FORENSIC CONSULTING 205

Forensic Engineering and Science: Your Passport to Consulting Practice 205 Credentials Required 207 Identifying Your Expertise 208 Marketing Your Specialty to Specialized Clients 209 The Forensic Engagement: Nature, Structure, Management, and Logistics 210 The Forensic Engagement: Lead Times and Fees 220 Twilight Zone 253

12 THE FORENSIC CONSULTING BATTLEGROUND 257

The "Battle of the Experts": Fact or Fiction? 257 The "Court-Wise Expert": Image versus Substance 262 What You Always Wanted to Know about Cross-Examination But Were Afraid to Ask (And Nobody Would Tell You) 266 The Expert Report: "Chamber of Horrors" for the Uninitiated 274 Retrospectus 286 A Word of Caution 287 Crossfires 291 Moving Ahead 294

13 A "CRASH" COURSE IN SAFETY ENGINEERING THAT CAN SAVE YOUR NECK 297

The Basic Problem 297 Safety Engineering 298 Practical Limitations of Safety Standards as Design Guidelines 299 Descriptive Catalog of Code and Standard Defects 300 The Universal Performance Standards for Safe Design 307 Defeatability Proneness of Product Safety and Functional Features 312 Defeatability Incentives 323 The Five Kinds of Engineered Protective, Safety, and Control

Features *326* An Epidemiological Approach to Safety Engineering *327*

14 SPECIALIZED APPROACHES IN SAFETY ENGINEERING *337*

Hazard and Risk Foreseeability *337* A Word about Warnings and the Unreasonably Dangerous
Product *341* Maintainability Safety and Safety through Maintenance *343* Applications Engineering *346* Human Factors Engineering *353* Some Additional Assistance for the Scientist and Design Engineer-Turned-Consultant *357* The Rest Depends on You *365*

15 HOW TO EXPAND YOUR PROFIT OPPORTUNITIES IN CONSULTING *367*

Self-actualization through Expansion *367* It's a Telephone Business *368* Expansion and Diversification *369* Other Avenues for Growth *370* Don't Forget How You
Started *371* Decontrol the Trivia *371* Raising Your
Fees *371* Multiple Fee Rates *373* Getting
Published *373* The Consultant's Consultant *374* Testing Laboratories *374* Your Consulting Associates *375* Practice within Your Specialty *376* The Multidisciplinary
Practice *376* Mergers and Acquisitions *377* Do Your Own Thing . . . and Dream *378* Partners *378* Stay
Solo *380* Unless . . . *380* Franchising *381*

16 ETHICS AND ISSUES *383*

The Limits of Negotiation *383* Avoid Conflicts of
Interest *388* Professional Ethics versus Professional Flexibility *390*

INDEX *393*

NOTE

Portions of Chapters 11 through 14 have been adapted from *Product Safety Engineering for Managers: A Practical Handbook and Guide* (Prentice-Hall, Inc., 1984, 438 pps.) by the author and also from various monographs published by The Seiden Group, Inc. Chapters 7 and 8 have been adapted from an unpublished monograph titled: "The Profitography System of Financial Analysis & Capital Budgeting," also by the author.

CHAPTER 1

NOW IS THE BEST OF TIMES TO BE A CONSULTING ENGINEER OR SCIENTIST

THE NEED

It's not hard to find convincing evidence that technologically based careers are at the heart of our expanding society. Because of this, specialized knowledge and practical experience of engineers and scientists are in greater demand than they have ever been. Society obtains great leverage from its technically trained people. One engineer designing a single prototype product or product feature can control the content of thousands or even millions of product units and earn the praise or censure of equal numbers of customers, thereby affecting the destiny of his or her employer.

There are great commercial risks in designing products that are not reasonably safe; not functionally fit; poorly designed for maintainability, efficiency, and economy; unreliable; shoddy, and so on.

Unfortunately, there are not enough well-trained engineering professionals to go around. That is unfortunate for society but fortunate for you. Therefore, many companies hire technical personnel with inadequate skills and then try to "home-grow" the expertise they need. In some instances, this works. In others, such "trainees" ultimately become as myopic and hidebound as their employers due to a stifling intellectual and professional climate within the company. What is worse, of course, is that this short-cut, cut-rate process can produce nothing more than incompetent technicians rather than top-notch engineers and scientists.

Some engineers and scientists ultimately lose their drive, creativity, and sense of self-esteem due to their working environment. It's easy to become too dependent upon a weekly or monthly paycheck that sometimes gives only an illusion of security and the "good life," which can sour.

On the other hand, engineers who are with top-notch firms may have unusual potential and opportunities if they decide upon a consulting career. The quality of their experience and even, perhaps, in-house training (being careful, of course, of the potential for conflicts of interest due to exposure to proprietary technology) may be very marketable on the outside. Thus, the more extensive your experience with leading companies, and the higher the quality of this experience, the more you will be in demand in your field and specialty.

But scientists, science, engineers, and engineering have acquired a new visibility in the public eye. Several major disasters over the past few years can be traced to inadequate engineering attention and imprudent compromises, in turn the result of political and economic pressures in some notable cases. Engineers played second fiddle to lawyers, politicians, and big business executives for so long that the profession had begun to develop an identity and image crisis. But this situation is now changing rapidly. The engineering profession is finally coming into its own as a credible, constructive, and critical force in society. It may yet become master of its own destiny and fulfill its mandate from society to promote and hold paramount the public safety, health, and welfare. And scientists have emerged from their proverbial ivory towers and experimental laboratories to provide competent, refreshing, innovative solutions to practical problems with dedication, energy, wisdom, and heart.

You should consider going into consulting as soon as you have something to sell. You should never view a high-salaried job or an executive position as your ultimate goal. Your goal should always be to go into business for yourself as an expert in some aspect of your chosen field. If, along the way, you happen to have the opportunity to stay in industry, become an executive, and are lucky enough to go down to the wire and retire in style, then by all means take advantage of this opportunity. However, always assume that your present job will be no more than a way station, and a training and proving ground on the road to individual success in your own professional practice or business.

If you do this, you will never stop planning ahead. You will never grow complacent, and therefore vulnerable. And you will never become obsolete. Continually improve yourself either through formal education or private study. You can even be young enough in your fifties to embark on part-time advanced-degree programs. More and more educational institutions have evening programs, including summer courses, leading to M.S. and Ph.D. degrees on a "residence" basis. However, beware the nontraditional, nonresidential, "life experience credit"-granting "university." There are many abuses in this particular area.

The point is that you must never let your employer call all the shots. What you *are* must ultimately be what you make of yourself, rather than what someone else decides you are worth based on his or her own motives. The two can be vastly different. In fact, you should prepare yourself to be ready, willing, and able to part company with your employer at almost any time.

If you live in some nice little "company town," seriously consider pulling up stakes and moving on if you can't identify any opportunities to do your own thing. You may eventually suffocate in such an environment if your professional growth is stifled.

THE CONSULTING PROFESSION

Engineering, like medicine, accounting, and law, is really consulting-based. These are all knowledge-based fields. Practitioners in all these fields are better controlled, in the public interest, with respect to the character of the consulting relationship. They are more formalized, more traditional, more circumscribed, and subject to professional boards of practice and conduct than many other types of technical consulting. However, the full-time, successful, independent technical consultant in almost any field can become financially successful. It is certainly a goal worth shooting for.

Structural changes in a field can create or change the market for specific types of services and specialties. There may be opportunities here. Or there may be indications that redirection and rethinking are necessary. In either case, there may be identifiable and important opportunities for the prospective consultant.

INCOME IN PRIVATE PRACTICE

If you are wondering what kind of income engineers in private practice make, Tables 1-1 and 1-2 will show you. They contain data abstracted from the "Income and Salary Survey" conducted annually by the National Society of Professional Engineers (NSPE). Information has been drawn from surveys for the years 1982 through 1985.

Notice the consistently very high income performance of practitioners falling within the private practice category, which includes consultants. Take particular note of the fact that the private practice category ranks toward the top, even considering the fact that the manufacturing and extractive industries figures (in the top half of each table) include executive and administrative job functions that skew data toward the upper salary levels.

Income of private practitioners is matched or exceeded by only a few categories, including petroleum and coal products, pipeline utilities, and construction and real estate development.

These figures clearly demonstrate that not only is there an identifiable contingent of engineering consultants, but that they do exceedingly well when measured against peer income. Naturally, there are many factors that enter into the construction of these figures, such as degree level, age, years in practice, level of professional responsibility, and so on. However, the figures are very revealing on an overall basis.

The survey data consist solely of incomes of Registered Professional Engineers. P.E.s are, generally speaking, at the tops of their fields with regard to income. On January 1, 1985, there were about 12,800 respondents to the survey out of some 60,000 total NSPE members. But 21,500 P.E.s reported that they were in private practice on the basis of general membership reporting. In the survey, about 3,000 members or about 25 percent of the total, stated that they were in private practice.

Briefly summarizing the 1985 NSPE data (Table 1-1), of approximately 60,000 total NSPE members, about one third stated that they were in private practice. Of respondents to the 1985 NSPE Income and Salary Survey, the median income for professional engineers in private practice was $57,459, with a median of $48,000 and a range from $30,000 at the first decile to $93,000 at the ninth decile. Where do you fall within these data? How does your income compare with the rest of your industry as well as with consultants?

Other data are also revealing. Notable are results of a study conducted by the Engineering Manpower Commission of the American Association of Engineering Societies (Table 1-2). As of February

Now Is the Best of Times to Be a Consulting Engineer

Industry or Service of Employer	All Members Responding						Full-Time Salaried Employees Only	
	Number Reported	Mean	First Decile	First Quartile	Median	Third Quartile	Ninth Decile	Number Reported	Median
All Manufacturing/Extractive Employers	3,034	$53,568	$31,980	$37,500	$46,545	$60,000	$79,800	2,743	$45,400
Aerospace & Aircraft Products	281	$48,155	$31,182	$36,519	$43,700	$54,000	$63,200	263	$43,264
Chemical, Pharmaceutical & Allied Products	332	$56,078	$33,952	$40,000	$49,050	$64,000	$82,950	314	$48,025
Electrical & Electronic Equipment	487	$53,252	$30,717	$36,885	$46,450	$59,789	$75,821	441	$45,552
Fabricated Metal Products	342	$50,575	$30,352	$35,320	$43,029	$55,000	$77,300	302	$42,000
Food, Beverage & Tobacco Products	141	$53,046	$34,030	$41,250	$48,000	$58,335	$75,520	129	$47,625
Machinery (except electrical)	196	$52,158	$30,543	$38,232	$45,300	$58,448	$83,250	172	$43,918
Petroleum & Coal Products	448	$63,023	$34,760	$40,466	$54,000	$72,050	$102,000	377	$50,500
Primary Metal Industries	123	$49,466	$32,376	$38,700	$45,000	$55,000	$67,480	116	$44,550
Rubber & Plastic Products	59	$52,080	$35,172	$37,578	$44,947	$60,000	$79,000	56	$43,968
Stone, Clay, Glass & Concrete Products	121	$52,534	$32,360	$36,780	$45,909	$59,700	$78,920	110	$44,700
Transportation Equipment	159	$50,419	$31,440	$37,433	$45,000	$53,475	$70,462	151	$44,000
All Non-Manufacturing/Extractive Employers	9,747	$52,013	$30,139	$36,500	$45,360	$58,200	$78,000	7,030	$43,000
Colleges & Universities	516	$51,330	$32,076	$39,925	$49,382	$60,000	$74,790	77	$46,400
Communications Services (radio, TV, telephone & telegraph)	201	$52,669	$33,395	$39,595	$46,300	$59,000	$81,485	183	$46,000
Construction & Real Estate Development	644	$62,264	$31,210	$39,000	$50,000	$72,000	$103,438	434	$45,500
Engineering Services	1,601	$53,265	$30,000	$37,000	$47,500	$60,925	$80,200	1,266	$45,000
Government—Federal	670	$43,694	$30,589	$35,639	$42,548	$50,000	$59,266	662	$42,548
Government—State	542	$38,571	$26,346	$32,000	$37,826	$43,848	$51,000	521	$37,707
Government—Local (cities, counties, etc.)	832	$40,631	$28,646	$33,555	$39,326	$46,400	$53,901	816	$39,308
Private Practice	2,903	$57,459	$30,000	$37,000	$48,000	$65,000	$93,000	1,369	$41,600
Research Organizations & Laboratories	119	$54,220	$32,900	$42,513	$50,400	$60,375	$77,360	110	$49,895
Transportation Services	72	$57,796	$34,710	$40,000	$49,288	$60,130	$75,060	65	$45,929
Utilities—Electric	932	$49,484	$33,270	$38,513	$45,954	$55,375	$67,283	907	$45,700
Utilities—Gas	151	$48,835	$32,150	$36,853	$44,000	$54,369	$65,129	145	$43,750
Utilities—Pipelines	81	$56,895	$34,596	$39,975	$52,000	$64,850	$83,660	75	$51,000
Utilities—Other or Mixed	228	$51,945	$31,930	$38,970	$47,171	$58,717	$75,000	213	$45,800

Industry or Service of Employer	All Members Responding						Full-Time Salaried Employees Only	
	Number Reported	Mean	First Decile	First Quartile	Median	Third Quartile	Ninth Decile	Number Reported	Median
All Manufacturing/Extractive Employers	3,453	$49,816	$30,000	$36,000	$44,800	$56,000	$74,500	3,226	$44,000
Aerospace & Aircraft Products	282	$45,063	$29,514	$35,691	$43,500	$51,900	$60,000	272	$43,500
Chemical, Pharmaceutical & Allied Products	397	$51,552	$32,510	$37,695	$47,000	$59,450	$76,752	390	$47,000
Electrical & Electronic Equipment	564	$49,467	$30,000	$36,430	$45,000	$55,000	$72,000	524	$44,000
Fabricated Metal Products	372	$47,861	$29,000	$33,600	$41,000	$53,050	$74,300	335	$40,080
Food, Beverage & Tobacco Products	155	$50,676	$31,000	$39,161	$46,000	$57,840	$73,900	151	$46,000
Machinery (except electrical)	240	$47,468	$27,600	$33,270	$42,430	$55,900	$74,750	218	$42,000
Petroleum & Coal Products	530	$58,546	$33,904	$40,200	$50,675	$69,000	$90,900	489	$50,000
Primary Metal Industries	121	$45,887	$29,986	$33,005	$43,000	$51,153	$66,340	118	$42,250
Rubber & Plastic Products	71	$45,370	$32,100	$35,625	$43,000	$53,000	$64,520	68	$43,233
Stone, Clay, Glass & Concrete Products	146	$47,969	$30,474	$35,940	$42,930	$54,200	$67,840	132	$41,925
Transportation Equipment	177	$45,043	$28,900	$34,875	$41,211	$50,670	$58,460	169	$41,000
All Non-Manufacturing/Extractive Employers	10,630	$48,026	$29,000	$34,866	$42,700	$54,121	$71,000	8,086	$41,000
Colleges & Universities	542	$46,692	$29,722	$36,000	$45,000	$55,000	$66,105	99	$44,190
Communications Services (radio, TV, telephone & telegraph)	247	$51,859	$34,888	$40,000	$47,000	$57,375	$77,760	236	$46,713
Construction & Real Estate Development	728	$56,508	$30,000	$36,350	$46,586	$65,000	$95,700	509	$44,000
Engineering Services	1,714	$48,631	$28,181	$34,820	$44,400	$56,000	$73,000	1,402	$43,257
Government—Federal (including Armed Forces)	774	$42,001	$29,500	$34,312	$40,869	$48,418	$55,899	764	$40,759
Government—State	628	$37,759	$25,000	$30,906	$37,000	$42,723	$50,887	615	$36,960
Government—Local (cities, counties, etc.)	884	$38,819	$27,500	$32,000	$37,681	$43,737	$50,032	862	$37,500
Private Practice	2,959	$51,853	$28,200	$35,000	$45,000	$60,000	$83,500	1,529	$40,000
Research Organizations & Laboratories	190	$47,964	$31,390	$37,100	$45,758	$53,000	$67,075	182	$45,625
Transportation Services	79	$48,468	$30,004	$35,175	$43,992	$53,594	$67,848	73	$43,992
Utilities — Electric	1,092	$46,507	$31,189	$36,135	$42,500	$52,000	$64,937	1,079	$42,500
Utilities — Gas	154	$47,909	$31,272	$34,560	$44,236	$53,700	$64,016	150	$43,683
Utilities — Pipelines	92	$53,825	$34,574	$39,690	$47,950	$63,360	$83,426	90	$48,225
Utilities — Other or Mixed	253	$47,267	$31,364	$36,600	$43,500	$53,725	$64,016	249	$43,100

Industry or Service of Employer	Number Reported	All Members Responding					Full-Time Salaried Employees Only	
		Mean	First Decile	First Quartile	Median	Third Quartile	Ninth Decile	Number Reported*	Median
All Manufacturing/Extractive Employers	4,045	$50,023	$29,151	$34,575	$42,590	$53,802	$72,000	3,774	$42,000
Aerospace & Aircraft Products	308	$44,349	$28,986	$34,100	$42,000	$50,000	$59,227	292	$41,519
Chemical, Pharmaceutical & Allied Products	466	$52,395	$31,326	$37,800	$45,526	$56,147	$76,790	452	$45,526
Electrical & Electronic Equipment	665	$48,091	$28,800	$34,193	$42,100	$52,140	$65,000	608	$41,621
Fabricated Metal Products	491	$50,041	$27,010	$32,205	$40,010	$53,025	$74,400	441	$39,900
Food, Beverage & Tobacco Products	164	$50,178	$29,940	$35,150	$42,750	$53,500	$65,338	161	$43,000
Machinery (except electrical)	281	$49,800	$27,960	$32,693	$41,000	$53,250	$76,774	251	$40,000
Petroleum & Coal Products	571	$60,146	$32,322	$38,500	$48,500	$63,575	$85,000	525	$47,640
Primary Metal Industries	162	$45,923	$28,673	$35,725	$42,421	$50,000	$62,526	159	$42,342
Rubber & Plastic Products	92	$46,729	$30,480	$33,708	$41,229	$48,650	$68,970	85	$40,394
Stone, Clay, Glass & Concrete Products	170	$46,540	$28,150	$33,300	$40,263	$50,000	$71,625	162	$40,084
Transportation Equipment	213	$44,899	$29,416	$33,745	$39,800	$48,000	$60,000	207	$39,710
All Non-Manufacturing/Extractive Employers	11,959	$46,930	$27,925	$33,000	$40,800	$52,000	$69,080	9,027	$39,168
Colleges & Universities	586	$44,651	$28,410	$34,560	$42,950	$53,200	$62,897	95	$39,300
Communications Services (radio, TV, telephone & telegraph)	296	$47,703	$32,400	$37,200	$42,950	$51,500	$67,900	282	$42,950
Construction & Real Estate Development	958	$57,618	$28,600	$35,500	$47,500	$66,500	$95,000	650	$43,960
Engineering Services	1,989	$48,301	$27,640	$33,515	$42,500	$55,000	$70,910	1,604	$40,677
Government — Federal (including Armed Forces)	812	$40,218	$28,598	$32,761	$38,476	$45,533	$54,024	803	$38,422
Government — State	683	$35,648	$25,000	$30,000	$35,000	$40,000	$46,420	676	$34,996
Government — Local (cities, counties, etc.)	968	$36,871	$26,415	$30,500	$35,700	$42,000	$48,800	952	$35,700
Private Practice	3,350	$50,882	$27,200	$33,000	$42,600	$57,000	$80,100	1,776	$37,620
Research Organizations & Laboratories	185	$45,168	$31,200	$36,000	$42,600	$50,850	$60,000	164	$42,750
Transportation Services	90	$46,447	$29,024	$33,800	$39,070	$49,956	$58,609	86	$39,304
Utilities — Electric	1,158	$43,741	$29,653	$33,876	$40,500	$50,000	$60,000	1,150	$40,500
Utilities — Gas	184	$47,501	$29,010	$33,990	$40,958	$53,000	$68,180	178	$40,358
Utilities — Pipelines	111	$50,225	$31,528	$37,710	$45,000	$55,000	$79,720	106	$45,000
Utilities — Other or Mixed	278	$46,241	$29,520	$35,280	$42,090	$51,100	$64,970	263	$41,500

*Excludes faculty members, self-employed owners and principals, and those not responding to Question A.

Industry or Service of Employer	Number Reported	All Members Responding						Full-Time Salaried Employees Only	
		Mean	First Decile	First Quartile	Median	Third Quartile	Ninth Decile	Number Reported*	Median
All Manufacturing/Extractive Employers	4,237	$45,498	$27,284	$32,200	$39,756	$50,000	$65,000	3,930	$39,000
Aerospace & Aircraft Products	317	$39,569	$26,780	$31,299	$38,000	$45,000	$52,000	310	$37,972
Chemical, Pharmaceutical & Allied Products	538	$48,640	$28,500	$33,600	$40,700	$52,500	$72,860	511	$40,560
Electrical & Electronic Equipment	675	$43,593	$26,400	$31,640	$38,000	$48,000	$60,700	613	$37,826
Fabricated Metal Products	486	$46,371	$26,459	$32,031	$40,004	$50,000	$67,026	440	$38,790
Food, Beverage & Tobacco Products	168	$42,641	$28,585	$33,120	$39,625	$50,000	$61,410	165	$39,600
Machinery (except electrical)	305	$44,999	$26,250	$31,537	$40,000	$50,000	$68,000	269	$40,000
Petroleum & Coal Products	590	$54,950	$29,500	$35,300	$45,060	$60,500	$80,874	531	$44,000
Primary Metal Industries	180	$44,951	$27,179	$32,300	$40,050	$50,000	$68,850	166	$40,150
Rubber & Plastic Products	93	$37,485	$26,880	$31,049	$36,968	$41,413	$49,712	89	$37,980
Stone, Clay, Glass & Concrete Products	165	$47,358	$25,133	$31,477	$38,400	$49,670	$71,749	153	$38,000
Transportation Equipment	241	$39,481	$28,048	$31,914	$36,250	$44,000	$52,742	231	$36,000
All Non-Manufacturing/Extractive Employers	11,972	$44,445	$26,157	$31,200	$38,978	$50,000	$65,000	9,048	$36,900
Colleges & Universities	612	$41,371	$25,570	$32,310	$40,950	$48,950	$57,388	103	$37,250
Communications Services (radio, TV, telephone & telegraph)	293	$45,452	$29,700	$33,984	$40,000	$50,000	$65,100	275	$39,900
Construction & Real Estate Development	927	$53,339	$27,000	$33,928	$44,600	$60,000	$90,000	671	$41,000
Engineering Services	2,147	$45,428	$26,200	$31,600	$40,000	$51,000	$67,500	1,752	$37,880
Government—Federal (including Armed Forces)	900	$38,945	$27,743	$32,013	$38,000	$44,991	$50,400	882	$38,000
Government—State	659	$33,677	$23,713	$28,400	$33,000	$38,168	$43,388	647	$33,000
Government—Local (cities, counties, etc.)	965	$34,761	$24,919	$28,575	$33,336	$39,944	$46,050	946	$33,338
Private Practice	3,177	$49,333	$26,000	$31,938	$41,000	$55,945	$79,445	1,608	$36,000
Research Organizations & Laboratories	201	$42,798	$28,489	$33,000	$38,900	$46,575	$55,800	189	$38,580
Transportation Services	109	$43,862	$27,744	$32,970	$38,770	$48,095	$57,776	102	$38,597
Utilities—Electric	1,119	$40,487	$27,000	$31,000	$37,500	$45,498	$56,000	1,098	$37,400
Utilities—Gas	164	$47,083	$28,230	$32,847	$40,908	$52,000	$72,000	163	$40,815
Utilities—Pipelines	110	$44,238	$28,470	$32,200	$40,256	$50,800	$67,000	106	$40,256
Utilities—Other or Mixed	251	$42,603	$28,354	$32,452	$39,050	$48,000	$57,000	240	$38,640

TABLE 1-2

EMC 1985 SURVEY SELECTED EMPLOYMENT GROUPS: MEDIAN SALARIES, ALL INDUSTRIES, FOR ENGINEERS WITH 0, 10, AND 25 YEARS OF EXPERIENCE

Industrial group:	Years of experience		
	0	10	25
Petroleum	$30,150	$51,950	$71,900
Research Organizations	29,150	45,350	55,300
Chemicals	28,350	42,800	53,950
Engineering Services	29,800	42,100	52,700
Electric/Gas Utilities	28,200	42,050	52,450
Consulting	25,150	39,900	50,800
Communications	28,400	40,400	46,850
Electronics/Computers	27,950	43,350	50,100
Aerospace	26,900	40,050	49,800
Shipbuilding	24,850	37,800	45,200
Metal Products	25,700	37,300	46,400
Construction	23,650	37,700	47,050
Electric Equipment	27,000	37,950	43,800
State/Local Government	20,250	30,100	37,300
All Industries	$27,400	$40,200	$49,750

1985, median salaries for all consulting engineers (i.e., not only Licensed or Registered Professional Engineers) with 0, 10, and 25 years of experience were, respectively, $25,150; $39,900; and $50,800. Mean data are not available. Notice the substantial skewing of the overall P.E. consulting median, at $45,000, toward the high end of the experience range for the total consulting segment shown by EMC data, which are based upon salary reports for over 75,000 engineers.

"All Engineers" data for an EMC sample of 6,477 engineers are quite striking. They show a range from $37,900 at the lower decile to $70,000 at the upper decile for twenty-five years of experience since attaining a B.S. degree, with a median of $50,800.

There are some measurement variables such as inclusion or exclusion of overtime, bonuses, fees, stock options, and other irregular payments. However, we are primarily interested in gross magnitudes for present purposes. Table 1-2 shows "Selected Employment Groups: Median Salaries, All Industries, for Engineers with 0, 10, and 25 Years of Experience."

It should be noted that private consulting practice costs and capital expenditures can include, in addition to salary, very significant pension plan contributions and property, plant, and equipment expenditures. However, the consultant is the direct beneficiary of such expenditures. Thus, there is a significant entrepreneurial or ownership interest in, for example, computers, instrumentation, real estate, property improvements, automobiles, and other earning assets acquired by the individual consultant that are essential to successful conduct of the business. Real estate can significantly appreciate and certain types of consulting practices can be sold as "going businesses."

The bottom line is that even successful industrial employment typically cannot begin to match the benefits of successful private practice. The true value of such benefits is not reflected in any of the available statistics relating to consulting "income." Actually, a more correct comparison would have to include consultant salary or income, fringe benefits, and the many entrepreneurial values that are developed in private practice. Even a part-time engineering consulting practice can offer important opportunities for financial growth not immediately evident if only salary or equivalent is considered.

The engineer or scientist embarking on a part-time consulting practice as a secondary income source can ease into consulting activity in a natural and progressive way. The consultant quickly learns how to deal with opportunities available from the standpoint of entrepreneurial financial planning and management. He or she will gain firsthand experience in the activities and potentials that lie behind the foregoing statistics, which do not tell the whole story and, in fact, severely understate the total compensation and allied value "package" available to the consultant as entrepreneur, including entrepreneurial profits over and above solely salary.

TEST THE MARKET AND THE POSSIBILITIES

Try to meet engineers and scientists who are in consulting, even if in specialties different from your own. Chances are you can meet at professional society meetings. As a matter of fact, your company probably encourages you to go to such meetings and keep abreast of what its competitors are doing. You might as well enjoy it and use it to your advantage. If your company *discourages* such contacts and will not pay for technical or professional society memberships, you may have a clue as to how your management looks at its professionals.

Keep a current list of subjects that could be specialties of yours. In which ones do you know consultants? In which ones do you know for sure that your employer goes to the outside for consulting help? Why? If you don't know, find out.

We're not trying to foment a revolution here, but just gain some perspective. As a matter of fact, if you do your homework right, you should be so well prepared that, if you should get fired or laid off, you will already have done your planning so you know exactly what your next step is. You will probably still look for another full-time job—this is too strongly ingrained in you. However, your experience may also be a blessing in disguise. Sometimes you have to be shaken up a little or even a lot before you realize who you are and who you can be. Engineers are trusting souls. Don't permit yourself to exist blithely from day to day, totally oblivious to the realities of working for someone else versus working for yourself.

AGE IS NEITHER BARRIER NOR CRITERION

How old are you? Thirty? Forty? Fifty? Your life expectancy is seventy-three to seventy-eight years. If you're not already doing it, pretty soon you will be living in that very "future" you used to think was going to be so bright.

In consulting, it is more the rule that the older the better. You have been through the mill. You have time-tested expertise.

As an engineer, you will probably have a nest egg of significance. You are not so obviously in a hurry to change the world. You may still be champing at the bit below the surface, but you function in a more socially acceptable manner. To quote an old Chinese proverb, you are like the duck: You have to learn to be cool on the surface but paddle like crazy below. You have learned not to be intimidated. This is especially important if you have been laid off. Besides the frustration and aggravation of job hunting and belt tightening, there is some inevitable loss of self-esteem. This is because our society often defines individuals by their jobs and titles.

For the engineer and executive, always remember that getting fired or laid off is nothing more than an occupational hazard. In certain instances, it is even a badge of honor. You might as well plan for it right now. The time to do planning is when things are going well. Be prepared for this common occurrence whether you are a Ph.D. or president of your company. You may only discover your true professional strengths and interests—indeed, passions—when you are forced to reexamine your career path. Chances are money will not be the only issue. It is merely one more way of keeping score.

SOME SAD CASES

Many of us have been witness to great personal tragedies. Outstanding engineers and executives are fired or laid off for reasons outside their control. They may have been fine and dedicated professionals in their fields and devoted to their companies. No doubt some of them were making in excess of $100,000 a year. They were too trusting, too complacent, and totally unprepared. They were team players who also thought they knew how to play the corporate political game, but to no avail.

Don't fall into this trap. Gird yourself for battle. It will be stimulating, exciting, professionally advantageous even if you stay with your firm, and potentially lucrative if you make the break—or are forced to. If you find the prospect frightening, lay your plans and then, if and when necessary, make your own move.

ENGINEERS AND SCIENTISTS ARE NOT TRAINED TO BE ENTREPRENEURS

There are three structural aspects of engineering practice that you should recognize:

1. Unlike lawyers, physicians, and other types of professionals, engineers are not trained to be in business for themselves. Engineers are basically trained to be and to remain employees. Consequently, many engineers are inexperienced when it comes to starting, operating, and developing their own professional practices.
2. By and large, engineers do not—indeed cannot possibly—own the resources and factors of production that they need to ply their trade.
3. Engineers predominantly go to work where the work is located. Usually the work does not—cannot—travel to the engineer.

When you are in consulting engineering practice, (1) you are in business for yourself, (2) you still work with client properties, whether physical or intellectual, to discharge your professional responsibilities, and (3) you may find it necessary to do some work on client or related premises, or in the field in some other way. However, the work you do in all cases is for your own account, significantly under your professional control, and now you are giving yourself the leverage of using somebody else's considerable resources to make money. Make other people's money work for you.

WHAT DOES IT TAKE?

The characteristics you need to possess or cultivate to be successful as a consultant and, generally speaking, as a small businessperson include integrity, reliability, intelligence, personality, ambition, effectiveness, creativity, energy, self-confidence, tenacity, desire to succeed, willingness to take on responsibility and take the initiative, self-discipline, knowledge, experience, and empathy.

However, first and foremost you must ultimately come to the realization, whether through first-, second- or even third-hand experience, that you have a driving, burning desire to be in business for yourself and work for your own account. You are willing to take a "businessman's risk."

The "what-have-you-done-for-me-lately" syndrome looms larger than life for every employee type, whether janitor, engineer, lawyer, physician on full-time corporate staff, secretary, manager, salesman, or even company president. It could be the positive desire simply to be your own boss (of course, your client is also your "boss," at least up to the point where no ethical compromises are necessary) or the more negative aversion to reporting to somebody above you. It's probably a combination of both that ultimately converges and jolts you right out of that "employee" mindset.

Of course, it may take a layoff to give you newfound self-understanding and motivation. Whatever it is, the important thing is that when it happens you recognize it and capitalize upon it.

Being convinced that you can do things better than your boss or your company, or that your contribution and potential haven't been adequately recognized, can also be important motivators. In addition, you will already have observed how much more a successful consultant can make compared with a salaried professional, even including what may well be generous perks, at least by industry standards.

The entrepreneurial mind and spirit are more "grass-roots" focused, however, than the outlook of the superficially well-established organization man. The entrepreneur will look for opportunities to go into business even if it means leaving his or her original field. Naturally, the preferred alternative will be to go into some private venture allied with one's interests, training, or experience so that it is possible to build upon strength.

IS "INTRAPRENEURING" FOR YOU?

A distinction has recently been made between entrepreneuring and "intrapreneuring." The latter term and concept was invented to describe a management attempt to retain innovative personnel by fostering internal "entrepreneuring" so the innovator doesn't feel it necessary to quit and strike out for him- or herself.

The "intrapreneuring" position is that only a large business may have the marketing, financial, and technological base and reach or clout to best pull off an innovative coup envisioned by a valuable employee. For an employee to leave the corporate fold and attempt to raise private outside capital, establish marketing capabilities, perhaps have the need to build upon proprietary company technology, retain "desirable" company ties, build upon or improve a specific company business, and so on, may be impractical or impossible to achieve.

The "intrapreneurial" argument further asserts that the desire of the employee to retain the friendship, security, and trust of the company can be made stronger than the mere desire for a crack at great wealth, especially in the face of the great odds against success due to the formidable risks of starting up your own business.

However, "intrapreneuring" is often misleading and gives a lot of good people false hopes of great in-company success, independence, and so on. It is often an illusion. None of these arguments even begins to address what are probably four of the strongest reasons that the entrepreneur is an entrepreneur in the first place, while the "intrapreneur" remains an employee. They describe two totally different kinds of people with divergent needs and objectives.

The entrepreneur is intolerably self-confident; uses failure to succeed; desperately wants to do his or her own thing; and has dreams of financial independence. However, it doesn't really even matter if income isn't as great as might have been the case had he or she stayed put. It is the desire to be free and independent that fundamentally drives the entrepreneur. It is the demon within that makes the difference. It is this that the corporate establishment and professional management do not understand or are not willing to acknowledge in public.

SET YOUR SIGHTS

When you begin to think about consulting, set yourself an income goal. Let it be modest for the first two or three years until you get your feet on the ground—perhaps $5,000 average per year. The most important thing is to get started *now*. Once you have gotten some engagements under your belt—even after your very first one—try to identify the amount of work that might be out there. What do you have to do to capture $10,000 to $15,000 a year? $25,000 a year or more? At what consulting income level are the demands on you likely to exceed your capabilities if you continue to hold down a full-time salaried position? Can you visualize building up a practice that you can manage part-time, one that you can nurture without having to give up your job unless and until your consulting income begins to approach your full-time, primary income? Can you manage a consulting practice where you, in turn, engage subcontractors who are experts in their own specialties, and whom you call upon when you get an engagement? Do you have friends who are engineers, scientists, or other types of technologists, and whom you can assemble into a multidisciplinary team capable of suiting the demands of the engagement? There are all kinds of possibilities. You might even be able to make *all* of these alternatives click.

You may identify consulting opportunities that are numerous but small. Or perhaps the engagements tend to be few in number but large. You must determine what the characteristics are of the specialties and opportunities with which you feel comfortable and that are open to you. It shouldn't be very hard, for example, to get a line on your own company's habits and patterns in purchasing technical consulting services on the outside. And try to find out who is supplying them. How are these outside service firms or individual consultants set up? What kinds of reports, services, results, and so on, do they ultimately provide? Get their brochures. Maybe you can even find out what they charge and how—by the hour, day, or job. We will look at this later on in a little more detail.

Most professional people tend to be reasonably open in discussing their specialties and work. This is because they are constantly seeking new opportunities and fields to conquer. Perhaps, in speaking with them, you will open some additional doors for them— and maybe even for yourself.

You might even start in the consulting business by doing some work for someone already established as a consultant. Consultants use consultants all the time. Engineering and nonengineering consultants use both engineering and nonengineering consultants. There are vast and lucrative opportunities here.

DEFINE YOUR PROBLEM AND FOCUS ON IT

Remember that planning to be a consultant is just another exercise in problem identification and problem solving. You're already accustomed to this process by virtue of rigorous training and practical experience in engineering. However, it could be the most important effort of your career. Simply view it as another problem in "applications engineering."

Think in terms of problem defining and problem solving as an exercise in "commercial intelligence." It is a practical approach to thinking about being successful as a consultant. It immediately establishes and focuses upon the financial, economic, and systems aspects of utilizing one's technical training and experience as the basis for the professional practice type of business.

Practices often involve a significant number of consulting associates. These are professionals, each of whom is specialized in a different field. Every one of their specialties, which call for high credentials, is demanding, lucrative, and even exciting in its own right. Nevertheless, such associates, each of whom may even be employed full-time in high-powered capacities, frequently discover that the engagements they work on are among the most satisfying, challenging, and professionally stimulating activities they confront. This is quite aside from the fees they earn, which, on an annualized basis, are commensurate with their high professional proficiency (and, on an annualized basis, amount to a lot more than their employers are paying them in salaries and perks combined).

So don't regard this book merely as an interesting curiosity, some unattainable idea, or too demanding because of current job responsibilities. Rather, view it as an incentive to attain your full potential as a professional person—as an engineer or scientist. The opportunities are out there. With a little help and prodding from this book, they are yours for the taking.

There is a very practical viewpoint that should make you look into consulting at multiple points in your career, no matter how successful and happy you may be in your present position, and no matter how bright your future appears. When you retire, which you will do someday, you will undoubtedly look at postretirement opportunities. You know you won't be happy being inactive. And you will probably be unhappy if you leave engineering altogether.

But why wait until you retire to consult, when you can do it today? Why wait until somebody tells you that you *have* to retire and it is forced on you? You may be unprepared for the change if you wait until the last minute to plan properly and set things in motion to permit a smooth transition. Make the change earlier—or at least pave the way for yourself. And why not reap the substantial rewards of a successful consulting practice when you still have a significant number of years ahead to embark upon a second career on a truly professional level at a relatively younger age?

CHAPTER 2

BUILD UPON THE COMMON BODY OF KNOWLEDGE FOR CONSULTANTS

THE TECHNICAL CONTENT OF CONSULTING

In this chapter, we will talk about a number of technical but nontechnological subjects that you should know something about. We will also suggest a number of technical but nontechnological steps you should take to get off to a solid start in your consulting practice.

Remember that running a business in your specialty is still "running a business." Therefore, your expertise is embedded in a larger structure that has diverse and even technical demands and characteristics. They may not be technologically oriented and may not require engineering- or scientific-type problem solving, but they do demand your undivided attention and minimal knowledge of other fields such as marketing, accounting, human relations, and management.

It is only when you understand how to run a business that you can give your primary attention to the practice of your specialty. If you don't acquire this understanding, problems will arise that will get in the way of your primary focus as a consulting engineer. In fact, if you don't understand what a business is all about, such problems can result in business failure no matter how good an engineer or scientist you are. Study after study shows that more businesses fail due to poor management than because of a poor product or service that is offered for sale.

This chapter will raise your level of consciousness and conscientiousness about the kinds of things you should know and do in order to be successful as a businessperson who just happens to be a consulting engineer. The emphasis is absolutely crucial to assure your success.

Many engineers like to believe that there are two kinds of consulting: technology-driven and nontechnical. Our interests revolve around science and engineering. *Technology* and *technical* mean, in fact, scientifically based consulting in the classical sense that refers to the study and application of natural phenomena. Maybe you think it is foolish to make an observation like this. It seems so obvious to engineers and scientists. However, nowadays many people in nonengineering, nonscientific fields characterize their activities as "technical" and even "scientific." In fact, a quick reference to any standard dictionary may surprise you. Notice how the terms *science* and *scientific* are defined. Nonengineers and nonscientists just haven't been as aggressive in the past in how they looked at themselves, or with respect to how they described "bodies of knowledge" in their fields and expected to be treated by the rest of us.

Competent practitioners and consultants in almost any field are organized and methodical in relation to their specialty. And all fields, not only science and engineering, have their own bodies of knowledge and practice. Many fields require more or less extensive study and method. All fields require relevant experience to a greater or lesser extent.

For example, if you will take the time to look around you, and if you become sensitive to it, you will discover whole fields out there you never even realized existed. And you will come to appreciate the nature and depth of knowledge and experience, theory and practice, of subjects and businesses you used to take for granted. Each is a world of its own, complete and demanding. It is a realization that comes slowly upon us, maybe even painfully at first, as we start to outgrow the narrowness of our "professional" course of study—even in engineering.

The technologically driven fields do require the acquisition of highly specialized quantitative *and* qualitative or descriptive knowledge, and usually require years of study, experience, and practice in problem defining and problem solving. Yet, the differences in length of time to acquire requisite capabilities may not be very great between the scientific and nonscientific fields. For example, for a pianist to become proficient in technique and interpretation, and develop a wide repertoire, could take ten or twenty years or more. The same is true for an engineer.

Some differences in kind are striking. The quantitative intellectual side of the sciences differ markedly from the qualitative aspects of the arts and near-arts. It does *not* make them "better," however. But the rules and "laws" of the social sciences, current-day psychology, and so on, are not immutable such as the laws of nature. Of course, each of us has to have a flair for whatever we do. The person who is skilled in languages and becomes fluent in ten or more before the age of forty is something special. The economist and professional appraiser also have special talents and skills. However, so do engineers or physicians who study their own "arts" and "sciences" and acquire the requisite skills to pass professional board exams, or lawyers who develop the high skill of using language instead of mathematical symbols and symbolic logic for arguments. And the sciences are not even necessarily more heavily theory-oriented than the nonsciences. The best we can say is that theoretical content is "different" and tends to be more related to the physical universe than to relationships among people. But the artistic content is pervasive. For example, architecture has been called "frozen music," and music has been termed "sublimated mathematics."

MORE SIMILARITIES THAN DIFFERENCES AMONG CONSULTING TYPES

When you come right down to it, except for the fact that an engineer does "engineering," the similarities among many fields of consulting are probably greater than the differences. When you think about it, with their specific technical skills, both lawyers and physicians really are consultants also. So are members of the clergy. The main difference in the consulting done by lawyers, physicians, and engineers probably is that the public at large does not deal directly with engineers the way it does with lawyers and doctors. Few consumers or members of the general public ever have direct contact with engineers except, perhaps, when they engage them to conduct prepurchase home or building inspections relating to real estate transactions. So where are the noncontent differences in consulting?

In fact, every consulting practice consists of two sides. There is the technical side, whatever it may entail in the way of specialized expertise, and the business side. On the business side, there are consultants ready, willing, and able to expertly assist their "consultant" clients. Such consultants' consultants include accountants, lawyers, advertising and public relations specialists, graphic designers, telecommunications experts, pension experts and actuaries, sales and marketing consultants, architects and interior designers, health maintenance organizations, and even engineers.

Thus, for present purposes, the fact that we are engineers practicing engineering is essentially irrelevant. We know our field. We understand its applications in our respective specialties. The important thing is for us to be able to practice our high art and science in a businesslike manner just as other professionals such as lawyers, accountants, physicians, dentists, and so on. To do this, we must become as versed in what it takes to run a business as the most successful "regular" businessperson.

THE COMMON BODY OF KNOWLEDGE AND YOUR LIBRARY

Therefore, the "common body of knowledge" for engineering consultants consists of a minimal understanding of such things as the principles and practices of business management, human relations and negotiating, writing, accounting and finance, advertising and public relations, personal time management, stress management, and accident prevention.

If this sounds like a tall order, it is, particularly if you immediately throw up your hands in despair before you attempt to find the shortcuts. Many of the best shortcuts are to be found in books. Among other things, we engineers are expert at learning through books. Engineers can learn the rudiments of many nonengineering subjects from books and, obviously, even engineering-related subjects. It is entirely possible to learn more than enough to provide information, methodology, and confidence to apply such information to your practice. In some instances, where a particular subject becomes a working tool, such as writing and publishing, the purchase of numerous books on the subject will cover it from as many, if not all, angles. You don't have to be a fish to understand how one swims, breathes, and functions. Or, if you want to learn how someone deeply involved in a field thinks, worries, and reacts, there will always be some author who is expert in the field, willing to share his or her knowledge and experience with you.

There are books on virtually every subject one can imagine. The trick is to keep away from the most exhaustive treatises and the overly compact and useless materials. In the middle will be numerous books that are conversational, informational, literate, and easy to read. They offer tips, shortcuts, lessons from experience, guidelines, checklists, selected references for further study, and other actionable content that will prove invaluable to you when you need some expert insight. They will bring you up to a threshold level of knowledge that will permit you to think correctly, intelligently, and even creatively about the particular problem even if of a nonengineering nature. Use this method to build a library that is useful rather than just for "show."

YOUR TECHNICAL LIBRARY

You obviously need a well-stocked library, especially in your chosen specialties, in order to do engineering consulting. Find yourself a local used-book store. Scour the shelves of both the technical and business sections. Telephone all of the major publishers and get their latest catalogs. Check your trade journals for announcements of new books. Visit the nearest large city and visit two or three of the largest bookstores and used-book shops.

You must keep abreast of *all* of the technical and trade journals in your field and be informed of new books. These days, keeping track of new books is not too much of a problem. Most major publishers send out numerous direct-mail card decks. No doubt you are already receiving several of these and probably get them at least once a month.

There's probably no physical aspect of your practice that so closely reflects your interests, capabilities, objectives, and accomplishments as your library. You can use it for information, inspiration, enjoyment, escape, and profit.

When you finally put together a library in your specialty, you will be able to identify errors in reasoning and judgment by certain writers, as well as acquire a sense of history of the body of knowledge and progress in your subject. When you reach this point in your mastery of your subject, you will feel as though a high-voltage electric charge has just gone through your entire body. You will feel exhilarated. You are now at a threshold, a takeoff point. You are now in a position to make real contributions to your field as well as serve your clients in a totally competent manner. You are *in control*.

The "older" part of your library may contain materials now technically obsolete due to the cumulative nature of scientific progress. But it will never be obsolete in terms of your perception of how your subject started, grew, expanded, and matured. And don't sell short the wisdom contained in older or classic works in your field. In fact, you may well find that much of the basic philosophy in your subject was much more comprehensively and eloquently developed and articulated in some of these older volumes than in newer ones.

THE "BUSINESS" SIDE OF YOUR LIBRARY

Certain books on sales and marketing, human relations, running a business, and so on are almost more inspirational than informational. However, these can be invaluable as well. As a practical matter, it is just as important to be able to get yourself into a problem-solving mood or frame of mind in relation to a particular situation as it is to have the requisite background information that allows you to think through specific problem solutions.

The information content of many problem solutions in business is trivial compared with the types of problem solving that we engineers do daily. It is more our resolve and commitment to solving problems, learning or knowing where to find solutions and how to apply specialized information that is readily available that makes the real difference. On the other hand, don't minimize for one minute the importance of learning how to solve the more generalized business management problems that you will invariably and frequently encounter in your practice. Lack of attention to this type of detail can get you into big trouble . . . irretrievably.

Again, the subjects with which you need to develop some familiarity are as follows:

1. Public speaking
2. Writing
3. General management and organization theory
4. Bookkeeping and financial management
5. Human relations and personnel management
6. Personal time management
7. Personal stress management
8. Negotiating skills
9. Advertising and public relations

10. Marketing and sales
11. Practical Law (so you know when you really do need a lawyer)

THE FIRST TEN STEPS YOU SHOULD TAKE

Once you have made the decision to go into part- or full-time consulting, immediately do the following things:

1. Telephone (don't just write) the secretary of state in your own state and get incorporation forms. Decide what you are going to call yourself, fill out the forms, and send in the typically nominal fee. Before you know it, you will get your incorporation papers back in the mail. You are officially in business. Incorporate to limit your liability to the extent possible. An attorney is not necessary just yet.
2. Get yourself an accountant. Shop around a bit. When you are first starting out, you don't need a firm with seventeen names on its letterhead. A CPA who has a part-time practice can be just the ticket. But be sure he or she actually knows what the small businessperson needs in the way of services. Check out fees and find out what first-year accounting costs will probably look like. Spend an hour or so interviewing your prospective accountant to get some idea of the services you will need and the kinds of books you will have to set up. Then engage him (or her), at his usual hourly rate, for an hour or so to give you some advice about pension plans. The law keeps changing, so it is vital that you can get current advice on this important subject. You may need other professionals in this field, such as an actuary and an attorney, as your practice grows.
3. Set up your own general ledger. It does not have to be of the double-entry type. Initially, use a tab-divided loose-leaf binder with about 250 pages in it, with the following sections, for example:

   - Operating revenues (use separate sections for each different type of revenue you can anticipate in your practice)
   - Office expenses (a catchall until you get rolling)
   - Salaries
   - Books
   - Dues, fees, and subscriptions
   - Out-of-pocket expenses and petty cash
   - Insurance

- Telephone
- Entertainment expense
- Accounting and legal fees
- Tax payments
- Subcontractor expenses
- Other revenues
- Other expenses

Initially, this will do nicely for your "chart of accounts," unless your special practice has other aspects that should be incorporated into ledger sections.

Eventually, however, your accountant will probably steer you in the direction of an enlarged ledger account lineup that might look more like the following:

- Fees—self-generated
- Fees—consulting associates/professional assistance
- Books and other professional literature expense
- Payroll—officers
- Payroll—others
- Payroll—taxes
- Advertising expense
- Bank charges
- Casual labor
- Contributions
- Conventions and seminars
- Depreciation expense (Noncash item)
- Dues
- Employee benefits
- Insurance—general
- Office expense
- Out-of-pocket expense
- Pension expense
- Postage
- Professional fees expense for services rendered (accounting, actuarial, legal, etc.)
- Repairs and maintenance
- Sales promotion expense
- Sundry expense
- Supplies
- Telephone
- Travel and entertainment
- Utilities
- Vehicle

- Taxes—other
- Other Income—interest
- Transfers (from one investment vehicle to another, from checking account to pension fund, etc.)

Your accountant will eventually generate a Profit & Loss Statement (Exhibit 2-1) that will look something like the following. Don't worry about a balance sheet for a while. But we'll talk more about financial planning and analysis in Chapter 7.

EXHIBIT 2-1

SMITH CONSULTING ENGINEERS, P.A.
STATEMENT OF INCOME
FOR THE MONTH OF _____ 198X
AND THE TWELVE MONTHS THEN ENDED

	Current period		Year to date	
	Amount ($)	Pct. (%)	Amount ($)	Pct. (%)
Sales				
Fees				
Cost of Sales				
Professional Assistance				
Professional Literature				
Total Cost of Sales				
Gross Profit				
Selling, General and Administrative Expense				
Payroll—Officers				
Payroll—Others				
Payroll—Taxes				
Advertising				
Bank Charges				
Casual Labor				
Contributions				
Convention and Seminars				
Depreciation				
Dues				
Employee Benefits				

EXHIBIT 2-1 (continued)

SMITH CONSULTING ENGINEERS, P.A.
STATEMENT OF INCOME
FOR THE MONTH OF _____ 198X
AND THE TWELVE MONTHS THEN ENDED

	Current period		Year to date	
	Amount ($)	Pct. (%)	Amount ($)	Pct. (%)
Insurance—General				
Office Expense				
Out-of-Pocket				
Pension Expense				
Postage				
Professional Fees				
Repairs and Maintenance				
Sales Promotion				
Sundry				
Supplies				
Telephone				
Travel and Entertainment				
Utilities				
Vehicle				
Taxes—Other				
Total Selling, and G&A				
Profit (Loss) From Operations				
Other Income				
Interest Income				
Profit (Loss) Before Taxes				
State Income Tax				
Federal Income Tax				
Total Taxes				
Net Profit (Loss)				

In the next chapter, we will attempt to attach some numbers to first-year expense and investment items. But now let's go on to some other "must do" actions.

4. If you or your spouse type, fine. If not, look around for a part-time typist and determine what the going rate is in your area. Preferably, anyone you hire should do the work on your own premise. Be sure to invest in a good typewriter. Keep names, addresses, and phone numbers on tap for typists.
5. Open a business bank account and have checks printed with your business name on them.
6. Design some stationery and business cards. We'll discuss this in greater detail later.
7. Get a large-sized post office box. Don't permit personal mail delivered to your home to mix with your business mail if you can help it. Use your post office box number on stationery.
8. Obtain and maintain insurance coverage including life, accident and health, general liability, fire and theft, auto, and workman's compensation. Analyze your need for professional liability insurance. You may or may not require it. It can be very costly, although you may be able to purchase it more economically through one of your professional, technical, or trade associations.
9. Select a banker who can advise you on loans for business working capital and capital investments. Your banker may or may not be the institution where you maintain your checking account. Eventually, you may want the two banks to be the same so that they have a stake in financing you, besides merely making money by reinvesting your deposits. And never let a check of yours bounce.
10. Develop a business plan for yourself, at least for your first year. Keep reviewing it so it remains current, realistic, and flexible. A one-year plan can be quite specific and useful, and is more amenable to "managing by objective" than three- or five-year plans, which, particularly for the smaller business, are more educational than practical. We will talk about planning in a later chapter.

YOU ARE READY TO ROLL

You are now ready to launch synergistically your fledgling business. You have entered the ranks of the small business start-up—with one vital difference. Your knowledge and services are valuable and your business stands a much better chance of succeeding than just any old small business. Your profession has been proven over time. You are not functioning in a twilight zone, the way so many start-ups do. As a professional person, your aspirations are achievable, unlike the dreams of too many new businesspeople who start with high hopes but not much else.

At this point, don't worry about any local prohibitions on conducting a business from your home. For the moment, your business is more a gleam in your eye than a reality. There will be time enough to worry about this problem, if there is one at all. When you reach the point where you have a real zoning problem, you will be delighted to confront it, without doubt.

Except for the bookkeeping items mentioned above, the rest of your accounting requirements will be delegated to your accountant. Just sign on the dotted line whenever he tells you to—not blindly, of course. Even if you have no revenues for a while, you will be able to deduct start-up expenses and begin deducting the expenses of operating your office at home, auto costs, and so on. Find out from your accountant what he recommends that you start expensing. And find out what assets you should allocate to your business, such as typewriter, filing cabinets, desk and chairs, your present library, auto, engineering equipment, computer, calculators, and so on. These are all now legitimately depreciable items as business assets and, if you wind up with a business loss the first year or so, you can have a loss carryforward to apply to later years.

YOU ARE WEARING ALL THE HATS

As a combination engineering consultant and small businessperson, you also are your own systems manager. Believe it or not, your small consulting practice is a miniature big business. Although you are personally wearing all the management hats, you have to deal with all of the same functional areas, to a lesser extent, as the largest firm in your field. You are a giant corporation in microcosm. And remember: Every giant company used to be a little company.

You have to be both a specialist and a generalist. You are a specialist in your field of technical expertise. But you must be generalist enough to manage your specialty within the framework of the much broader dimensions of the business you run. Like any business executive, which you really are, when you have to function out of your primary area of expertise you may need help. You bring in consultants—your accountant, attorney, technical consulting associates to give you depth in multidisciplinary engagements and so on.

Eventually, running your little big business will become second nature to you. You will grow into it. There's plenty of time. Start now studying some of the more unfamiliar, or what may even be distasteful, subjects. For example, to many people writing and public speaking are anathema.

CHARACTERISTICS OF THE SMALL BUSINESS

Your body of knowledge must also include an understanding of what the small businessperson is really like. The economics of the small business and the personality of its owner(s) are revealing. They set the small business apart from the rest of the commercial community. The small business entrepreneur is not well understood by engineers either, since most engineers are professional employees, but employees nonetheless. However, if you understand the small businessperson, you will be in a better position to understand the consulting business—and yourself as an engineering consultant—and use this knowledge to your advantage. At a minimum, you will be alert to the many traps that await you if you are not careful. Following is an abbreviated list of small business economic and personal characteristics.

1. The small business is less likely to survive for any significant period of time than its larger counterpart.
2. The small business has poor access to the capital markets with respect to both working capital and investment capital, and the lack of capital in the small business community is legion. The owner or ownership group supplies needed investment capital.
3. The small business is less able to compete effectively than its large counterpart, with some exceptions.
4. The small business is more likely to be a "mom and pop" endeavor with several, if not all, family members active and involved.
5. The small business is notorious for its lack of business skills and know-how, particularly with respect to management acumen and expertise, although its owner may well be quite knowledgeable in the purely technical aspects of the product or service. The small businessperson typically survives through tactical (shorter term), rather than strategic (longer term) commercial maneuvering. Survival is the result of "mountain goat" or "crisis" management (i.e., shifting or managing from crisis to crisis).
6. The small business is more likely to service very local markets and trade.
7. The small business is more likely to be owner-operator managed. Ownership and management are identical.

8. The small business is almost never dominant either in its own industry or its community; it has unidentifiably small impact on local, regional, national, and international economies.
9. In the small business, there is seldom a discernible separation of individual operating functions and, therefore, there exist attendant multiple vulnerabilities with respect to the depth of technical knowledge available to it.
10. The small business is under great pressure to maintain price levels rather than discount merchandise or services; it must make its profits on smaller volumes and higher margins, whereas the larger business enjoys purchase discounts due to price breaks and can marshal higher volumes with smaller profit margins to its net advantage.
11. The small businessperson is under continual pressure to "cut corners," "buy cheap," and "sell dear"; survival is seldom feasible through "competitive pricing."
12. The small businessperson is susceptible to flexible working hours, resulting in relatively poorer discipline and adverse managerial results, although, typically, long hours are worked whenever required.
13. The small businessperson is prone to develop and maintain faulty working habits. There is no such thing as performance review and guidance by more experienced and successful superiors. There is total and utter dependence upon one's own resources, and being left to one's own devices in running the business. Decision making, if poorly thought out and implemented, is more likely to have catastrophic consequences for the small businessperson. There is a greater vulnerability to be "OBE" (Overtaken By Events). Most regrettably, the small business owner-operator gets precious little business management advice and is not usually in the fortunate financial posture to be able to pay what it costs to obtain it from competent professional management consultants.

CHAPTER 3

PREPARATION, INVOLVEMENT, AND GOOD HABITS THAT ASSURE SUCCESS

THE FIRST ORDER OF BUSINESS

Prospective candidates for admission to engineering school, as well as recent graduates, have one particularly vital factor to consider for career planning purposes that isn't obvious unless it is brought to their attention. Young people should analyze themselves with as much objectivity as is possible to try and decide whether they will be happier spending the rest of their lives working for somebody else or for themselves.

All professionals possess specialized knowledge that is valuable both to themselves and to society. There are only two ways to apply that knowledge—either as an employee or as their own boss. If one has an "employee mentality" and finds a durable, career-length position with some large corporation, that's fine. But for someone who has a freer, entrepreneurial spirit and recognizes it, then go into industry and acquire enough knowledge and experience so you can ultimately strike out on your own or else go on to higher levels of education. Get advanced degrees and your P.E. license so you can go into consulting as soon as possible on a meaningful level. Unfortunately, however, one does not learn how to be a consultant in college. Consequently, for most engineers consulting is not an option they consider right out of school, even if they have attained their doctorates. However, consulting is considered to be a viable initial career track in a number of other fields right from the start.

The important thing is to know yourself, and then plan your future (as well as anyone can plan the future, of course) based on who you really are. Importantly, professionalism starts at home. You and you alone control it.

SUCCESSFUL IS AS SUCCESSFUL DOES

Not every engineer in industry or in consulting will make it big or even be content. And not every engineer is a genius or has to be one. But believe it or not, not every lawyer, physician, dentist, accountant, or psychologist is going to be at the very top of his or her field either professionally or financially. However, you are missing the whole point if you think that becoming a part- or full-time consultant is desirable only because of the potential financial rewards or that it is for every single engineer or that every engineer who tries it will like it or be a success at it. However, you can be a consultant if you want to be. Financial success will find you if you function on the highest professional level—by design.

The engineer is capable of starting up a manufacturing business, becoming a manufacturer's representative, becoming a successful technical author, running seminars, publishing newsletters, and so on. And every engineer is inherently capable of having a private consulting practice. Unfortunately, the typical engineer is like a fish out of water because of the pressure to work for an employer that comes very early in one's career. It seems the easiest way—although not necessarily the best—for an engineer to find gainful work. Becoming a consultant is more difficult. It takes more creativity and initiative. It takes greater dedication. But only to the uninitiated does a job give the illusion of a "steady income," at least for as long as you get to keep it.

You don't have to be totally unique to be successful. Just be a hard-working, dedicated, motivated professional who is still learning, striving, and planning. As a high-paid employee, perhaps with great responsibility and even authority but also possibly discontent, the only thing that makes you different from the independent consultant is that, for whatever reason, he or she broke loose. It doesn't matter how or why. Perhaps you're lucky if it's done for you. That way, you don't have any choice and you don't have to agonize over leaving that good job.

Just do your own thing in your own way, with your own style, skills, interests, passions, objectives, and all the rest. You could be a highly successful consultant. If you are a graduate engineer with high credentials in your field, you can make it happen. And you can start today.

Whether at age twenty-five or fifty-five, any engineer worth his or her salt can make the transition. But you must have the courage to do so. If you prepare yourself properly for the consulting scene, you may almost be better off if you get fired or laid off. Otherwise, it may be impossible for you to break loose by your own decision, even if you are basically an entrepreneurial type with the highest professional credentials and a readily identifiable and marketable specialty. You may overestimate the risks in your own mind and do nothing. And if you discuss the prospects with trusted friends who are not entrepreneurial in nature, you probably will find them trying to discourage you.

BREAK THREE BAD HABITS

So drastic a remedy is not necessary, however. Start consulting part-time. The worst habit you can have that will hold you back from being a successful consultant—whether part- or full-time—is having an "employee mentality." Get rid of it. Daydream about being a consultant; about being your own boss. What would you do? How would you do it? Whom would you consult for? What are your technical interests? What do you need to get going? Fulfill your dreams. It can be done. Dreams are the stuff from which successes are wrung.

Having an employee mentality is not an inborn characteristic. It is something we acquire because nobody tells us there is another way. Some people do not learn that there is an option until too late in life. Some people do not learn that success can be attained in any other way. So the employee mentality becomes a way of life, and they never experience the exhilaration of forging ahead on their own, even though they may well have what it takes to do so.

The second worst habit is being too meek and submissive. We engineers are a curious lot. We are highly skilled at what we do. Our expertise is legion. And catastrophe frequently strikes with certainty whenever our best counsel is brushed aside or compromised. But many of us are not assertive. We react. We are unsure of ourselves because our employers won't always let us make the final decisions. We usually are in the unenviable position of being able to say "I told you so" when our advice is not taken—a hollow victory when lives and/or profits have already been sacrificed to nonengineering decision making and risk taking. We shut up. We are afraid for our jobs. Of course, we often become the whipping boy anyway and sometimes lose those jobs. And who among us is brave enough to be a "whistle blower"?

The consultant is the opinion holder of last resort, in a manner of speaking. While his or her best technical counsel may not be followed by the client, the engineering consultant, like the medical or legal consultant, maintains his integrity throughout. And for the most part, clients will indeed follow consulting advice. In fact, they rely upon it. So it better be good and it better be right the first time.

The consultant is in a position to provide "best efforts" solutions and opinions. The employee too often has to compromise his professional and personal integrity to humor an uncompromising employer who "owns" the bottom line. On balance, it is the employed engineer who must leave a job to salvage his or her professional integrity.

After many years in industry, if you are still idealistic enough to feel that the world can be made a little better place by your efforts, and if you feel that your particular employer doesn't know—and/or doesn't care—what he is doing to his customers or to you, maybe this is the time for you to resolve to go into consulting.

Break a third bad habit that typically afflicts new consultants. Most of us are very organized in our business lives, but much less so in our private lives. As an independent engineering consultant, you must now pay as much attention to the details of running a professional practice as to solving the technical problems that clients will ask you to address.

This may even mean that your business life will intrude on your personal life. So what else is new? As an employee, you are probably bringing work home from the office anyway. But now your office may even be in a part of your own home. And what does "intrusion" mean? Right now, the hour or more that you spend commuting to your job each way is creating a barrier between work and family. Is not the time you commute totally wasted?

The logistics of running a business from your home can be delightful—as many consultants will tell you. You can wake up at 5 or 6 A.M. and do a few hours' work while everyone else is still asleep.

Then you can have breakfast with your family and go back to work. You can even have lunch with your spouse occasionally. You can work into the early hours of the morning without necessarily being away from home. And your commuting is easy, inexpensive, and 100 percent efficient. When you have to commute from home to office, it takes exactly one minute or less per day.

How would you like to have two or more hours a day—every day of your working life from now on—to earn professional-level income dollars? $100 an hour? $150 an hour? $200 an hour? What is your professional time worth on the open market? What are you making now? Once you get established as a consultant, you can make an excellent income on the commuting time alone that you save every year: two hours a day, five days a week, forty or fifty weeks a year times $100 minimum.

Is it real? You bet. And you, too, can do it. Is it demanding? Does it have its moments? Of course, but so what? When you become saturated, you'll feel weary, numb, maybe even depressed. But most of the time you will be wildly excited, exhilarated; your "engine" won't stop turning and your "computer" won't want to quit. In fact, going back to working for somebody else, even in a high-paying job, will never even enter your mind again. There is absolutely nothing like the feeling you get working successfully for yourself.

Talented consultants on their way to significant successes in their fields say that they would rather make less money in their own businesses and practices than work for somebody else for a lot more. They mean it. Of course, you can also become your own worst enemy. You will be a taskmaster as far as your own work and schedule are concerned. You will demand perfection or close to it. You will have *control* over the quality of your work.

Since consulting is basically an intellectual pursuit, and because you are your own salesperson, production and project manager, engineer, and almost everything else (we'll talk more about this later), you will not be able to control when jobs come in and when they don't. You will no longer lead a "balanced" life (if there ever was such a thing). Nobody is going to hand you three jobs a day, every day, with a half day off on Saturday. It doesn't work like that.

DETAILS ARE CRITICAL

Most engineers are so intense about the technical aspects of their profession, and so intent on accomplishing a job and solving a problem that they consider it a waste of time to have to worry about the many details of running a business. Unfortunately, these details are what will make or break you if you stake out on your own. Don't let the details of running a business or consulting practice throw you. Everything is a compromise. Having to become a businessperson and spend whatever time it takes to do it successfully is well worth the effort. Don't look at it as a sacrifice but rather as another investment in your future.

Do whatever it takes to be successful in consulting. Once you get used to it, and once you begin to taste the rewards of private practice, not only will you never want to return to working for somebody else but you will also find that, while the business side of your practice may be time-consuming and take you away from your first love—the technical aspects of engineering—it will actually help you become liberated. This is because running a successful practice means running a successful business. The more successful you are at running the business side of your practice, the more successful you will be in having a practice to run for the duration.

BE BUSY EVEN WHEN YOU'RE NOT BUSY

Consulting can be a feast-or-famine business even for the most successful practitioner. On the other hand, the successful consultant knows how to make the best use of slack periods. And he or she always has many professional irons in the fire. You work off your backlog. You do research and study. You renew your contacts with old and valued clients. You do client development. You write books or articles. You attend seminars in Hawaii or Florida. You take your spouse to Paris. Maybe you'll even take him or her out to dinner occasionally. You do everything *except* worry. There's no need for it; just understand how the business works.

There are lots of things you can do during slack periods. As a matter of fact, you may well opt *not* to go on vacation. Chances are you will continue to get telephone calls and letters. You want to be around. Of course, these kinds of things apply primarily to the full-time consultant. The part-time consultant will always be busy for the predominant part of the week because some employer feeds a steady flow of work to do.

For the full-time consultant, slack periods can be rewarding in a number of ways. Once you are totally immersed in your full-time consulting practice, you may even welcome those occasions during your own business cycle when things lighten up a bit and you can do some heavy thinking about the future course of the practice and new fields to conquer.

In this connection, too, if you should become somewhat established in a full-time consulting practice, don't let your first light period devastate you. The relatively new consultant, if he doesn't know better, will undoubtedly feel that he or she has been doing the whole thing wrong and panic. But this is just the nature of the consulting business.

Even successful attorneys complain that sometimes they sit in their offices and think that nobody likes them any more. Dentists and other types of professionals have been known to say the same thing. So don't fret. Every business and every style of doing business ultimately has its own pattern of peaks and valleys. Believe it or not, July and part of August can be the busiest times of the consulting year, depending upon the type of practice you have. On the other hand, September can be a light month—sometimes. You have to discover for yourself what the business pattern is. As we will see later, this is also important in terms of cash flow planning, analysis, and control.

The point is, don't permit yourself to go into a state of shock every time you have a lean period and little backlog. Keep the juices flowing. Keep up client development. Do all the things that should lead to inquiries sooner or later—they will.

The trick is always to keep client development activity at a high pitch. Marketing your consulting services is no different in this respect from being an insurance, real estate, or securities salesperson. Again, even lawyers, accountants, and other types of professional people—as well as all other types of consultants—have peaks and valleys in their lines of work. Physicians are in a different category for the most part, since the medical profession has been able to successfully limit the number of practitioners in business—so far.

Time management is thus a necessary skill to acquire and cultivate. A good simple rule is "First things first." And don't put off until tomorrow what you can do today. This may sound terribly trite, but it works every time. You might as well utilize that storehouse of energy and stamina you still have left even after working all day. You will be surprised how invigorated you suddenly become when you set up shop for yourself, even on a part-time basis, and do a little bit of planning whenever you can. Manage time by organizing yourself.

Another bad habit to break is "giving up." Just keep up your client development activity until it begins to work for you. Play the percentages. The more contacts you make, the more people you talk to, the more times you run your ad (we'll talk about this later), and the more PR packages you send out (more about this later also), the more chances you'll have to do business. Be patient. Be cool. Be tenacious. It will work for you.

Wouldn't it be interesting to see the entire engineering profession become a community of high-powered, high-earning, highly dedicated, highly responsible, highly regarded consultants? This would not occur by artificially keeping engineering school enrollment low so demand exceeds supply like in some other professions. Instead, it would take place through the realization by all engineers that their skills are valuable and vital, and that the only way they can retain their professional integrity and fulfill their mandate to protect and promote the public safety, health, and welfare is to become independent.

BREAK MORE BAD HABITS

Besides having an employee mentality and some of the other bad habits we have talked about, here are some other things to avoid that could keep you at ground zero:

1. Sitting in your office all day waiting for the phone to ring
2. Being "me"-oriented instead of focusing on client needs in an empathetic way
3. Not treating consulting as a business but as a pleasant and stimulating, noneconomic hobby or as an escape
4. Not managing by objectives (MBO)
5. Giving away free advice and having free meetings as the general rule
6. Being afraid to discuss money and fees for fear you will not get an assignment or engagement
7. Being afraid to ask for fee balances due
8. Renting the biggest office you can find, more computer capability than you need, three full-time secretaries, and trying to create too much impression too soon
9. Practicing out of your specialty
10. Sending a hastily and emotionally composed letter on a troublesome subject without first "sleeping on it"
11. Failing to follow through
12. Throwing everything away every two years (you must keep everything forever!)
13. Investing both time and money in a project or engagement (invest time or money, but not both)
14. Failing to return telephone calls on a timely basis

Preparation, Involvement, and Good Habits that Assure Success   37

15. Procrastinating, procrastinating, procrastinating
16. Taking an apodictic approach to things (i.e., being a perfectionist to a fault)
17. Being afraid to take even nominal risks
18. Not learning from your mistakes
19. Failing to take stock of things periodically to perceive patterns and trends
20. Resisting change
21. Permitting a distaste for travel to interfere with consulting opportunities
22. Remaining overly introverted
23. Persisting in the belief that you don't have to be nice to people on the way up because *you* are not coming back down
24. Being too timid to take the initiative when it is precisely your initiative that the client wants
25. Failing to listen continually and sincerely to both verbal and nonverbal messages being sent out by clients on the phone, in meetings, and so on
26. Permitting yourself to bend to unethical demands of clients, for whatever reason, so that you compromise your professional integrity
27. Expecting that you can work "banker's hours" (the impossible dream)
28. Not being willing to take full responsibility for the quality and responsiveness of your work
29. Failing to study the dynamics of your own business and those of your clients
30. Putting your personal interests above the needs and reasonable expectations of your clients
31. Failing to become involved in your technical and trade associations on a meaningful basis
32. Failing to be a self-starter
33. Failing to know when to forgo an advantage as well as grasp an opportunity
34. Being too greedy
35. Losing your grip on yourself and running scared when you are merely going through a phase of your business cycle, thereby not being adequately selective in accepting engagements, not screening prospective clients adequately, permitting clients to intimidate you into quoting low fees, and so on

CIRCULATE, LISTEN, AND LEARN

As you begin to circulate in your professional and trade associations with your new incentive to keep your eyes and ears open for consulting opportunities, identify professional colleagues who are already in the consulting business. Tactfully discuss with them the bad habits that have to be overcome, perhaps that they themselves overcame, and the skills that have to be developed. Chances are that every consultant will tell you some of the same things. However, there is also a good chance that each will have his or her own pet list of dos and don'ts. As mentioned in Chapter 1, even these brief and chance meetings can open your eyes wide, as well as let people know that you may be available for occasional consulting assignments in your specialty. And remember, all of us have a list of don'ts that we have done and maybe even still do. Learn from us. You know the old saw: "Do as I say, not as I do."

Don't become unstrung or disenchanted just because some colleague in consulting seems to be unhappy. Listen to what people say. Make notes later. Analyze. Discuss apparent problem areas with other consultants. It's all in the mind. Think positively, constantly. There's a difference between acknowledging an occasional setback and having a totally negative outlook. Don't be negative. Think positive. Think success all the time.

Productivity is extremely important in consulting. However, you are an engineer. Nobody should insult you by telling you how to be productive. If, as an engineer, you don't know by now, then nobody can tell you. Likewise, as an engineer you have been trained to be creative. In fact, the engineer with even ordinary skills is capable of being highly creative in a technological sense. In engineering, novelty is the norm . . . by definition and by design.

Engineers frequently feel deficient in people skills, communication and marketing skills, and negotiating skills. Engineers often tend to be introverted. This is less true now, since engineers more frequently are becoming drawn into areas of business and situations where increasing interactions with diverse types of people are vital. In addition, the engineer who comes out of the technical shell will automatically become more communicative. Interactions with people demand that we communicate. The biggest obstacle to communicating is an unwillingness to try.

Many formerly introverted engineers handle themselves well when they become active in community affairs. Sometimes this even leads to political involvement. There is nothing magical about communicating or becoming communicative. Just do it. Don't be afraid of it. Don't be afraid of people. We engineers have as much, if not more, to contribute to the goings-on about us than many. Communication includes public speaking and writing. Don't be intimidated into being silent.

YOUR SUPPORT SYSTEMS

Your most important support system is your family. Even if you have your office in the house, which is very likely if you consult on a part-time basis, take your spouse into your confidence. It will be an adventure for both of you. You may even get some new ideas out of it. If nothing else, you will have someone else to talk to. This will relieve the inevitable stress that occurs in a consulting practice for any one of a number of reasons. Even if everything is going great, you may find yourself so tense and stressed that you need an escape. Your spouse offers the best emotional support system to help you relax. Even a talkative teenager or a trip to the supermarket can have a beneficial effect. Take delight in both. It gives you back your "grass roots."

You should also find a mentor. This can be very important because such an individual will be in the same general field, will have been through the mill, will probably have been reasonably successful, and knows his or her way around. You can discuss the more delicate problems with such a person. Try to identify such an individual.

It probably isn't a good idea to expect to find a mentor among potential competitors. It can happen this way, but it is unlikely and you probably shouldn't seek this kind of relationship. It will become strained very fast if your own practice begins to take off because of the inevitable competition for clients.

YOUR HOME OFFICE

You need an office. It can be in your home. After fifteen years of successful engineering consulting practice, mine still is. But it must be your personal domain, large or small. Privacy is one absolutely essential ingredient for conducting a successful consulting practice—or any other business, for that matter. Appropriate a den or get a movable partition and use part of your family room. Working out of an apartment can be a little more difficult. But when you are working, the rest of your family must know it is serious business and to leave you alone.

As you graduate to the higher levels of consulting income, you will need more space. I moved from a very comfortable but moderately sized home to a significantly larger one in the same town when I started to stumble around my office because of the clutter. It is now starting to happen again.

It is financially feasible for me to move my office out of my residence and/or move again to a still larger home. But I don't want to. A consulting practice is exceptionally demanding—as is being in any kind of business for yourself. Working out of your home as long as you can, or even in a converted garage or porch, can be delightful and provide just the margin of relaxed atmosphere that you need on those nonstop days that every consultant approaches with some trepidation. Nonstop days have a way of turning into nonstop weeks or even months. Once your practice takes off, those nonstop months may turn into nonstop years. It can and does happen.

Of course, if your practice is equipment-intensive (such as if you need a lot of computer gear or you are into some laboratory-based activity, or you need a large supporting staff physically located in or about your office, including technicians, secretaries, clerks, draftsmen, etc.), you may have no choice but to move into an outside space relatively quickly.

In my case, my only employees are secretaries who also work out of my office-residence. My consulting associates are all part-time subcontractors who also hold responsible and lucrative full-time professional or executive positions in industry. Some of them even have their own businesses but enjoy the challenge of an occasional consulting engagement.

PIPELINES AND WINDOWS FOR SUCCESS

Whether you work out of an office-residence or an outside place, you must quickly establish disciplined ties to the outside world—that "real world." You need "pipelines," and you need "windows." Up to the time you become interested in consulting, you may not have been active in your technical, professional, or trade organizations, but this is absolutely vital now. And don't stop with one or two organizations. Join or rejoin every single organization that can add to your contacts, knowledge of your specialties, and knowledge of what is happening out there. It is no longer sufficient to lead a cloistered, sheltered life that too many employers encourage in both obvious and subtle ways so that employees live, breathe, and die with the company name on their lips. Was this living? Was this being an independent professional?

OFFICE HARDWARE AND SOFTWARE

Get yourself a businesslike desk and chair. Don't just use some makeshift table or shelf. You'll need at least the equivalent of one five-drawer filing cabinet to start. You can buy one used. Be sure you have a good typewriter with a businesslike type face. Don't get fancy with type faces. Even a new IBM electronic is now under $1,000.

You will want to keep separate filing systems on the technology in your field, your personal contacts, actual engagements, administrative and accounting matters, and so on. Don't lump them all together. Organize yourself right from the start. Just think it through. It's easy, but you have to plan and follow through. Get lots and lots of file folders.

You'll need some storage for stationery, supplies, measuring and photographic equipment, other field-related apparatus, and so on. Buy yourself a small copier. It is deductible. And don't forget to get a business telephone number and listing. *Don't* let your residence telephone also be your business number. And get an answering service. *Don't* use an answering machine. Many people hate them, especially prospective clients.

You may already have a PC. If that's all you need, fine. If you need greater computer capability, try to rent time from someone or some company nearby where there is an installation that meets your needs. Keep recordkeeping simple.

Keep a telephone log in the form of a 250-page spiral-bound notebook. *Don't* use little slips of paper or index cards or bits and scraps of anything you can find to take down telephone messages. Your telephone log is one of your most important tools. I have been keeping one faithfully for fifteen years. It is worth its weight in gold. Retain the log book permanently.

For every message from your answering service, telephone call you take yourself, or message a secretary records, make a separate entry. Always start every new day at the top of a new page and record the time, the name of the caller, and a brief description of the inquiry. Record fee quotations.

You can't think straight when you are faced with a bunch of little scraps of paper. But with a permanent log book system, you are treating telephone calls as important items of business data. They are very efficient for note taking, analysis, priority setting, reviewing to find a certain caller, or to determine the date of a call. There will be no more lost phone numbers or contacts. Although you might forget to make a record of a phone call when you answer the phone yourself, if you train your secretary to use the log book faithfully, it will at least be 95 percent effective.

BE A "CLIPPER"

Clip magazines, newspapers, newsletters, mailing pieces, and everything else that is relevant to your work that comes across your desk.

Clipping can get to be an inexpensive hobby of sorts. All you need is a sharp pair of scissors and a whole bunch of files. Reading, scanning, and then clipping articles on subjects of interest to you will help you feel involved and current. You'll feel that you're not losing anything of value that has hit the media—of value, that is, in relation to your own particular interests and needs.

Clipping also extends your reach. It forces you to think. It forces you to evaluate. It's a way of keeping your mind whirring. It's a way of keeping interested. It's a way of cross-indexing everything that happens in the whole world that is related to your own little world of consulting, as well as personal interests.

Professional people tend to be born collectors. Collecting has its good points and its bad points, but it's always interesting and relevant.

A word of caution, however. Your library and clippings files, utilitarian as they will prove to be, can grow, and grow . . . and grow.

KEEPING TRACK

Many types of consulting practices are one- and two-person businesses. They are somewhat akin to "cottage industries." Therefore, although the work content of the business may be highly sophisticated, the business as a business is usually relatively small as enterprises go—well under, say, $1,000,000 a year in gross revenues, and most of the time, under $500,000.

However, you still have to keep track of what is happening. And you have to manage your resources. Actually, the most important and difficult thing you have to manage is *you*. You may not have to run elaborate statistical analyses of performance, but you will develop a series of files that represent the lifeblood of your practice. Just as revision blocks on engineering drawings represent and document cumulative know-how, the cumulation of casework reports is one of your most valuable resources and assets. It reflects progressive development of your thinking in a multitude of problem areas; it is an inventory and catalog of problem solutions; it houses individual collections of research studies on specific subjects and projects; it represents the distillation of the literature on numerous topics; it is a resource you are constantly applying, reapplying, and building upon; it is rich in both accomplishment and potential. It nourishes you.

Therefore, you should be as careful as possible in how you set up and cross-reference your files. Cross-referencing can be very troublesome. The major problem is that you are only one person and have only so many hours in a day. There are some things you can't delegate very well even to a skilled secretary.

For example, I have many file drawers full of machinery-type accident investigation reports. There are numerous subdivisions within this single broad category. They cover practically every type of equipment you can think of. I have another file section on cases relating to slips, trips, and falls. Again, this is divided into numerous subcategories. And there are file sections on numerous other major casework types, with section classifications including automotive, fires and explosions, electrocutions, toxic and flammable substances, economic and feasibility analyses, and psychological studies.

Within the total filing system, and within each subject category, cases are physically filed by client name. But exactly where do you file a report relating to a stairway trip and fall where the "stairway" is on a large rolling mill, truck-tractor, cargo ship, construction machine, swimming pool, amusement ride, or the like? There are separate file sections in each category.

Or suppose that an in-ground swimming pool collapses due to a construction or maintenance defect and inundates the property next door plus the adjacent street. A child riding a bicycle skids, falls, and is hurt. We get called in by the child's attorney. Where should this be filed? Under bicycles, swimming pools, or construction? Again, there are file drawers for each of these categories, with subdivisions.

Suppose a child is killed while riding a tricycle. The case involves an automobile accident. Where should the report be filed? Under bicycles and tricycles, or under automotive accident reconstruction?

Or an industrial worker suffers severe toxic effects from inhalation of chemicals used in a process plant. The involved equipment is part of a printing press system. During the course of the accident there is also a fire, since the substance is flammable. Where would you file this information? Under fires and explosions, some machinery category, or industrial hygiene casework? And so it goes.

Since you don't hire either an engineer or another secretary to do cross-indexing in the smaller practice, you simply go on to the next problem. The important thing is that there is a method underlying my filing that has served reasonably well and is affordable. I "lose" about a dozen files a year and sometimes spend up to an hour trying to find one of them. But since I never throw anything out, I know it can only be in five or six places. It may be misfiled by category or out of alphabetical order, but it always eventually gets found.

Keep one filing system on engagement reports; another on engineering, scientific, and management topics; and a third on matters relating solely to the business side of your practice, including taxes, expenses, office equipment, bills received, invoices rendered, insurance, and trial and deposition fee letters.

One convenient file drawer should be for maps of states, counties, and cities visited often, with several copies of each. Setting up an itinerary is much easier if you work off detailed area maps. Surprising as it may seem, not everyone does this.

PROJECT INDEXING

Project or casework files can also be indexed by job or project number. However, this may be an inefficient way to file if you have to access your accumulated "wisdom" in a particular area of problem solving frequently and if you also have lots of files. If I have a case involving plastics injection molding machinery, I may want to review previous casework on this subject. If files are indexed and physically filed by project number alone, I would have to spend too much time retrieving individual files on the particular equipment type. Since I have thousands of files and literally a whole roomful of file cabinets, this would simply be too much of a chore. Hiring a file clerk would be counterproductive from a financial standpoint.

For your particular needs, the best compromise may be to have complete physical control over the filing system and forgo a numerical filing system that would immediately require meticulous cross-indexing. When you first start out, of course, you may be able to use a numerical indexing system since you have relatively few files. You'll have to make a transition later, however.

One of the fascinating things about running a small business is that, as complicated or sophisticated as it may be in the engineering fields, if you are alert to the things you should be doing as "chief cook and bottle washer," you will sometimes find yourself thinking along thirty-seven different avenues all at once. This can be invigorating if you don't get numb first. What does this problem mean to your overall grasp of the subject? What does it mean to the client? What does it mean to you in terms of follow-up business?

If you take this engagement, what investment does it require in time and money? Does it fall within your capabilities? Does it offer the potential for growth? Will you have to call upon consulting associates? How did this new client come to you? Was he referred? By whom? Is your advertising still effective? Should you expand or modify it? If you set up this itinerary, what opportunity do you have to visit that client, or this prospect, or some site or plant where you have done work and want to see job progress firsthand or the results of your recommendations?

If you bring a new consulting associate aboard, what are his or her strengths and deficits? Can you do some additional type of work? What is the target market universe? How do you get to it? What and how do you charge? What kind of sales, advertising, and marketing efforts will be necessary? Can there be real synergy? Do you need even more expanded or diversified expertise to service the new market opportunities suggested by his or her presence as an associate? What will it take in dollars and cents? In cash flow? Can the associate travel? Does he or she have a full-time job? How much work can he or she actually do before it starts to interfere with his or her regular full-time job? What will be the demands on your own time? Is your technical expertise required or just engagement management time? Who would be most effective in interfacing with the client? What are the obstacles? The pitfalls? How do you overcome them? And so on.

How do you contact this or that potential client since he or she is known to be very difficult to reach through normal channels? Do you have contacts who can help you? How can you improve your "networking"? Can you fit this or that conference, seminar, or convention into your schedule? Who will probably be there? Can you do a little bit of client development? What should you take with you? Should you design some special material? Should you take a booth at a convention? How much will it cost?

Or maybe every one of these questions will be going around in your mind at the same time as you skip from one thought to another. Solutions begin to appear. Keep a tape recorder with you at all times. Should you have a telephone installed in your car? Should you get a pocket pager?

You will not find a need for luxuries such as a private pager or automobile telephone at the beginning, and you may return to your office and find extremely important and urgent telephone calls waiting for you. But when you are in your car, you can do a great deal of important and sometimes creative thinking. It is about the only time you will be alone except for weekends. In any event, you will almost constantly have numerous interdependent problems—and solutions—on your mind.

TAKING STOCK

Your filing system constitutes your own private archives. Besides using it for business development, you can use it for self-analysis and evaluation of your progress. You must periodically take stock of where you are. "Inventory" your resources and assets, new and old capabilities, your sheer capacity to handle existing and anticipated casework loads, your potentials in terms of practice development, and soft spots. Create your own "red flag" managerial control system. You must make a disciplined effort to eliminate soft spots and capitalize upon strengths. It is easy to bend with the wind and wait for opportunity to come knocking. Surprising as it may seem, it won't.

Determine what your own business cycle is. When are you the busiest? The slowest? Where do your clients go on vacation? When? When do they go to conventions? Where? Should you go to some of them? What time of day do you receive the most phone calls and, therefore, when should you be sitting at your desk to receive them? What day of the week is heaviest for you? Are these patterns constant from year to year? What day is slowest during the week? When is the best time for you to go on vacation? To take off a day during the week? Build all of these seasonal and other patterns and cycles into your planning, including periodic tracking of cash flow needs, even if only informally. Bank deposit slips can be used for this purpose.

REVELATION

A traffic engineering consultant who now has a very successful practice says when he was just another engineering employee with a large consulting firm he almost never interacted with anybody except other engineers, and then generally only professionals on his own level. Once he broke loose and set up shop for himself, he was talking to business owners, executives, governmental officials, executive engineers, and other more highly placed and successful people. To him, this was a revelation. It was mind-expanding, exhilarating, and made his life fascinating and worthwhile. It exposed him to numerous other opportunities. He was able to exploit his considerable professional expertise and obtain many new and profitable assignments. And he has made lots of money and has a sizable organization now. He started out working for a medium-sized consulting firm.

EFFICIENT IS NOT EFFECTIVE

We move ahead by keeping track and taking stock on an ongoing and disciplined basis. Build solutions, rather than problems, into your plans. Again, remember that being efficient is not the same thing as being effective. If you incorrectly define a problem, you will come up with an ineffective, although possibly very efficient, solution. Always make certain that you are identifying the real problem.

As you grow, you will find that you are also improving your judgment. In fact, the three most important things in professional life are "judgment, judgment, and judgment."

You will have to discriminate among apparent opportunities that only look good superficially, and those that have real substance and potential. Engineers in private practice are forced to consider the financial and economic consequences of their every move.

The engineer is part scientist, part businessperson, part humanitarian, and part artist. The engineer must learn to juggle efficiently multiple demands made upon his or her skills. Maybe you are already doing problemsolving on a very high level. Consulting gives you an unparalleled opportunity to take on as much high-level work as you can handle and of which you are capable.

In fact, as you progress in consulting you will find that you progressively move on to higher and higher level engagement quality. The scope of assignments you feel comfortable accepting will expand. Of course, you may need the assistance of consulting associates, but this is a true sign of progress. Your effectiveness will be expanded as your leverage increases through use of associates and employees. However, always beware of efficiency without effectiveness. This is a much more insidious problem than effectiveness without efficiency.

CHAPTER 4

HOW TO LOCATE, SIZE UP, SELL, AND KEEP CLIENTS

ATTRACTING CLIENTS

Finding clients is a high-priority task, but then you have to get them to sit still long enough to listen to you and think of you in connection with their own problems. It's not difficult, but it takes time and perseverance.

You have a specialty. As a consultant, you're working for yourself. Not every company wants—or can afford—a full-time specialist on staff. There are prospective clients out there who will be delighted to pick your brain on a part-time basis for a fee, and hopefully more than once.

If you have unique or highly specialized knowledge or experience, or if you have an angle that makes your knowledge or service novel, you can probably command a premium for your time. However, if, as is more likely, you have a specialty that other professionals also have, you must differentiate your service from that of your competitors. Of course, aggressive client development may get you to prospective clients before your competition. This, too, is a plus, but it takes planning and dedication on your part.

Marketing consists of packaging yourself and your services in an attractive, professional way that will make a prospect want to contact you and find out more about you. Typically, the prospect will just pick up the telephone and call. So be there, even if it is through a telephone answering service. *Don't* use an answering machine. This turns off a lot of people. Monitor the performance of your answering service. Such services periodically don't pick up in time and let callers ring nine or ten times before they finally get frustrated and give up...and maybe call another consultant.

Do these things immediately:

1. Prepare a standard resumé just as you would for a headhunter and general job search. This is your basic document. If you have one now, update and refine it.
2. Review, rethink, and recast your resumé into a 300–350-word, single-paragraph narrative that will fit on one page.
3. Next, prepare a blurb about your overall capabilities and activities that is approximately 500 words long and will fit on a single piece of paper.
4. Prepare a separate summary for each subsidiary specialty or activity in which you can competently and competitively engage, each about 500 words long.
5. Design professional-looking stationery and business cards. Design a logo for yourself. (you may want to copyright it). For second sheets and report paper, design a ruled page with margins, plus a cover or title sheet for reports, brochures, and so on. A local printer can help you if you are not artistic.

Whether you have it run off on a copier and stapled or done up professionally, this is what you need to put together a capabilities brochure. If you don't type, hire a typist. Have it typed on an IBM typewriter or equivalent with an Elite typing ball, not fancy type face styles, and with *no mistakes*. Run off 200 to 300 sets for starters. This preliminary package is something you can send or hand prospective clients. You'll probably be modifying it and polishing it up as you go along. After over fifteen years of consulting practice, I am still changing my resumé and am ready to redo an expensive brochure I had printed not so long ago.

The foregoing exercise, if you've never been through it, forces you to look hard at yourself, decide who you are both personally and professionally, and reduce this to writing in such a way that your prospective client universe can decide if you will be of any help.

Many years ago, I developed just the kind of package described above. It works. I continue to use it, update it, and rely on it. Probably no two consultants practice in an identical manner even with respect to what outwardly might appear to be similar, if not identical, specialties and client universes. Therefore, this is an opportunity to differentiate yourself from your competition. The materials I developed and utilize should be instructive to somebody just starting out and perhaps even to anyone who already does a significant amount of consulting but has never perceived the need, or taken the time, to create a public relations package. It will help you start up in

**Biographical Profile
R. Matthiew Seiden**

R. Matthiew Seiden, PE, CPSM, CSP, CHCM is a principal of The Seiden Group, Inc., product safety engineering, management and economic consultants. Mr. Seiden has had wide experience in private industry in engineering, planning, financial and executive posts. He holds the National Engineering Certification and is a graduate of Columbia University with degrees in Mechanical, Industrial and Management Engineering. He is a Registered Professional Engineer, Registered Safety Engineer, Certified Safety Professional, Certified Product Safety Manager, Certified Hazard Control Manager, Certified Standards Engineer and a Designated Appraiser in the field of human capital valuation. A member of numerous professional, trade and technical organizations, Mr. Seiden has lectured in managerial economics, forensic engineering, product safety and safety engineering. He is a founder and Director of the International Product Safety Management Certification Board, a past member of the Product Safety Management Academy Advisory Board, Past Director of the International Hazard Control Management Certification Board and has served on the State of New Jersey Carnival & Amusement Ride Safety Advisory Board. Past President of the N.J. Chapter of the American Society of Safety Engineers and Past Chairman of the Joint Committee on Product Safety & Liability Prevention of the American Society of Mechanical Engineers, he is also Chairman of the Product Safety Committee and a Trustee of the Essex County, N.J. Chapter of the National Society of Professional Engineers, and Chairman of the Engineering Division Product Safety Committee of the American Society of Safety Engineers. Mr. Seiden was also the recipient of a National Merit Award in the 1980 Hartford Insurance Group Loss Prevention Awards Competition. He is the author of: *Product Safety Engineering for Managers—A Practical Handbook & Guide*, published by Prentice-Hall, Inc., Englewood Cliffs, N.J. (1984, 438 pps.), which has been offered as a featured selection of its Engineers and Designers Book Service. Other book projects are in progress.

P.O. Box 99
Livingston
N.J. 07039
201-992-4788

ACTIVITIES PROFILE

the consulting business, or else move to a higher plateau in your existing practice. The image you project should be real, or at least you should strive to attain the image you want to project.

When you have developed your capabilities brochure, always keep it readily at hand. Redesign it, change it a little or a lot, keep it current, and keep it fresh. Try to make it innovative. Evaluate it in the light of each new engagement. Don't skimp on cost. It can be the single most important document you produce in your practice. Don't forget that this presentation package is *you*. Figures 4-1 through 4-5 show various materials I designed for my own use. You might find them helpful in custom designing your own brochure and report formats.

Activities Profile
The Seiden Group, Inc.

Since 1973, The Seiden Group, Inc. (TSGI) has conducted thousands of engagements related to negligence and products liability litigation for defense and plaintiff counsels, insurance carriers and claims adjusters, and governmental agencies. These have included building inspections, machinery and equipment guarding investigations, industrial workplace and construction jobsite occupational injuries and toxic exposures, consumer product safety and health, fires and explosions, electrocutions, economic analyses, human factors engineering, psychological assessments, and a wide variety of other safety, health and environment-related problems in industrial, commercial, residential, construction, institutional, agricultural, transportation, military, maritime and infrastructure settings.

TSGI activity has remained predominantly forensic in character. TSGI Consulting Associates are specialists in their respective fields and professional disciplines. They possess appropriate registrations, certifications, designations, academic credentials and/or other qualifications. Engagement management in the more complex cases is based upon a multi-disciplinary team or systems approach to scientific accident reconstruction, failure analysis and technical problem-solving.

In the selected fields of engineering, science, technology, management, economics, psychology, architecture and commerce most frequently encountered in its forensic casework, TSGI maintains one of the largest private libraries of older safety-related books and other reference materials. Many of these volumes date back to the turn of this century. They constitute an invaluable resource and asset in forensic practice for purposes of establishing state-of-the-art vs. state-of-the-trade on an authoritative, retrospective basis. TSGI is also expert in the application and interpretation of safety, health and environmental codes, standards, regulations, practices and procedures.

Other services for lawyers, insurers and manufacturers include seminar and training program offerings, technical monographs on safety, health and litigation related topics, and design and development of corporate product safety management systems, procedures and organizations. A major textbook titled: *Product Safety Engineering for Managers: A Practical Handbook and Guide* by R. Matthiew Seiden, President of TSGI, is available from Prentice-Hall, Inc., Englewood Cliffs, New Jersey.

We look forward to the opportunity to be of service to your organization in forensics, education, and/or management consulting in the aforementioned areas of specialization.

PROSPECTING

For the moment, assume that you have thought through what it is you are capable of offering in the marketplace, the kinds of needs that exist, the kinds of prospective clients who have these needs, and the alternative ways to reach them.

Once you have designed your consulting brochure, you are in a position to get out there and pound pavement. This, too, is not difficult, but it is time-consuming. You know you have something to sell. Don't be afraid. Don't be timid. Knock on doors. Place inserts in newsletters published by your trade or professional associations. You can advertise in dozens of ways, so *invest*. But don't be penny-wise and pound-foolish. The bottom line is: Talk to people and get them to talk to you.

Early in your programming, you must take the time to pick up the telephone and/or make cold calls. Don't procrastinate. It is an

Risk-utility & cost-benefit analysis
Rope, cable & chain failure

Safety codes, regulations, practices,
 procedures & standards
Safety equipment & personal protective
 gear
Schoolroom & shop equipment
Scientific accident reconstruction
Security equipment & systems,
 vandalism & intrusion
Service industries machinery & equipment
Specialties include engineering,
 electronics, metallurgy, psychology,
 economics, engineering economics,
 architecture, chemistry, industrial
 hygiene, management science, real
 estate, computer science, others

Toxic & hazardous substances
Toys & novelties

User error, failure & oversight

Vendor failure
Vocational, employability, rehabilitation
 & psychological assessment

Warnings, signs, labels, instructions,
 manuals, pictorials, multi-linguals

Your Inquiries Are Invited in Strictest
Confidence
Comprehensive 10-Year Index Available
upon Request

THE
SEIDEN
GROUP
INC.

Consulting
Engineers

THE
SEIDEN
GROUP
INC.

Consulting
Engineers

P.O. Box 99
Livingston
N.J. 07039
201-992-4788

CONDENSED INDEX

exciting and potentially profitable experience. Find out whom to talk with in the organizations you contact, if you don't already know. If you are courteous to a receptionist, telephone operator, or other employee, he or she can be tremendously helpful in guiding you to the right person or department. Speak with your peers in different organizations. Most of the time, they want to see a consultant succeed. Each fancies him- or herself a successful consultant, too. Therefore, if they can help *you* make it, there might be hope for

TSGI Condensed Index to Selected Product, Subject & Professional Specialty Casework Categories

Agricultural equipment
Automotive & other transportation equipment, systems, products & accessories

Bathroom, kitchen & other household appliances, equipment & fixtures
Bicycles, tricycles, kiddie vehicles
Burns and scaldings

Carnival & amusement rides
Chemical process equipment
Clothing & textile products
Construction defects & failures
Construction machinery
Consumer products
Controls & instrumentation

Design & manufacturing defects
Dusts, fumes, gases, mists & vapors

Elderly & handicapped safety hazards
Electrical appliances & equipment
Electrocutions
Elevated workplaces
Elevators & escalators
Engineering economics
Environmental hazards, pollution control

Failure & fracture analysis
Fasteners & adhesives
Fire, smoke, water & transportation damaged merchandise & equipment
Fires & explosions
Flammable substances
Floors, stairs, ramps, landings & other walking & working surfaces

Heating, ventilating, air conditioning & other building equipment
Homebuilding materials & products
Home workshop hand & power tools & equipment
Household products & equipment
Human factors analysis

Indoor & outdoor maintenance equipment & products
Industrial, consumer & commercial guns
Industrial workplace & construction jobsite safety & health
Interlocks, quick-releases, captive devices

Juvenile furniture & other products

Laboratory apparatus
Ladders & scaffolds
Lost earnings & wrongful birth & death economic valuation

Machinery guarding & failure
Management error, failure & oversight
Materials & cargo handling equipment
Materials & workmanship
Misuse & modification
Multi-disciplinary engagements

Office furniture & equipment

Packaging, containers & closures
Patent infringement, trade secrets & unfair competition
Playground, gymnasium, exercise & recreational equipment & facilities
Product misuse & abuse
Product safety management
Product safety training programs & seminars
Product servicing failure
Prosthetic devices & rehabilitation products & equipment

---

them. Also, if you are really good, they can learn from you. They might even be heroes. Don't be afraid to give away a little bit of useful information. It's a sort of "loss leader" type of public relations, but it is PR nonetheless.

Right from the start, be totally open in your relationships with prospective clients. There are other schools of thought on this subject, of course. The dominant one probably is that you shouldn't give away something of value for nothing. As a practical matter, however, reference materials. Many of these volumes date back to the turn of this century. They constitute an invaluable resource and asset in forensic practice for purposes of establishing state-of-the-art vs. state-of-the-trade on an authoritative, retrospective basis.

TSGI also offers in-house training seminars on product safety engineering and management for insurers and their assureds, and corporate, trial and patent counsels.

Your confidential inquiries are invited at no obligation.

P.O. Box 99
Livingston
N.J. 07039
201-992-4788

FORENSIC CONSULTING

giving away a little bit of knowledge or experience can begin to make you useful to a prospective client. Importantly, the client may begin to depend on you and to consult you.

In my own practice, when prospective clients call, for example, I am happy to screen problems at no charge, which usually requires no more than five or ten minutes of my time. This is a fantastic type of ongoing client development. Over the course of an average week, for example, I might spend many hours in such discussion. I also get

Forensic Consulting Services
For Lawyers & Insurers

Since 1973, The Seiden Group, Inc. has conducted thousands of negligence and products liability investigations for lawyers and insurers. TSGI analyses and opinions have been instrumental in proving damages and obtaining favorable jury verdicts and out-of-court settlements of many millions of dollars in meritorious plaintiff lawsuits, and in precluding or mitigating potential damages in viable defense matters.

Cases have involved industrial workplace and construction jobsite occupational injuries and toxic exposures, machinery and equipment guarding, household and other consumer products safety and health, fires, explosions and electrocutions. They have also included economic analysis of lost earnings and wrongful birth and death, human factors engineering, psychological assessment, slips and falls, building defects and failures, sewerage and drainage systems, and fire, smoke, water and transportation damaged merchandise and equipment. Casework has also covered scientific accident reconstruction, failure and fracture analysis, automotive accidents, materials handling, power tools and outdoor power equipment, carnival and amusement rides, material defects and poor workmanship, building equipment and other construction products, vendor failure, product misuse and modification, juvenile products, warnings and instructions, patent infringement, trade secrets and unfair competition, and numerous additional product and subject categories. TSGI is also expert in the application and interpretation of safety, health and environmental codes, standards, regulations, practices and procedures.

In more complex cases, multi-disciplinary teams and diverse technological specialties and professional personnel are brought to bear. Case-specific problem-solving needs alone control the nature and magnitude of TSGI forensic efforts, consistent with client resources.

TSGI Consulting Associates possess appropriate licenses, registrations, certifications, designations, academic credentials and/or other qualifications in their respective fields and professional disciplines. Expert testimony has been provided to clients in courtrooms across the country. TSGI Consulting Associates have been qualified in Municipal, County, Superior, Federal and Chancery proceedings, and arbitrations. Deposition testimony is routinely given in connection with forensic engagements.

In the selected fields of engineering, science, technology, management, economics, psychology, architecture and commerce most frequently encountered in its forensic casework, TSGI maintains one of the largest private libraries of older safety-related books and other

---

exposed to what's going on out there, where the action is, and how to get a piece of it.

These are screening conversations. Believe it or not, many such contacts can turn into cases or engagements, and more and more as you start circulating. Maybe it won't be the case you were called about, but the prospect interacted with you. It is hoped that rapport has been created and that he or she will call you again. Maybe someone will even be referred to you.

areas of executive control have been intensively and extensively explored and developed, generally speaking. But it is less widely recognized that a product fully complying with engineering specifications, and industry and governmental safety codes, standards and regulations, nevertheless may be unreasonably unsafe and, therefore, economically unsound. Thus, responsible, competent, professional product safety management can help make the difference between corporate success and commercial suicide.

Your confidential inquiries are invited at no obligation.

Consulting Engineers

P.O. Box 99
Livingston
N.J. 07039
201-992-4788

PRODUCT SAFTY

Among other things, learn to identify engagements that are doomed to failure. This can happen for a number of reasons such as lack of data, constraints placed on the assignment, and so on. But failure can lead to an unhappy client or even a lawsuit, so don't get sucked in.

Once in a while you will have someone call you who's just "picking your brains" and really doesn't intend to engage or pay you at all. Such an individual may even have already decided to use

Product Safety Management Consulting Services for Manufacturers

For over a decade The Seiden Group, Inc. has specialized in forensic engineering relating to negligence and products liability litigation. Clients have included lawyers, insurers and governmental agencies. Forensic engagements have numbered in the thousands. TSGI opinions have been instrumental in obtaining favorable jury verdicts and out-of-court settlements of many millions of dollars in meritorious plaintiff lawsuits, and in precluding or mitigating potential damages in viable defense matters.

TSGI now offers its broad expertise in product safety directly to manufacturers. Scientific product safety management is one of the last great unexplored frontiers for creative and meaningful profit planning. Design, development and implementation of cost-effective, results-oriented product safety management systems, procedures and organizations are both technically and economically feasible if there is concomitant, bona fide top management commitment in this neglected area of executive control.

The Seiden Product Safety Management (PSM) System, tailored to the needs of your company, will facilitate efficient, economical, responsible and responsive corporate-wide product safety planning, analysis, control and education (PACE). Treating product safety as the specialized management science it is, the Seiden PSM System helps you build upon a basic blueprint for designing and marketing safer products. Your entire organization will be sensitized to product safety costs and benefits. Useful analytical tools and formats for engineering and managerial control will be incorporated into your customized program. TSGI will also assist you in staffing key PSM positions and integrating them into your overall business structure, as well as in effectively training key PSM personnel, whether promoted or reassigned from within, or newly hired. In-house seminars are available.

The Seiden PSM System provides you with a comprehensive approach to product safety-related loss control management. It balances product safety assurance and products liability prevention needs, thus addressing both preventive and prescriptive aspects of managerial control. Or, TSGI will review your existing approach to product safety assurance, if it is not achieving its goals, and make recommendations to correct its deficiencies. PSM system auditing can also prove beneficial.

The Seiden PSM System is DESIGN DEFECT oriented. While manufacturing quality assurance and product reliability engineering are of critical concern, the profit opportunities of these more traditional

---

somebody else, for whatever reason. Thus, he or she is stealing your expertise.

However, in over fifteen years of professional practice in my own field and specialty, only a handful of people have tried this, among literally thousands of bona fide clients. The gains have been far greater than the losses. I have always believed in being perfectly open with both present and prospective clients.

As a practical matter, all you are doing in any attempt to be secretive or overly protective of your expertise is frightening or intimidating a prospective client. There is absolutely no point to this, and the prospect doesn't deserve to be treated this way. In most consulting situations, the process is truly a partnership. The client has certain experience, expertise, and needs; you have certain experience, expertise, and needs. Preliminary discussions, especially if they take place on the telephone or during client development time that you have allotted anyway, if relatively limited in duration, can be one of the most productive parts of your marketing effort. You are presenting yourself, interacting with the client in a "real world" environment. Also, you are becoming exposed to your target client universe under the best conditions imaginable—the client is exposing you to his or her innermost agonizing over real life problems, the solutions to which mean money in his or her pocket. *Don't* use technical jargon if you can help it.

Don't confuse client development activities with casework. Very quickly, client development-type conversations and/or meetings should lead to a discussion of fees. Obviously, you're not going to spend a day on a prospective client's premise, giving him or her all kinds of useful and valuable information without being paid. You should always be on your way to the next appointment or telephone call within a short time before you get trapped into such a situation.

When visiting a client, always try to make some cold calls afterward, if it seems opportune. There may be prospects in the immediate vicinity of your client. No matter what your itinerary calls for, you should always be prepared for a little bit of "itinerary detouring" of this type. Always keep geography in mind and a whole bunch of maps in your hip pocket. And take a generous handful of your capabilities brochures and business cards to leave behind.

PART-TIME PROSPECTING

Since you are probably working as an employee in some full-time job, you won't have the time to search for clients. Your client development has to take other forms. This is another reason why well-thought-out advertising is important. You can always take time off for appointments with prospective clients who respond to your ads. And this is why a telephone answering service is so important in conjunction with your advertising. Then, with your PR package in addition, you can follow up in still another way. If you work in a highly industrialized area, you may even be able to use lunch hours, early mornings, or late afternoons or evenings—even Saturdays—to advantage in making appointments.

You probably have a few weeks of vacation time. If you don't use up all of this time vacationing, you can spread it out through the year. However, use the time efficiently. Have a program in mind before you take the time off, have appointments made, and have an itinerary.

It may be difficult to do client development while you are employed, but it can be done. It has been done. It has more to do with what you have to offer and your commitment than with mere time available to physically knock on doors. Consulting practices have been built from postcards and a single small advertisement well placed every week of the year. If there are 10,000 prospective clients, invest in postcards that present you professionally, or use some other type of mailing piece. But do it! Get started *now!*

Remember not to let anyone tell you that you don't know his or her business and that you're too theoretical because of this. A sound theory is the most practical thing in the world. Engineering science has been built upon this single fact. Comparative analysis and technology transfer have made more contributions to our lives than closed-mindedness.

CLIENT LISTS

Obtain lists of potential clients when you first start out and keep them current as your practice grows. Get copies of the various industrial directories for your state, surrounding states, and other types of markets you serve. Use the Thomas Register to identify prospect firms in your area. Clip announcements of promotions that appear in trade magazines and local newspapers that you receive. They may lead you to prospects. Also clip articles of interest that may suggest new business leads from other standpoints, such as company expansions, relocations, large orders and contracts received, new project descriptions, profiles of local executives and businessmen, ground breakings, new offices, and factory, mill and industrial center construction and openings.

ADVERTISING

Advertising is costly. A one-inch-wide column can cost from three dollars to five dollars per line or more. However, even a small ad can lead to big profits. You must find a way to advertise your talents and skills. You must not take the attitude that an occasional contact and your confidence in yourself alone will bring hundreds of clients beating a path to your door. Advertising dollars well spent can be the best investment you can make in consulting, as every place else. Once you have packaged yourself in a brochure or equivalent, design a couple of advertisements of different sizes for different purposes. Design a number of ads with different angles and content. Seek out trade papers, magazines, newsletters, and so on where you will get the best exposure.

You don't want to go broke advertising, but there is a middle ground—you have to decide what it is for yourself. The only guideline is to think of well-placed advertising as a cost-effective investment. You must wage a campaign. Don't place one or two ads and then stop. Run an ad someplace for a minimum of six months to a year to give your target client universe a chance to see your ad, remember it, react to it when there eventually comes a need, and to instill the comfortable feeling that you're in business to stay. It works. However, buy space judiciously. You want it to have reasonable visibility. A premium for such positioning can be worth its weight in gold. In addition, pay your advertising bills promptly—within ten days, if not sooner. You want the newspapers or magazines you advertise in to be on your side.

Due to the importance of advertising and public relations to any consulting practice, you should be certain that you subscribe to virtually all of the important publications in your field. Get sample copies of all the publications that your prospective clients read with rate cards and media packages. There are many sources and lists of magazines, newspapers, and so on at your public library.

Remember that you have to spend it to make it. Although it may be hard for you to forget that it is your own money you are spending, force yourself to treat the money you spend as an investment. If you were spending it for an employer, you would not think twice about applying cost-benefit analysis in making recommendations to advertise. Even though it's your own money, you have to think the same way. It's just good business.

Ethical advertising is the best advertising, especially in the professions. All other things being equal, if you advertise your capabilities and availability you are at least one step ahead of your very capable colleagues who sit at home on their hands waiting for the phone to ring but make no other discernible effort to develop clients. It is advertising that makes the difference. Advertising is an American tradition that works. It can be cost-effective. You may not want to go on TV, but you certainly can afford alternative types of high-impact precision advertising.

You might even be able to get into your local newspaper or trade publication, or a publication put out for your prospective clients, on the basis of some accomplishment or service that is newsworthy. Public relations is basically nothing more than advertising for which you don't have to pay. It increases your credibility. You can use it in your own advertising brochure or other promotional package. It doesn't even have to relate specifically to your specialty in its primary descriptive content, but it will probably mention who you are and what you do, including the name of your fledgling consulting firm and its specialty.

Send a good black-and-white photo along with any news release that you develop. Have professional photos taken of yourself in 3" × 5" format. Get several dozen copies and keep them handy. That way you will always have one when you need it. Wear a business suit, white shirt, and tie for these sittings. Look like a consultant rather than someone who just stepped off the tennis court or overslept.

INVESTMENTS THAT MEAN BUSINESS

Thus far we have not talked about the kind of investment you need to get into the consulting business. Chapter 3 did discuss resources and services, however. Part of your investment will go toward items discussed in Chapter 3.

Some consulting books and articles tell you that all you need will be from $1,000 to $3,000 to set up shop. However, this refers to setting yourself up in a small corner of your own in your residence, where you can work privately and intensively. It cannot include advertising, for example, or a good typewriter. Working from home is cost-effective, but $3,000 is too low.

If you're serious, do it right. Results don't come all at once. A few well-placed ads, even though small, more than pay for themselves eventually. If you're just going to dabble, forget it. By the way, a part-time consulting practice that grosses you $10,000 must be treated with the same respect and tender loving care as one that pulls in $100,000 or more full-time. This includes how you treat clients.

Since you are an engineer, scientist, or technologist with special training and skills, you probably have not done badly financially over the years. If you mean to get into the consulting business part- or full-time, you will have to make a firm decision to commit $6,000 to $12,000 or more of start-up money or "seed capital." You can borrow this from your bank. You should not plan on many meaningful engagements for about a year. Some sources say that six months should do it, but I would recommend that you think in terms of a year.

For someone looking forward to serious part-time consulting and serious money, the modest investment represented by $6,000 to $12,000 will be recouped. Anyway, it's a business expenditure, so Uncle Sam is your ready, able, and willing partner. You can always find an occasional client, but the "occasional" consultant or "dabbler" will probably not be charging professional-level fees and will probably not have thought through his or her presentation package. He or she will operate very informally. We will discuss more about "dabbling" and its negative effects in the next section.

Let's see what this $6,000 to $12,000 investment will get you.

Referring back to our Chapter 2 discussion of the chart of accounts and basic equipment needs, a new IBM electronic typewriter with some supplies and different type faces will run around $1,000.

A good used five-drawer filing cabinet could cost from $100 to $250. You will need both a 35mm and instant cameras. These could cost around $1,000, including a good zoom lens, flash equipment, a tripod, camera cases, and a starter supply of film. If you don't have a work desk or table, you can get something simple initially for under $250. So far, you've invested around $2,500.

You already have a car and the beginnings of a library. Feel your way on books. However, plan to spend another $500 for books the first year. Figure advertising at around $200 per month, or $2,000 a year, excluding July and August. Professional, technical, and trade association dues will run you perhaps $1,000 per year. Office expense and petty cash each will come to around $1,000 a year. Since we are itemizing various office items in our listing, the $1,000 mentioned here could give you a little bit of a cushion. Postage might run you $500 a year; the same is true for supplies ($500), telephone ($500), gasoline and tolls ($500). So far, we have another $7,500, or $10,000 including office and photographic equipment. As mentioned earlier, your office telephone should be a brand-new instrument rather than the same as your residence phone. Don't mix the two.

Travel and entertainment could run you around $1,000 a year to start. Sales promotion will be at least $1,000, including printing. So here you have another $2,000, for a total of around $12,000 to cover your first full year of start-up activity if you do it right, whether part- or full-time. The first $5,000 to $6,000 will be expended very quickly once you make the decision to get going and set up shop, even if only part-time. Plan your every move. The second $6,000 or so will be spent as you start functioning as a consulting business, whether your first engagement comes in the first or the ninth month.

Let us now list below everything we have discussed, for easy reference:

Investment or expense item	Amount
1. Typewriter & Supplies	$1,000
2. File Cabinet	250
3. Photographic Equipment	1,000
4. Desk & Chair	250
5. Books	500
6. Advertising	2,000
7. Dues	1,000
8. Office Expense (Excluding Supplies)	1,000
9. Office Supplies	500
10. Petty Cash	1,000
11. Postage	500
12. Telephone	500
13. Gasoline & Tolls	500
14. Travel & Entertainment	1,000
15. Sales Promotion	1,000
Total First Year Start-Up Costs	$12,000

Notice that we have *not* included certain expenses that are already being incurred in the normal running of your home, such as heat, light, gas, electricity, and so on. However, depreciation on your automobile is now assignable to the business, as well as depreciation on other fixed assets (such as furniture, fixtures, office and field equipment, etc.). Your accountant will help you assign assets and other expenses that will now be business-related. Don't worry about "rent." Home office "rent" would only become personal income. Concern yourself with "real" out-of-pocket dollars rather than non-cash transfers.

As your practice grows, you can expect your expenses to increase. Some will increase more or less in proportion to the amount of business you do, such as telephone, postage, and travel. Others will be essentially fixed, at least up to some incremental "scale of enterprise" that represents a change in kind or quality rather than merely in degree or quantity, such as secretarial expenses, utilities, and equipment service contract expenses.

In Chapter 7, we will talk more about Financial Engineering, including engineering economics, financial planning, analysis and control, flexible or variable budgeting, capital budgeting, lease versus buy, cash flow, discounted cash flow and present worth, and so on. Your start-up consulting practice probably will not require rigorous or extensive application of most of these financial-analytical tools, but you should become familiar with them so you can at least recognize emerging problems (i.e., opportunities) and patterns. You don't have to be an accountant or financial analyst to understand these techniques. As a matter of fact, as an engineer you will find them simple since they can be described and understood through the use of graphs, flow charts, formula charts, mathematical tables, growth curves and other kinds of technical concepts, visual aids and data bases with which you have long been familiar.

BEWARE "DABBLING"

The part-time consultant can make a significant percentage of salary in outside consulting income as he or she begins to circulate where prospective clients may be found, or otherwise working. But the "dabbler" can and will be taken advantage of by unscrupulous clients. This is because the dollars are not as meaningful as they are to the determined professional who is looking for an ongoing, significant, even if part-time, supplementary income source. A businesslike approach cannot be informal or casual with respect to the value of expertise. You can give away some knowledge as part of your PR program, but *don't* accept or ask for bargain fees when you do charge for your services. And remember, the best investment you can make is in yourself. So don't skimp. You are worth it.

Very few consultants have started out with full-blown practices; most began on a part-time basis. Most consulting practices do not get started as quickly or as surely as would probably be the case if the new consultant would make the decision immediately to invest properly in his or her business and self, even if the business is a part-time undertaking. Not doing this can be a great, if not fatal, mistake. This is because the chances are good that if you have a marketable specialty you will be able to do a significant amount of consulting, but you must attack the problem in the formal, serious manner it deserves.

Don't start spending your money until you have worked up a capabilities brochure and decided how and where to market it and how you can most efficiently and effectively reach and service your chosen market. Only then are you ready to make a financial commitment to do real client development. However, by then you are also in a much better position to move quickly, precisely, and successfully. This assumes, of course, that you have prepared yourself in other ways, as discussed in Chapter 3.

CREDIBILITY SELLS

The consultant is a problem definer and problem solver. Consequently, you are probably also expert in acquiring broader and/or deeper technical expertise. Such personal progress gives you not only greater confidence in yourself, but also greater credibility in the eyes of prospective clients. Keep several steps ahead of your client in your specialty. Your expertise will be visible and impressive. The client cannot help but notice this when you discuss problems intelligently, realistically, and with timely insights that can only be the result of conscientiously keeping up with your field.

But ultimately, whatever competitive advantage you possess will probably not relate to your technical expertise alone. How you present yourself and relate to your client will be just as important, if not more important. Be open, listen, and don't be afraid to "intellectualize" with the client about his or her problem. Always remember that your main competitive advantage is *you*. You are unique. You are different from all of your competitors. You can sell something that your competition can't. This is *not* technical knowledge but, rather, *you*. Try to identify what your competition can sell better than you . . . and why. It works both ways.

CLIENT RETENTION FACTORS

Keeping clients is not hard. Obviously, you have to deliver quality performance. You have to provide acceptably quick response and on-time service as promised or contracted for. You have to remain visible to clients at all times. If you don't, there will invariably be lots of competitors ready and willing to step into your shoes, even if they are less able. Don't give them a chance to gain a foothold.

You may have opportunities to provide follow-up or after-market services. You can keep a step ahead of your clients by periodically reviewing their files and/or problems. Determine whether there might be some opportunity to contact them again to see how things are going or what new problems may have cropped up that they may want to talk about with someone they can trust. Your client may find your follow-up contact as much of an opportunity as you do. Much of the time, you can bank on this. Experienced consultants will guarantee it.

Invite your clients, as well as prospects, to bounce ideas off you—and expose you to problems—any time they wish. Get them accustomed to thinking of you first when they have a problem or an idea that needs your kind of professional insight. When an engagement materializes, you can crank in the fee for a few productive telephone conferences. It will make up for the "missionary work."

In fact, you will find that a periodic personal follow-up contact with the client results in his or her wanting to discuss a new problem or case, or even several. Interestingly, the client sometimes won't let you hang up before he or she has described the latest problem. This is great for both of you. Don't charge for such conferences. Sometimes you will learn as much as the client. It's almost always a two-way street, and the client is appreciative. Again, if and when a new engagement develops, the fee will take care of itself. You really can't lose . . . and neither will your client.

One way or another, you're always going to have competition. It's normal, healthy, and even an opportunity. We'll talk about this later in greater depth. For now, just remember that if you do your homework and remain in the client eye, you have a comfortable edge because you got there first. Entrenchment is always an important plus, as long as you don't become complacent, take the client for granted, or take advantage of him or her.

KEEP THE CLIENT UP TO DATE

Stay close to your client with respect to progress you are making on the engagement. Don't forget about the client even if you are hard at work on your assignment. If an assignment is expected to be protracted and extend beyond a few weeks, make it a point to call him or her now and then—perhaps every two weeks—to report on the progress you are making and that you are, indeed, making progress. Don't force the client to guess where you are and what you are up to. You must never be too busy to hold a client's hand when you are spending his or her money at your usual rates. Of course, don't expect to be paid for such missionary work—it is just good business.

On the other hand, clients have a way of opening new doors on engagements during their course. An interim phone call to report status or progress can be invaluable. On occasion, it can change the entire complexion and focus of an engagement. It can be an opportunity to really do a job for the client. It may even lead to an expanded assignment.

What it also does, of course, is permit you to get closer to the client. Little by little, he will take you into his confidence. He will depend upon you more and more. He will discuss new problems with you, some of which may lead to still more work. He will come to think of you first, rather than one of your worthy competitors. Remember, the name of the game is client development, so develop.

Develop yourself, and develop your clients. In fact, you develop yourself by developing *them*. Be client needs-oriented and needs-sensitive at every stage of a consulting engagement.

FOLLOW-UP IS IMPORTANT

After you render your report, follow up to see how implementation is going. After your recommendations are implemented, if they are, follow up to keep track of results. If they are not, find out why. If you find that things are going as well as can be expected (and hopefully they will!), follow up to see if any fine-tuning is necessary. Do not charge for any of this as long as it doesn't entail serious casework.

In fact, once you are certain that good results have been obtained, ask your client if he or she would be willing to write you a letter of recommendation that you can use for client development and PR purposes, and if you can also use his or her name as a reference.

DON'T BE INTIMIDATING

Always try to find ways to continue to educate your clients, even in follow-up phone calls. Try to create a standard in your specialty, or at least keep your own services on a level that meets or exceeds existing standards and trade practices. Surprising as it may seem, your clients may even be a little bit afraid of you, as well as of your competitors. This is easy to understand. If you need a TV fixed, a car repaired, professional-level advertising, or a lawyer to defend you or sue for you, or medical or dental treatment, just recall your own feelings. When you have to depend on someone else to help you, and you also know that they are going to charge you a significant sum for their professional or other expert services, there is always a feeling of intimidation and being somewhat out of control. You don't always know exactly how much they're going to bill you, just what you're going to get for your money, or what kind of follow-up services they're going to tell you that you need. What can be more intimidating? You're in their hands.

On occasion, clients of mine have been most complimentary with respect to the open way I conduct my practice. The worst thing that a consultant can do is attempt to play God with a client in the consulting specialty. The client will resent it. The client will resist any recommendation that may cost him any more money than he or she originally planned on, and may even begin quietly searching around for a new consultant. A relationship like that can go on for some time, though strained, but at the first opportunity that client is going to shift allegiances in favor of a consultant who is more open, easier to work with, and, perhaps most importantly, who treats him or her fairly and as a full partner in the venture.

Being open with clients permits and encourages early discussions and relationships to develop that provide important insights into client method of operation. With enough experience, a consultant can size up a client sufficiently well to decide whether or not to become involved with that particular client. Over the course of consulting practice, you will learn to identify prospective clients successfully who would have just been nothing but trouble, and for no good reason or fault of yours. In some instances, you may decide to take a calculated risk. In others, you simply decline politely and tactfully. Once in a while, of course, you wind up getting burned. Try to be on your guard at all times to preclude such a problem.

"QUALIFYING" CLIENTS AND CASES

It is also important to attempt to identify incompatibilities that you may have with prospective clients. Quite aside from the matter of fees, you might find that a client is simply very hard to work with. He doesn't trust anybody; he wants to remain in control; he wants to be the boss; or, perhaps, he tends to be very patronizing. He may think you're on his payroll; he may tend to curse, yell, and scream a lot; or he may threaten you during your first conversation, and for no good reason. You have to know when to back the client off, reassure him, or accommodate or otherwise compensate for such incompatibilities. And you have to do this before it's too late. Think through the consequences of each and every engagement in terms of personal relationships, before you accept it. Otherwise, you could have nothing but trouble on your hands through absolutely no fault of your own except misplaced good faith.

It is also important to be certain that the needs of the case or engagement can be matched to an effort that will produce acceptable results within a time and cost or fee frame that the engagement, and the client, can sustain. There is no point in moving ahead with a client if the problem to be solved requires a magnitude of effort that would clearly be impractical for the client to finance. Lay it all out at the outset to the extent that you can. Let the client know to what extent the engagement is "open-ended," or the probability of success.

The client must be aware, of course, that all activities are on a "best efforts" basis. This is *very important*. There are no guarantees—and don't be intimidated.

Also, size up the prospective client to be certain that he or she can pay you. Don't permit yourself to be sold a bill of goods. Try to "qualify" your client in a manner analogous to that used in a credit check. Who is the prospect? What have you heard about him or his company? Ask around if you have to. Once in a while, you may decide to take a calculated risk. This is fine, but always expect the worst in such instances. You should always have a substantial retainer in hand and be in a position to put a hold on follow-up activity if fees are not forthcoming. One sign of a possible problem engagement and/or relationship is when a prospective client comes to you in a panic with a super-rush job that won't wait and that requires you to make a significant short-term commitment of time, but where you are told there is not enough time to get your retainer to you before you do a substantial amount of work, even all of it. Also, beware anyone who keeps saying "I'm an honest man."

Beware of a sudden change in client style, such as how the client wants to pay you. It may signal a problem engagement or it may reflect some problem the client has. In any case, you may be in for big trouble if it seems to you that you are dealing with a "new client" or a drastically different relationship.

LIMITED ENGAGEMENTS

A client may specifically engage you to help him "put out some fires," so to speak. Your solutions to such local or limited problems may be only temporary and, perhaps, not very effective or efficient. But to the client you may have done precisely what he wanted and he may be tickled pink.

On the other hand, you may realize that addressing such problems will not solve his or her fundamental problem. In fact, the client may also realize this. But whether or not this is true, you are in a position to discuss the broader situation with him or her even though the client may still have a reason for wanting you to do the limited work, whether for a good or bad reason in your own mind. The lesson here is to learn as much about the client and his business as possible at every point in your relationship. Your objective should be to become his alter ego in the area of your specialty.

Incidentally, this type of situation occurs quite frequently. One of the reasons the client will insist on what you consider to be a limited, ineffective, or inefficient problem solution often will be due to lack of time, money, or other resources to solve the larger or tougher problem in an adequate manner. Of course, sometimes you have to mind your own business.

In some instances, your recognition of the limitations of the engagement can lead to follow-up work in the future. So work with the client and address the problem he or she needs solved at the moment, keeping in mind the broader picture. As long as your professional integrity is not compromised, there is no harm done. Where ethical considerations and/or compromises enter the picture, you will have to play it by ear, and be certain that you are not caught in a trap from which you will not be able to extricate yourself reasonably.

BEWARE THE LONG-TERM OFFER

You must also evaluate carefully an engagement where the prospective client wants to put you on the payroll full-time for the duration of your consulting assignment, looking forward to some longer-term relationship. This is obviously not for you if you already have a fulltime job that is reasonably good and secure. If you are attempting to build up a consulting practice, being on someone's payroll can cause you to lose your flexibility and independence in seeking out other clients. You may be committed to be on your newfound employer's premises at specified times or for a greater part of every week than will permit you to develop your own practice seriously.

Getting on the payroll might be tolerable in itself and even afford you some security, but it will also probably mean you are being paid less than a regular consulting fee. Your newfound employer naturally is bearing expenses that you would otherwise have incurred as an independent businessperson, and which you automatically would have built into your consulting fee by nature of the consulting relationship. A consulting fee includes a profit . . . *your profit*. Don't make it *your employer's* profit.

Also, he or she is getting a bargain. If you are in business for yourself, your consulting fee should be at least three or more times a salary, on whatever basis, since you have all expenses to bear, including client development, and must also turn a profit on your activity in addition to taking a salary. Anyway, do you really want just another job? Can you afford it? You want to benefit from the leverage on your knowledge. A client can "rent" your know-how, but do you want to let him "buy" you? You may be trying to cut loose from a job-based income. Why take another one?

Don't be afraid to tread water for a few years, making contacts and consulting fees wherever you can legitimately and ethically find them. It's good training. Just be certain that you keep your antennae up. Focus on your specialty. Refine, redevelop, redirect, and expand it. Locate clients. Don't be afraid to be taken advantage of during these early, formative times, even if years. Consulting is a growing process, like anything else, and you cannot avoid experiencing the inevitable growing pains. Even later, when more experienced, you may be stung once in a while, but the overall benefits of consulting will far outweigh the negatives.

Remember that a job with a large consulting firm is *not* the same as independent consulting. It is just that . . . another job. A large consulting firm is no different from a large corporate employer. Of course, in either case you can develop a specialty that can conceivably set you up later.

In my own case, even having spent my early career with very large corporate employers, I still identified several consulting opportunities and made good fees on each one. By the way, I always told my employers about my outside work. Never try to do moonlighting without the full knowledge of your employer. Check company policy—you may be with some outfit that frowns on it and thinks it owns you, body and soul. Also check back to see what kinds of papers you signed when you were first hired. You might be surprised to find that you are specifically prohibited from consulting or moonlighting. Maybe you even signed a patent agreement. Maybe it's time to find a new employer.

You will ultimately identify a dynamic engineering specialty that permits you to apply full-time the full range of your training, experience, talents, skills, creativity, and inclinations. Because of this, you will be able to make contributions to your field, and your specialty will become exceedingly rewarding both financially and professionally.

CHAPTER 5

ECONOMICS AND MANAGEMENT OF THE CONSULTING ENGAGEMENT

TYPES OF ENGINEERING AND SCIENTIFIC CONSULTING

When you first enter consulting practice, keep in mind from the start that you are in a business. As a practical matter, however, you will be initially preoccupied with "managing" individual engagements. Taken together, they define your "practice" and the emerging dimensions of your "business," even though you are setting aside time to do client and resource development, set up your office, and all the rest of it.

There are three quite distinct types of engineering and scientific consulting. One revolves around civil engineering practice and building trades-related specialties. The second deals with nonconstruction-related general industrial engineering consulting. The third is the world of the consulting scientist, metallurgist, chemist, the health science professionals (including industrial hygienists, psychologists, etc.), and other technology-oriented consulting specialties.

Generally speaking, the first is the world of proposals and contracts. It is also the domain of lump-sum, cost-plus, percentage-of-construction-costs, and larger consulting firm fee arrangements such as salary plus percentage, standard time rate plus percentage, and so on. Other fee types are also used, of course, even in construction-related consulting, such as straight time charges plus expense reimbursement, retainers, and per diem rates. Any standard textbook on business aspects of engineering will treat you to elaborate and intricate discussions, complete with charts, graphs, tables, and so on, on all of these standard financial arrangements between engineer and client.

FEE ARRANGEMENTS

However, the focus of this book is on general industrial consulting as seen through the eyes of the one- or two-person practice. Here you should stick with hourly, per diem, or even half-day charges. Although fee arrangements based on direct salary or wages plus burden or overhead may be attractive to many industrial clients who are accustomed to thinking in such terms, for the individual consultant this is a potentially dangerous game and even a fiction. The engineering consultant, like the physician or lawyer in individual practice, is much better off recognizing salary, business development and other overhead costs, and profits in the form of a single hourly rate quote or a per diem rate or equivalent.

Furthermore, once you let a client begin to think of you in terms of an employee who gets a salary but who is providing a consulting service, you are absolutely finished. If you're going to be a professional, then act like one, charge like one, and provide services like one. You can't do this when you think in terms of salary. For example, any prospective client whom you allow to think of you as an employee will probably also believe that he or she should only pay you the equivalent of a salary that covers, say, fifty hours of work instead of forty since his other engineering employees all work fifty hours a week even though their weekly salary is presumably for forty hours.

You are a consultant. You have a consulting practice. You are in business for yourself. You have business expenses. Always remember it. Gracefully tell your client how you charge for your services, but *do not* negotiate. This is the wrong way to buy into consulting practice.

Of course, even in industrial consulting you will sometimes find it more reasonable, even to your benefit, to make a proposal or to sign a contract depending upon the type and/or extent of work to be performed. For example, if you give seminars to corporate clients, you may be delighted to sign a contract that you will have carefully reviewed and revised if necessary, of course. You will also be pleased to have a local meeting or two with a prospective client at no charge in order to present your story and brief him on proposed seminar content, format, and so on. You may even have prepared a separate seminar brochure.

However, for the bulk of your practice, charge by the hour, half day, or day. Also require a substantial up-front retainer. When this has been used up, provide the client with an accounting and go back for follow-up retainers, if warranted by the extent or nature of the engagement. From the start, clients are entitled to be briefed on the probable nature, scope, and extent of such follow-up work. Sometimes an engagement will be open-ended or will become a "can of worms." You can't always tell when this will happen or even prevent it from happening. Clients must understand this, as well as the fact that all of your professional endeavors are to be on a "best efforts" basis.

When you have expended time that begins to approach the amount of the retainer, simply put a hold on casework. Call the client and report on the milestone reached. However, do not exceed the amount of the retainer in terms of work performed without additional money in the house. Otherwise, you will eventually find yourself in the soup. You will wind up with thirty-seven clients who each owe you for five to ten hours. At $125 to $150 an hour or more, that could be over $25,000 in receivables. You'll probably spend another three to five hours per client trying to collect—plus your phone bill. This is an unenviable position to be in, unless your available hours per week grow on some tree in your backyard. If they do, I would love to buy some seeds.

FEE COMPONENTS

All of the time you spend on a job, including worrying over the problem while you're shaving, should theoretically be billable. This includes all attributable travel time that is required by the job or the client.

With respect to travel time, as a general rule you should charge for all such time that is spent at the convenience or demand of the client. However, you should swallow travel time that is at your own convenience. There is a twilight zone in the situation where your own scheduling problems result in increased travel time. You can't always charge extra where you have a deadline to meet and you wind up doing a field inspection on a Sunday. Such annoyances come with the territory.

In your own specialty, you already know the sequence of steps necessary to solve problems and prepare reports, but when you are consulting you should be much more alert to the patterns of time spent on each engagement phase for different types of jobs. From the very start, try to keep track of time details. This is important not only for billing purposes, but also for future fee and cost estimating. Some modern telephone systems, for example, have a display of call time that you can record.

As you do more and more consulting, the patterns of time you experience on different job segments and different job types will permit you to talk much more intelligently to clients about probable dimensions of an engagement, both time- and fee-wise. Clients like to speak with professionals who not only know their business technically, but who are also cognizant of the commercial ramifications of problems in terms of logistics, fee requirements, and so on. This will become second nature to you the more consulting you do.

If your field of expertise requires that you do significant experimental or laboratory work, try to align yourself at the outset with some local laboratories. Develop a preliminary relationship with the technical directors of your laboratories of choice. Obtain their fee schedules and determine lead times for typical lab routines that you can foresee needing.

With respect to lab work, do *not* add any profit for yourself onto lab fees. This should remain a pure pass-through to your clients. As a matter of fact, any actual lab work should be the result of an agreement on fees between the lab and your client. Laboratory invoices should be sent directly to your client. You should charge only for your actual time interacting with the lab, determining what test routines you need, consulting with lab personnel, interpreting experimental results for your reporting purposes, and so on.

Never pay a laboratory and then bill the client for the total job. Sooner or later you will get stuck. Also, a client might feel that you are getting a kickback. Keep all financial arrangements with laboratories on a visibly arm's-length basis at all times. Otherwise, you might also get some surprising bills from the lab. Your client might decide to order more lab work directly and have the lab bill you for incremental services rendered, representing, naturally, that he will ultimately be responsible for the total bill. Make clear to both lab and client what the interrelationships are to be.

SALARY, FEE, AND PROFIT LEVELS

According to the engineering income and salary surveys cited in Chapter 1, recent median primary income for Registered Professional Engineers in the United States has ranged roughly from $40,000 to $50,000 annually, and it is rising.

Now, if you assume that engineers work a forty-hour week and then go home—which they don't—this amounts to around $800 to $1000 per week or $20 to $25 per hour plus fringe benefits. Such salary levels may sound reasonably good, even though they are "medians." However, when you realize what the implied hourly rate is, and when you consider that most engineers probably work at least fifty hours a week, that $20 to $25 an hour decreases to more like $16 to $20 an hour or even less, depending upon how faithfully and conscientiously we attend to professional problem-solving needs.

Compare this with hourly engineering consulting fees of anywhere from $100 to $150 and more, depending upon the specialty. Of course, these fee levels assume that you have all the expenses of a consulting practice. Even if you have a part-time practice, you have your business expenses to contend with. This is serious money. Conservatively speaking, figure that only one-third of your gross hourly rate will filter down to before-tax income, including profit and salary. Remember that a consulting practice, like any other business, not only must cover expenses and salary, but it should also throw off a respectable profit for your entrepreneurial troubles.

For a part-time practice, you may only be thinking in terms of consulting income as a supplement to your full-time salary, but this is an incorrect viewpoint that can cost you dearly. You should also be thinking in terms of some profit. As an example, suppose you want to make a minimal supplemental income of $5,000 a year just to get your feet wet. Here's what you need:

Gross Consulting Revenue	$15,000
Client Development	$ 5,000
Other Expenses	5,000
Your Salary	5,000
	$15,000

This is not a realistic approach from the viewpoint of good business thinking. Your salary is for your pure direct technical expertise and labors. How about your entrepreneurial profits? Here's where a lot of consultants, new and old, go wrong. The numbers should look more like this:

Gross Consulting Revenue	$20,000
Client Development	5,000
Other Expenses	5,000
Your Salary	5,000
Total Expenses	$15,000
Profit Before Taxes	$ 5,000

Notice that in the first case cited you have inadvertently set yourself up in such a way, through erroneous thinking, that your "profit" is actually pegged at zero. Correspondingly, there is a big difference in the hourly rates reflected by the two cases. Assuming you work an average of five hours a week for forty weeks on a part-time consulting basis, in the first case you would charge $15,000 ÷ 200 hours = $75 per hour. If you are a businessperson, you would want an entrepreneurial profit on your laboring as well as a mere salary. After all, you are entitled to it. Therefore, for the same 200 hours per year, including a small profit, you would bill out at $20,000 ÷ 200 = $100 per hour.

The difference may sound somewhat academic to you if your salary is only worth $15 or $20 an hour, but as your practice expands you will be more and more sensitive to your entrepreneurial needs in order to continue to make running your own business an attraction. After all, isn't this what it is all about, whether you are an engineer, physician, lawyer, accountant, or other kind of small business owner?

You will soon grow very jealous of your entrepreneurial posture and prerogatives. You will make more demands on yourself and your practice—and you should. Don't let *anyone* tell you that you don't deserve or haven't *earned* a profit. Anyone who says so is thinking just like an employee. And most employees don't realize, or don't want to realize, that their boss or company is making a sizable and tremendously leveraged entrepreneurial profit off every hour of employee labor, including time spent by "hired hands," whether professional or not. This includes *you.*

One consultant I know put it this way: Your salary is for your technical knowledge, education, and experience. Your profit is for aggravation past, present, and future.

WHAT WILL THE TRAFFIC BEAR?

You may or may not know what the traffic will bear in your own field on the consulting end, but at this writing, an hourly rate of $100 to $150 for a Professional Engineer is certainly not out of order even if it is not as high as it is possible to command in some instances, depending upon a number of factors. Again, this assumes that you are running your own business rather than providing purely technical input to a consulting firm that is set up to obtain all the engagements or casework for its consulting associates. More about this later.

Don't be greedy when it comes to fees. You don't have to get the absolute top dollar in your field in order to either feel professional or begin to liberate yourself financially. Just try to charge a fee that is more in line with the top of your field than the median or in the lower end of the range. You should have enough pride and confidence in yourself to expect something approaching the top of the market.

After all, if you feel confident enough that you can provide professional, competent, and valuable technical counsel to prospective clients, and the top of the market is $150 an hour, then there is absolutely no point in charging $100 an hour. You may consider charging $125 an hour. However, bear in mind that buying in at a low rate, aside from possibly locking you in with clients for a good long time, will only backfire. This is because, try as you may, there is absolutely no way you will be able to charge clients for every single minute or hour you spend on an engagement. You may think that good record keeping will do the trick, and you may think that you will be able to get paid for the actual time you spend. But you will soon find out that many jobs have a way of taking longer than you anticipate, and that it is simply imprudent, impractical, or inopportune to attempt to get additional retainers, send follow-up invoices, and so on. Consequently, if your services are valuable, and you are good in your specialty, you might as well charge like a professional from the very beginning. While it is illegal to collaborate with your colleagues and/or competitors in fee setting, you will soon gain some idea of what the going rates are. You may even want to charge more if you have some novel twist in the way you solve client problems, follow through, and make services available to them.

YOU CANNOT AFFORD TO WORK "CHEAP"

Don't let prospective clients beat you over the head on fees. In many instances, the reason a prospect has leverage is because engineers seem to be so willing to work cheap. Too many engineers think like employees. And we have let purchasers of our valuable services think of us in the same way. Therefore, many prospective clients expect to pay no more for a consultant than it costs them to hire a salaried professional for the same job: $15 to $20 an hour. They forget the overhead they apply to each professional hour, fringe benefits, their entrepreneurial profits being reallocated, bonuses, and so on. They also forget that if they do *not* have a full-time need for professional-level expertise, even a high-priced consultant may be one of the best investments they can make.

Most engineers don't realize how much leverage a client enjoys when he or she is hired or engaged. Part of it is due to the fact that most engineers have no management training and do not understand the accounting and financial aspects of business management. They do not realize how much money is being made on their efforts.

Look at it from another standpoint. The consulting "firm"— and there are many large, respectable, and successful ones—charge handsome professional fees for the valuable services of their own engineering consulting associates and/or employees. So why should you charge less if you are in business for yourself? You, too, have significant overhead (if you practice properly). You, too, have to do client development on an ongoing, expensive basis. You, too, have to buy or maintain capital equipment, an adequate library, telephone services, supplies, advertising, postage, secretarial help, insurance, gasoline, periodicals, expensive photographic film, and so on. You, too, have to pay license fees, dues, accounting and legal fees, convention travel expenses, and room and board. You, too, have legitimate entertainment expenses. And a consulting firm, large or small, is entitled to a profit to permit reinvestment in its business to allow it to grow better and stronger, and even to have a cushion in the form of earned surplus for that rainy day.

You are also entitled to be able to put away something for yourself for your own retirement, such as into a pension plan. You'll need a new car every once in a while. Some consultants are more like salespeople. They travel 50,000 to 75,000 miles a year. Where is that going to come from? To the extent that your needs derive from your consulting practice, it is your consulting practice that should be able to pay for them. An automobile that you use intensively in your consulting practice is a legitimate item of "Plant and Equipment" if you use it daily to travel to field sites, client premises, and the like.

So build well-earned entrepreneurial profits into your thinking about fee rates.

RETAINERS

Do not do any extended work without an advance retainer. Whatever you estimate the job to be worth at your hourly rate, get a retainer of the full amount under $1,500. Get one-third (or $1,500, whichever is more) of the estimated fee for jobs worth over $1,500, and then *put a hold* on the engagement when you have reached the level of the retainer. Go back to the client and get a follow-up retainer for the next milestone or segment of the engagement. Always get retainers up front for each succeeding milestone as you come to it.

The best practice is to avoid entirely the necessity of billing for fee balances due to the extent feasible. The best way to be successful as a consultant is to get paid up front.

Naturally, you should also attempt to give the client some idea of what kind of time will be required on the engagement, or explain that it may be open-ended, right at the outset if possible.

Always think "retainer" for each and every engagement milestone—up front. There may be instances when you feel it is more appropriate to bill for services rendered. That's all right, but remember that *you bill for services at your own peril.*

More often than you like, you will be contacted by a client who is in an absolute panic. He or she got your name from someplace—doesn't remember where or when. An inspection has to be made no later than next Sunday. Can you do it?

Of course you can do it. There is nothing else you would like to do more. As a matter of fact, at least 5 to 10 percent of your practice will probably consist of compressed lead-time work. Just be sure that you get a retainer up front, even if only to cover the cost of the field work, with the additional retainer amounts to follow when it is mutually agreeable, except maybe for "old" clients. But be sure to outline your complete operational style to the client, together with fee requirements. In situations like this, incidentally, do not make the client feel that you are trying to take advantage. This is the worst thing you can do. Don't charge a premium for Saturday or Sunday field work, unless some very unusual circumstance warrants it. If no retainer is forthcoming or the client calls and says he or she cannot possibly raise the money on such short notice, forget the whole thing, although tactfully. Things are tough all over.

The rush job is a real challenge. It is invariably an unusual or significant problem, and you usually have a client who is desperate for professional input. You can get new clients this way when their regular consultants are away, sick, or otherwise unavailable. It's foolish to be proud. After all, your practice is like any professional practice. There are still physicians and lawyers who will service patients and clients on weekends, holidays, and evenings. Your business is a professional service.

Be available for consultation whenever a client feels like picking up the telephone—or must do so. If you are out, your answering service will take the message and also some private number where you can reach the client as soon as possible.

Here again, part of your success will be due to playing it straight and making yourself available when you are needed. Sure, sometimes it will interfere with your family life, but you are a professional. You have a mandate to serve the public faithfully. Your ultimate client *is* the public.

Aside from everything else, rush jobs, weekend jobs, holiday jobs, evening jobs, and so on can add significantly to your income. Surprisingly, you will find that such emergencies don't interfere markedly with your personal life. Here and there it can become sticky, of course; but you just have to cope. Being attentive to clients when they need help the most will cement a good relationship and bring additional business.

THE PROPOSAL, CONTRACT, AND NONDISCLOSURE MORASS

The individual consultant—especially the new consultant—should steer clear of those types of consulting opportunities where proposals and contracts are the rule or are required by clients. You usually don't get paid for sweating the details of proposals, and contracts do not offer you adequate protection, no matter what anyone says. Don't forget that if you decide to sue someone on the basis of a broken contract, no one wins. Many lawsuits won are, on balance, but Pyrrhic victories, and frankly most lawyers are too busy with serious suits to worry about the relatively small amounts that are likely to be involved at your financial level. They probably won't even want your case.

When somebody buys an automobile or a house, or a contract is let for a building or piece of machinery, the contract serves a valuable purpose for all parties involved. However, contracts are not always useful in the consulting relationship. Contracts can be valueless to the small consultant if you ever have to sue somebody. Engaging a lawyer, going to court, taking the time to prepare the case, and spending the money that is necessary to sue a nonpaying client is hardly worth your time and trouble since most of the amounts will not be that large. And clients know that you will have to go to a great deal of time and expense—far more than it is worth—to collect that $250, $1,000, or even $2,500. Collection agencies don't always solve the problem and their cut can be one-third of the proceeds recovered. The best policy is to replace the need for a contract with up-front retainers.

Use of lawyers is invaluable under the right circumstances, but the problem of bad faith cannot be overcome merely because you have a piece of paper with somebody's signature on it.

In engineering and general scientific work, both the quantity and quality of information are vital backdrops for competent problem defining and problem solving. Where you must have access to proprietary client data, the client may request that you sign a nondisclosure agreement. Be careful. What a client may consider to be proprietary may well be technically trivial, but if you sign a nondisclosure agreement you may be sorry later.

To illustrate the absurdity of some corporate policies on this subject, here are two examples. In one case, the consultant had to do some work relating to an accident investigation on the premise of one of the world's leading pharmaceutical manufacturers. It was a very serious accident, involving major injuries and loss of body function. No photographs were permitted and the company corporate counsel was on hand looking over the consultant's shoulder every inch of the way.

Mind you, this is a billion-dollar-plus company. The accident took place in the maintenance carpentry shop. And do you know what the "proprietary" machine was? It was a table saw cutting wood for some maintenance job.

Another interesting example was an inspection on the premise of a large chemical manufacturer. There had been a serious explosion in a piece of machinery during a servicing operation. The plaintiff survived, but was bandaged like an Egyptian mummy. The inspection proceeded without incident. The employer was most cooperative.

About a year later, the consultant received a communication from his client transmitting a nondisclosure agreement from the equipment manufacturer, who was a third-party defendant, requiring him not to disclose equipment or process details. Guess where *that* went?

Here, again, there was absolutely nothing about the machine—or process—that deserved to be called proprietary. In fact, the consultant's original forensic report analyzing product defects was more damaging to the manufacturer than any possible detailed disclosure of equipment or process details. If the case had ever gone to trial, all courtroom testimony would have been in the public domain anyway.

INVOICING FOR BALANCES DUE

If and when you do have to invoice for fee balances due, try to make sure that they are relatively small. Don't be bashful about calling a client, and as many times as you feel it is necessary, in order to inquire about the status of the invoice and when you can expect payment. Don't be intimidated by people who sound insulted because you are asking about money they owe *you*. An overdue balance means that somebody is borrowing money from you and that you are forgoing interest. You are financing them. It's very difficult to charge interest on fee balances due in the consulting business.

Theoretically, it is true that you are entitled to it, but it just doesn't sit well with clients. When fee balances are due on the basis of invoices rendered, expect payment within thirty to forty-five days. In fact, where invoices have remained unpaid over forty-five or sixty days or more, satisfy yourself that it is for a good reason. Most of the time it will involve administrative holdups, but check it out to be certain so you don't encounter any surprises later.

CONTINGENCY FEES

As an engineer, especially if you are a licensed or Registered Professional Engineer, you know that contingency fees are not only frowned upon but may also be unethical. In fact, contingent fees usually are viewed as unethical by our own Engineering Code of Ethics due to the undesirable potential for conflicts of interest to develop if they are dependent upon outcome or size of job, and so on.

Some clients may try to play a nasty game with you. When fees are due as progress payments or as installments, the client will make every payment but the final one. This is never collected or collectible. It usually turns out to be too expensive to sue. The solution is to charge a larger retainer up front that is equivalent to the last one or two installments—sort of like requiring a few months' rent in advance, which you then hold for security.

Thus, when you are using a progress payment financial arrangement, don't put yourself in the position of expecting that last payment or installment to come in on a timely basis. You may never see it. There probably will be enough problems along the way without looking for trouble. But such a client will probably have forgotten that he gave you the initial amount and, therefore, continue to look forward to doing you out of the last payment. However, the joke is on him. He will never receive an invoice for the last work installment because he has already paid the fee. In effect, what he doesn't realize is that each "progress payment" has been an advance retainer.

RELATIONSHIPS

Once you have retainers to work with, you'll be surprised at how much smoother your client relationships become. You and your client know exactly where you stand relative to one another. Also, your client has an investment in you, so he or she is more likely to treat you professionally. You will feel more of a responsibility toward your client, since you are going to try to give true value for that investment and, of course, you would love follow-up business.

When you are just starting up in practice, prospective clients want to know who you are. They want to see you, shake your hand, study your face and the way you dress, and learn about you on a firsthand basis. This is what you want—it means that your client development efforts are beginning to pay off.

Thus, at the beginning, be "physical." In fact, the only time you should not make extreme efforts to "reach out and touch" clients, so to speak, is when you are just so busy, happily, that you don't have the time.

You will enjoy meeting clients, but you may not have the time to invite them to lunch. Of course, they are mostly too busy for this anyway. This is but one of the sacrifices you may have to make as you become busier and busier. Your professional life tends to become more efficient. The pure technological content of your time becomes higher and higher. You may even tend to be somewhat out of reach occasionally. Watch out for this one.

Never be so busy that you are out of reach for long periods of time. Even though your practice may become efficient and very busy, make the policy decision to try to be around most of the time to service your client universe. Do not actively solicit remote engagements from remote clients. A certain percentage of your work will unavoidably come from out of state or across the country, but make it a practice to hold yourself available for clients in the two- or three-state area that you primarily service—unless, of course, your practice is highly specialized and clients are scattered geographically.

Have meetings. Go to lunch. I am not going to presume to tell you who should pay, but it can work both ways, depending upon circumstances. Do whatever seems most natural. Here and there, you might feel that you are being taken advantage of; that somebody is trying to "pick your brains." However, personal meetings are important for a number of reasons. They get you circulating physically, they get you out of your shell, and they keep you from talking to yourself.

SCOPE OF WORK

You will outline your services and make a proposal—even if orally—for the work to be accomplished. Okay, the client buys it. You're in. Now you have to start taking notes. Just what is the scope of work? What will it take? Will it include on-site visits? Interviews or inspections? How many? Where? How long? When? Overnight? Drive? Fly? Room and board? Taxi? Is there a deadline? Any lab work? Do you need any expertise beyond your own? How about consulting associates? Can you estimate the amounts of retainers you will need at each stage, even if only roughly? Can you provide the client with a ballpark estimate as to the order of magnitude of fees and expenses? Write it all down. Put it in a letter. Keep records of everything.

MILESTONE MANAGEMENT

In your own specialty, manage your engagements milestone by milestone. Think in these terms. While you may have the dimensions of the total engagement firmly in mind, do first things first and complete each step as it demands attention. Any particular segment of any particular engagement may be interrupted two dozen times. Attaining the sought-after milestone may seem to take forever, but keep that objective uppermost in your mind. Force each engagement step to its conclusion and then go on to the next. If you need help from the client, don't hesitate to call him. If you encounter administrative or logistics problems, call him. If you meet unexpected obstacles or require additional resources, call him. Don't let the client remain in limbo while you put off difficult tasks. They have a way of not getting any easier and lingering on and on and on. And then you really have explaining to do.

Since you are your own boss, you have all of the authority and responsibility you can handle on every engagement. You are your staff and line. However, in your mind, separate all of the parts so that you understand, at every step of the way, how you have to function, what is required, what type of problem identification, decision making, evaluating, controlling, and so on, is required. Segregate and formalize the tiniest element of each engagement. Don't think of an engagement in a "broad brush" way. This is the best way to forget details and go off on a tangent, to neglect schedules, do the wrong things at the wrong time, and generally lose your grip on the requirements of the engagement. Once you do this, of course, your practice also goes to pieces since you are not managing it. Always remember the axiom that "Management must manage."

Once you learn to separate each engagement and your total business into its parts and learn to keep track of each separate element as a discrete organism with a life of its own for all practical purposes, you will find it a breeze managing twenty, thirty, or even a hundred engagements at a time. However, you have to start thinking like this very early in your consulting career. As an employee, everything was done for you except the task at hand. Even if you are an executive or manager, you have a relatively smaller domain to deal with compared with the entire business, no matter how broad-gauged you may be. And, of course, if you are already thinking in this way about the highly responsible job, organization, and all those people you are managing, you might as well be doing it all for yourself and reaping the financial rewards.

Each individual engagement has a schedule to meet, even if it is set by you. Therefore, you are your own production control manager. As a professional, you set your own standards and comply with those of your specialty and/or industry. You will also utilize established, if not innovative, methods and procedures. Production must be on a timely basis. Again, you don't want to ever leave a client dangling until he or she forgets about you, and the engagement and you both go out of style. The client may even want his or her money back.

ENGAGEMENT QUALITY CONTROL

And then there is quality assurance. You must exercise stringent quality control over every engagement and every segment thereof. You should only be so busy, of course, that quality control of engagements becomes a problem. But, you must be certain that your considerable expertise in your specialty is built into the engagement in a reliable and quality way. You must resist compromises. You must develop a kind of checklist mentality, using checklists for everything. Eventually, you may even be able to standardize some of the work you do through the use of checklists. This can save you time and also save your clients money. One of your competitive advantages may well be that you can respond faster and for less money, or else offer more services for the same fee as the competition.

For example, never charge a client for telephone consultations if the time is relatively short, say fifteen minutes or less. As a matter of fact, typically don't charge prospective clients at all for telephone consultations. You may spend a significant number of hours every week on the telephone screening cases, talking with old and new clients, and generally providing a great deal of free information. That's great. It will come back to you time and time again. Sometimes it may become burdensome. Do it anyway. Make up for it in the level of fees you charge. But this type of missionary work will pay off. Clients will come back for engagement-level consultation where regular paying casework is involved. Not even doctors and lawyers can account or charge for every single minute of the time they are providing potentially valuable advice to patients or clients. It is built into the overall fee structure.

A STRATEGY FOR BUILT-IN ENGAGEMENT VIABILITY

In setting the stage for consulting engineering engagements, it might seem realistic that the more formal and sophisticated your effort and interactions with your prospective client, the more solid the engagement will be. Those classes of engagements where proposals, contracts, fancy laboratory analysis, and computer simulations are necessary may be right up your alley. You might thrive on the high-powered, high-technology assignment.

However, in many areas of engineering the best way to build a successful consulting practice is to aim at the more "mature" and mundane segment of your prospective client universe. Here you may find that poor management decisions and problem solutions have been the rule over the years, and that it would take a hundred consultants to undo all of the damage that has been done. You may even find such assignments interesting and challenging. More important, you can make a lot more money solving the less sophisticated problems and build a more substantial consulting practice thereby.

In addition, you will probably find that consulting fees flow more freely in this field of application. This is because prospective clients have real needs to serve and real problems to solve that have been draining profit dollars for years. Now you come along and hold out the promise of correcting some of these situations.

Furthermore, if it runs true to form, you will begin to perceive patterns in the problems that need solving. You will be able to gear up for an entire class of engagements, even standardizing your approach to a greater or lesser degree and offering top value for your client's consulting dollar.

The one-of-a-kind high-technology engagement does not offer you any way to give yourself significant commercial leverage. Problem solving in such a situation is a fight every step of the way. And your client may ultimately get tired of paying you for follow-up hour after follow-up hour due to the uncertainties of the work.

Just consider for a moment what companies are most successful and have been able to grow the most. Are they the ones that sell only custom-tailored products or offer only customized services? Or are they the General Motors, Continental Cans, Polaroids, and General Electrics of industry? The company that discovers a standardized, quantity production niche for itself is frequently the one that survives, grows, and prospers. There are important lessons here for the service industries as well, including consulting engineering.

Of course, you can still pride yourself on your technical prowess and seek out a modicum of "high-end" business. A few of these can add nicely to your qualifications profile. But look hardest for relatively simple and mundane problems, and perhaps even the glaring examples of bad engineering. You may be surprised at how many there are that have simply not been remedied and that are eating up potential profits of prospective clients right and left by not being addressed at all, or if so, only inadequately.

If you have high expertise in a state-of-the-art specialty, then look for old problems to solve in new ways. *Innovate.* Do not try to invent. Innovation is the best source of consulting fees.

Actually, there are many examples of successful companies in every industry that innovate, even imitate, or both. But there are precious few that invent at every turn. We engineers are trained in all these problem-solving techniques, but as a general operational style, innovation will be a much more practical, dependable, and lucrative foundation for your practice.

CHAPTER 6

PLANNING, OPERATING, AND CONTROLLING YOUR CONSULTING PRACTICE

THE GRAND ILLUSION

All this time I'll bet you have been under the delusion that, since it's your business, you're the boss. Right?

Wrong. Almost everybody else is your boss *except* you. The fact that, to a greater or lesser degree, you set up your own schedules, do your own work, engage and pay your consulting associates, develop your own clients, personally receive all retainers and other payments, pay all of the bills and do all of the billing, try to plan for the future and implement your plans is irrelevant to a significant extent.

All of this merely gives you financial security and independence, professional satisfaction, and the potential for long-term growth that no one can deny you or take away from you.

However, the real bosses are your clients, your suppliers, your secretary, your associates, and anybody you depend on to keep things going smoothly, or even at all. This means that you have to be responsive to client requirements and highly sensitive to the needs of your employees, associates, suppliers, and so forth.

Sometimes you may seem to hit nothing but snags every time you turn around. Learn to shrug it off. There are far more times when things go right and everything falls into place.

WHEN YOU ARE THE BOSS

When you work for somebody else, the worst thing that can happen is that you get fired or laid off. When you are in business for yourself, you will go through periods when you are supposed to be in two or more places at the same time, when the critical supplies you ordered don't come in on time or arrive in the wrong size, when your copier serviceperson breaks down on the road while en route to your office, when your typewriter doesn't work, when the phone rings and a very critical situation arises just when you are leaving your office to take care of another very critical situation, when your brand-new luxury automobile needs a new electronic control module eighty miles from home just as it's starting to snow, when you can't find an important file in your meticulously cross-indexed filing system for two days, and when that rush job you were supposed to do gets postponed after you already cancelled your whole day for it.

When things go wrong, that is when everyone finally lets *you* be the boss. Be prepared for things to go wrong more than they go right—that way, you will never be disappointed. It's an extreme viewpoint that will serve you well. The scientific method you learned so well has lots of business applications, and more of the important things will go right rather than wrong.

State the problem and/or objective, formulate alternative hypotheses and plans, test the effectiveness of your plans through experimenting and learning by doing, analyze your successes and your failures, draw conclusions and fine-tune your efforts. Perhaps you'll have to reformulate your plans and start all over, but for every effect there is a cause. Work backwards from unwanted results, redesign your efforts and set things in motion again, as many times as you have to in order to achieve your objectives. It may take three months; it may take three years.

THE SCIENTIFIC METHOD

The process of management is just an extension of the scientific method. Establish your objectives by gathering information, synthesizing it, planning, and then making some decisions among alternatives. Direct the attainment of your objectives by organizing, communicating, keeping yourself motivated, and continually developing yourself.

Finally, evaluate and measure your results so you can control your practice. It is at the very beginning, when you are establishing your objectives, that you should try to be innovative in how to approach the start-up opportunity. Notice the emphasis on "opportunity" rather than treating it as a "problem." Problems require solutions. Every need for a solution represents an opportunity so that, in effect, a problem can be viewed as an opportunity. If you think that this is too simplistic a viewpoint, stay right where you are and do absolutely nothing.

THE THIRTEEN RULES FOR BUSINESS SURVIVAL

Remember the following thirteen rules for business survival:

1. Planning is the cheapest part of running a business.
2. Every decision is a financial decision.
3. Management must manage. That means you.
4. Work smart and not hard. Knowledge is leverage.
5. Always have storm cellar plans in reserve.
6. Know your customers and be sensitive to their needs.
7. Know your competitors and be sensitive to their moves.
8. You are your own most important asset (and maybe your own worst enemy).
9. Change and grow, or wither away and die.
10. Good management is hard work and positively *not* hassle-free, no matter what anybody tells you.
11. Don't talk in jargon, buzzwords, or faddish acronyms.
12. Be nice to everybody, but tactfully make it known that you don't just get mad . . . you get Even!
13. While you are starting up your consulting practice, try to keep your job for as long as possible. Make all your consulting mistakes while you're still employed full-time. Make your job work for *you* for a change.

THE FUNCTION OF MANAGEMENT

The classical functions of management are planning, organizing, directing, coordinating, controlling, and staffing. Of course, you are your own staff—all of it. The dynamics of management apply to the large business, the small business, and even your own little start-up. The process of management would be easy except that it is usually embedded in a morass of uncertainty, incomplete data, and inadequate measuring tools.

In getting started on your own, it will be useful to you to think of management as an exercise in commercial intelligence. Again, the intelligence process consists of three stages as follows:

1. *Collection.* This involves the acquisition of all information believed to pertinent, sometimes called raw intelligence data.
2. *Evaluation and production.* The sifting, sorting, and judging of the credibility and utility of assembled data, plus the drawing of pertinent inferences through data analysis, and the interpretation of such inferences consistent with your multiple requirements and objectives.
3. *Dissemination.* Communicating intelligence findings in forms most suitable to specific applications such as planning, and also including communication to all responsible for implementation of decisions. This includes your clients.

MANAGEMENT BY OBJECTIVES

It is only through the effective and efficient application of information that you can then establish your final goals. This is planning. Always manage by objectives. You have done this in your engineering work for your employers for years; now do it yourself. However, be certain that your objectives are realistic and attainable.

MANAGERIAL CONTROL

When it comes to control, remember that the standards of performance that you use and that have been established through your planning are not automatically self-fulfilling. They are means rather than ends. You have to work hard at control. Controls can be preventive or prescriptive. Performance variances from your targets must be scrutinized continuously. Corrections must be made as soon as feasible, *if* feasible. If variances cannot be corrected, find out where you went wrong and start over, just as in engineering.

Be sensitive to the concept of "control"—it is most meaningful only if you can exercise it while there is still time to do something about problems. So always be on top of what is going on in your business. Don't let it get away from you. If you can't control a problem this time, be certain you can the next time. Establish your own performance standards. Determine performance variances. Evaluate your performance and initiate remedial action. Control is ongoing and future-oriented. Planning accounts for the future results of present objective setting, decision making, and risk taking as well as the development of new or changed objectives and courses of action. Planning is a prerequisite in all other management functions.

My first secretary was a highly competent young woman. She typed extremely well, took shorthand, could use a dictating machine, and had incredible English skills. She was phenomenally effective and efficient. As the practice grew, she tended to take more and more responsibility on her own initiative. She scheduled appointments, interacted with clients, and ran the office. Even though it was a small office, there was a lot to do. An engineering practice is very intricate. Her salary also increased accordingly, of course. Things went great.

However, there came a point when I felt I was beginning to lose control. I was not talking directly with clients as much as before, and I depended upon my secretary to tell me what my schedule was. Whenever I returned to my office after being out for a few hours, she would brief me on what had gone on in my absence, and on what I had to do for the rest of the day.

This was fine. But little by little, I felt less involved in the management of the practice than before. She ultimately outgrew my little consulting practice and went to work for a giant financial corporation, where she is today a successful manager.

This also proved to be an important turning point for me. I got back into the management of the practice. I began to speak more and more with clients again. Control was returned to my own hands. When I realized what I had just gone through, I decided then and there never to let it occur again. Even to this day, with a highly successful practice and exceptionally capable secretaries, I still manage the business from start to finish.

Don't give up control by default. Don't let even an incredibly effective, efficient, and skilled secretary or other employee usurp your managerial prerogatives until you are ready to give them away by design.

CONFLICTS AND STYLES

Your entrepreneurial incentive will quickly become a managerial imperative. But don't confuse the two—they are quite different. If you give up the managerial imperative, you may inadvertently close entrepreneurial doors. For example, by not continuing to interact with clients, but letting my secretary do that, I'm certain I missed opportunities to discuss new cases with them. Even though I was doing work for them at the time, my involvement with clients became less intense.

Clients need intense relationships with consultants in areas of their expertise. You need continuous client development. You must not permit barriers to be raised between you and your clients. A good secretary may sometimes create as many problems as he or she solves, if you're not careful.

MOUNTAIN GOAT MANAGEMENT

Sometimes you will feel as though you're managing by jumping from crisis to crisis. This is officially termed "mountain goat management." Sometimes you can't avoid it; just don't make it your favorite operating style. And remember, although parallel management is a time-saver compared with series management, there are limits to when it can be used. For example, a man can't make a baby in one month by getting nine women pregnant.

In consulting work, you must expect your fair share of crises that may extend over a few days. Some of the worst ones have to do with super-rush jobs that occur all at once. But somehow you manage to get through it. Setting priorities is a juggling act 75 percent of the time. In my particular case, the schedules of my clients, most of whom are trial lawyers, are busy due to the demands made on them by the courts. Their schedules are subject to rapid change and are almost always up in the air. Consequently, even though I myself am not in court very much, and even though I happen to be a well-organized person, I am almost always on call for one case or another. So, since my schedule depends on client schedules, and their schedules are unpredictable, mine is even worse. But I've managed to survive it for over fifteen years.

Mountain goat management is certainly not "hassle-free" management. In fact, putting out fires constantly is symptomatic of more deeply rooted problems. And what may be erroneously regarded as problem solutions in reality merely address symptoms or syndromes. The small business owner-manager cannot afford the luxury of building cushions, buffers, and safety factors into his or her operation. Of course, if and when things turn sour, even the giant corporation can fall on hard times as the layers upon layers of fat and inefficiency are pared. When this happens, the bug business becomes leaner, but it can also become vulnerable like the small business, only on a larger scale.

THE MORAL

The moral is that you must make strenuous efforts to get to the bottom of every problem you encounter in your practice, and in time to do something about it. Identify. Evaluate. Control.

From one standpoint, the executive or professional who always has a clean desk is up to no good. It may not necessarily be a sign of good management—it may merely be a sign that he or she isn't doing anything.

This doesn't mean that your desk should always be littered with mountains of paper, reports, books, files, and miscellaneous assorted "stuff." But the consultant is in the "applied knowledge" business. If you have consulting problems on your mind, you're going to (or should) have working materials on your desk. The more problems you are involved with at the same time, the more "stuff" is going to cover your desk and your credenza.

Periodically, or when you simply can't function any more because of all that organized chaos on your desk, clean up. This should work pretty well for you most of the time.

"HASSLE-FREE" MANAGEMENT

By "hassle-free" management, one recent writer simply meant that management and employees are on the same side and behave that way. They are not in constant conflict or turmoil. Management manages professionally and does not view its work force as an adversary and vice versa. There is no discernible corporate combat; things get done reasonably smoothly.

But there is another side to "hassle-free" management. No matter how smoothly running your internal operations may be, you are always faced with hassles in the form of your external relationships. The big business has twelve publics, for example, as follows:

1. Middle management and first-line supervision
2. Employees and their families

3. Unions
4. Customers or consumers
5. Suppliers, vendors, and subcontractors
6. Dealers and distributors
7. Stockholders and/or bondholders
8. Bankers
9. Investment bankers
10. The competition
11. The federal government
12. The local community and/or general public

Even as a small businessperson, you may still have as many as ten publics with which to contend:

1. Your employees
2. Your customers or clients
3. Your suppliers
4. Your dealers and distributors
5. Your banker
6. Your competitors
7. The government
8. Your local community
9. Your consulting associates (who are subcontractors)
10. Your own family (if you work six or seven days a week, twelve hours a day or more on average, then watch out!)

Doing a balancing act with all or even some of these dynamic forces at work can create stresses and strains that pull you apart and in different directions. But as before, you somehow manage to survive. You just have to learn not to be intimidated to the point where you throw your hands up in despair.

"VOODOO" MANAGEMENT

There is another management style that you should avoid. This is commonly referred to as "voodoo management," management by "cult" or by "trance." It is a new-old management system based on an animistic, ritualistic belief that if you ignore problems long enough, complain about them enough, or place blame elsewhere as the official line, they will disappear. It utilizes various secret processes of logic, bypasses generally accepted ethical, scientific, and economic principles and considerations, and can be very effective in a sociocultural environment where there are no clients, bankers, competitors, suppliers, the IRS, or other real-world troublemakers to stir things up but only pure, unadulterated engineering problems to solve.

However, as usual, all is not lost just because you make a mistake. But at least learn from your mistakes, keep your guard up, and learn from other people's experiences.

WEEKEND PROJECTS

Once you are established, consulting is a seven-day-a-week commitment. If you are a "part-timer," the weekends are ideal for actually doing the detail work, but you should also spend some weekend time planning. Design and write your capabilities profiles and brochures. Study. Engage in some type of project or work connected with your consulting practice-to-be. It will make consulting seem more real to you. And you will be fresher than at the end of a long day at your regular job. You can also probably set up some appointments here and there. The types of clients you will have may well be working on Saturdays and may even look forward to the kind of break in their routine that meeting you will permit. Also, the phones won't be ringing and the atmosphere will be a little more relaxed.

Set yourself some projects that tie in with your prospective consulting practice. For example, you may be quite expert in specific areas of engineering and technical problem solving. If these are also the areas you are planning to develop, prepare some model solutions to typical problems complete with data sheet forms, analytical procedures, computer programs, and so on. Prepare "model solution packages" that you can even use to impress a client down the pike.

Or, if there are some gaps in your knowledge of your subject, fill them in quickly by studying on weekends. Take a night course during the week. Once your practice gets rolling, you may not have the chance to go back to school for a while—perhaps even not for years.

THINK AND REACH

Whatever your style and however your practice develops, never be so busy that you don't spend a few hours a week thinking about the future. Try to relate your specialized interests and capabilities to the broader world around you. Expand your horizons. Constantly be reaching. Always be on the move intellectually in relation to the broader implications of specific engagements or the needs of, and prospects for, your business.

"INTENSITY" FACTORS

Keep in mind the difference between capital-intensive versus cost-intensive decisions. For example, buying an automobile or production machinery is capital-intensive, and you have to make significant down payments. But renting or leasing is more cost-intensive. In the former case, you can deduct depreciation. In the latter case, your lease payments or rental expenses are tax-deductible. Leasing is more expensive than owning. But if you are faced with a large capital outlay to buy an asset, seriously consider the leasing alternative. You may have a better use for the capital that would have otherwise gone into an initial investment outlay.

Another example is the emerging trend to replace cost-intensive human labor with capital-intensive automated and robotic equipment, systems, and processes. There are many business advantages, as you well know as an engineer.

Still another is that although you should stay in your house as long as you can, as your practice grows you may toy with buying a small building for your office instead of renting. This is also capital-versus-cost-intensive, as well as a lease-versus-buy decision. Remember, however, that buying rather than leasing also entails increased responsibility to manage what you now own. This takes time, which is also valuable.

THE PLANNING PROCESS

The most logical plan for engineering consulting activity should be to (1) do some good, (2) enjoy oneself, and (3) make some money. But whether a business is small or large, managers have to plan. Planning is one of five or six generally recognized managerial factors. These typically include (1) planning, (2) organizing, (3) staffing, (4) leading, (5) coordinating, and (6) controlling.

Plans can be short- or long-range. They can be daily plans, five-year plans, or even ten-year plans. Naturally, the further out in time we go, the more fraught with uncertainty they are. But the function of plans is to minimize risk, identify alternative courses of action and their consequences, and facilitate rational selection from among alternatives for purposes of conducting the business. A properly designed and constructed plan permits the future to be as "surprise-free" as any future can be, recognizing that the process is unavoidably and interminably embedded in a morass of uncertainties at every turn.

Planning is a process. Set your objectives, both short- and long-term. The shorter-term they are, the more explicit and quantitative they may be and, probably, the more achievable. But they should have both generalized and specific content. The longer-term they are, the less specific they will be.

It is necessary to be careful that all assumptions built into the plan are explicitly recognized. Mere projections of past trends or current performance may not be relevant, realistic, or adequate for planning progress. It is also necessary to set performance standards in order to measure progress toward both short- and long-term goals.

Plans should be reasonably flexible. Alternative plans should be ready for implementation if the primary program proves unworkable. It is desirable to define a range of acceptable performance. What is the optimistic outcome? Pessimistic? Most likely? What conditions are implied by one or another alternative future? What performance standards should be used? Should they necessarily be the same for each plan? What outcome values will define optimistic, pessimistic, or most likely performance attainment? What goals can be quantified? What goals are qualitative?

PERFORMANCE MEASUREMENT

In the smaller consulting engineering practice—or any small business, for that matter—gross fees or revenues per week, month, or year will naturally be among the performance measures as well as before tax income or profit per year. For the engineering practice, number of engagements and number of new clients per year can also be performance measures, number of inquiries, and number of reports prepared. Number of billable hours is a measure, and so is number of hours expended per engagement. Client development and advertising effectiveness can also be measured, such as number of new clients per client development hour, advertising dollars spent per new case, ad dollars per prospective new client inquiry, and so on. Try to screen out those inquiries that are new versus those from old clients or referrals from old clients.

DON'T BE OVERTAKEN BY EVENTS

The pessimistic plan outcome alternative should be accompanied by a "storm cellar" plan. An optimistic outcome should trigger a range of actions to capitalize on successes on a timely basis. A highly successful ad might suggest that advertising be expanded, as well as suggesting market reach in terms of other places to advertise.

If you find that you are being overtaken by events, rethink your strategies, tactics, and performance. Most important, analyze your situation until you understand what went wrong. Try to make next time more surprise-free. Don't let lightning strike twice, or even three or four times. Too many businesspeople don't learn from experience.

Even a small ad can help build a sizable business if it strikes at the heart of the prospective market. One consultant placed a very small ad in a publication he knew his prospective client universe would read. He has been using variations on this ad every week for the past fifteen years. It worked. The cost has gotten much greater, but it has been worth it. At present, his practice is over 95 percent referral, but he continues to advertise because he wants his clients to know that he is in the engineering business for keeps. So many engineers go in and out of consulting that he doesn't want there to be any questions about his own intentions. He would run the ad even if it didn't draw at all.

THE FUTURE IS WHAT YOU MAKE IT

As you continue to agonize over the future of your consulting engineering practice, decide where you would like to be three, five, and ten years from now. What will it take? At what point do you want to have an office outside your house? What do you want to be the scope of your practice in three, five, and ten years? How many employees and professional associates do you think you will need—or aspire to—in your practice at these points? What are the alternative futures in terms of specialization? What are the likely client universes of interest? Do you envision full-time staff? To what extent? How many full-time and how many part-time subcontractor-type consulting associates? Do you want the business to grow beyond your personal span of control, do you want to continue to do purely technical work, or do you want a combination of both managing and technical engineering?

The nice thing about planning your own business is that it's an open game. Naturally, you start with what you are today, but when you consider the future you have many options. Your present specialty may really be just a little piece of a larger field or activity that you can slowly ease into over time. Your present practice would be but one of a diverse number of activities that, together, comprise an integrated spectrum of capabilities.

Once you get rolling, there's no stopping you. Any limitations you have will be significantly self-imposed. The "balance" in your life will be more a function of the extent to which you are willing to let the demands of your profession and practice be controlling. Technically speaking, almost all fields of engineering are experiencing explosive growth. Engineering, medicine, and law are not very different in these respects, nor is the income potential if you are in private practice.

PRACTICAL PLANNING

Big business management develops plans for every functional area, including R&D, engineering, manufacturing, finance and accounting, advertising, sales and marketing, employee and community relations, legal, and so on. Every department, plant, division, group, product line, and marketing channel has its own plan. Objectives are set, financials are developed to the extent feasible, action programs are formulated and set in motion, management audits are conducted to assess conformance with and variances from plans. Capital budgets and flexible or variable expense budgets are constructed to support goals and performance needs.

Overall, the organization is poised to move ahead on the basis of some "grand design." Of course, things don't always work out the way they are planned, but the planning process educates management and employees alike. The exercise does, in fact, engender greater awareness of what is going on and what corporate performance is likely, depending on which conditions prevail, as specified by the plan for alternative scenarios or futures. Indeed, the future is somewhat "surprise-free" and less "shocking" than it otherwise would be. A large company five-year plan can run hundreds of pages.

However, "planning" will not be so formal for the small engineering practice. It may even be mainly in your mind—and that's fine. But plan. Keep a file folder on "Planning." Make some notes from time to time. Make time to think about your future. Nobody else will. Otherwise, your future will be some residual due to default.

Look at your situation. On a scale of 0 to 100 percent, what is the likelihood that your company will survive the next three, five, or ten years? Surely you're with a company that will stay in business forever. Right? What is the likelihood that your company will make a respectable income-to-sales ratio and return on assets over the next three, five, and ten years? What is the likelihood that your employer will not experience a downtrend in sales and profits in three, five, and ten years? What are the chances that you will not be laid off or fired? What are the chances you will want to continue working for your company? What is the likelihood that commuting traffic will stay the same or increase intolerably?

What is the likelihood that you will be promoted? What is the likelihood your salary will be doubled, tripled, or better, if you stay with your employer?

What are the chances your boss will still be your boss three, five, and ten years from now? That you will take his place? That he will be promoted, laid off, retired, fired, transferred, dead? That you will be transferred? To where? And so on.

Try to answer these questions and assign best estimates to them. How does your future look from this perspective? Better or worse than you thought? Have you been honest or dishonest with yourself? How likely is it that you will become a manager or executive with your present firm? A competitor? What is the probability you will have to change companies? Industries?

Reevaluate yourself and your situation every five years starting *now*. Lay plans to advance in your job and improve your income. It is the conscious effort to plan and influence your destiny that will start you moving. You have to be a "self-starter."

Planning ultimately comes down to dollars and cents. Although the process of planning will cost you nothing, each alternative future entails a cost of accomplishment. You have to decide which scenerio is optimistic, which is pessimistic, and which is most likely. These alternative futures will change from time to time, as the future actually unfolds. At each five-year planning milestone, you may find that your current alternatives are quite different from the last ones. Actually, of course, you will be evaluating things continually, not just every five years. But major evaluation is important every five years at a minimum.

Look for both "chance" and "assignable" causes for scenario changes. Capitalize upon both. Expect that both will occur. Be alert for randomly occurring opportunities to grasp, but also attempt to "design" change.

CHAPTER 7

FINANCIAL ENGINEERING FOR EFFECTIVE BUSINESS MANAGEMENT

FINANCIAL AND PROFIT PLANNING

The first law of management is that *management must manage*. The second one is that *every decision is a financial decision*. However, one of the stumbling blocks to effective and efficient managerial control is failure to comprehend the interplay of all business elements.

For the engineer, therefore, the opportunity presents itself to model the business and study the interactions of major factors. Graphical and mathematical modeling approaches can assist in acquiring an overview of how the business behaves in its financial and economic aspects. Such models need not be overly complicated. Naturally, they will be oversimplifications of what is actually happening, but they can be invaluable in comprehending how management decision making, risk taking, and resource allocating affect business performance, and in understanding how business responses to management objective setting affect both shorter- and longer-term business structure, condition, competitiveness, and so on.

The techniques and analytical formats of these next two chapters comprise a useful foundation upon which to build a sound understanding of why a business behaves the way it does. They permit us to determine the constants of the business for purposes of estimating responses to varying pressures and parameters, and the ranges of probable performance.

Most importantly, they permit us to gain an essentially instantaneous yet integrated picture of the business in its major aspects. They constitute a threshold over which all other financial planning, analysis, and control tools and techniques must pass, and upon which all must stand if they are to be relevant and effective.

To talk about financial and profit planning for your start-up consulting practice may seem like overkill, but you should understand the financial anatomy of a business. Break-even points and cost-volume-profit relationships, for example, are very real, as well as volume and efficiency-type budget variances, cash flows, the time value of money (involving compound interest, present worth, future value, discounted cash flow, profitability indexing, annuities, and other actuarial-mathematical-statistical subject matter), capital or asset planning, ratio analysis, and so on.

However, this chapter and the next are *not* going to be a crash course in basic accounting, standard engineering economics, or even managerial finance, accounting, or budgetary control. Rather, we're going to look at a few key concepts and tools that are essential to understanding how a business actually works for purposes of managerial planning, analysis, and control, as suggested above.

Interestingly, financial management is a subject particularly amenable to the creative use of flow and formula charts, general graphic presentations, elementary calculus, mathematical models, and other common engineering tools. And since there is nothing as practical as a really good theory, although there are naturally some practical limitations to the utility of the "financial engineering" approach it facilitates incisive analysis of business "microeconomics." It also encompasses the design of actionable performance standards, pragmatic performance monitoring, and comparative analysis.

The bottom line is that financial and profit planning, analysis, and control are not the exclusive domain of accounting or finance majors. Engineers also have much to contribute, as well as learn, in this fundamental and critically important area of business management. In fact, a number of important financial and accounting techniques and subjects were "invented" and developed by engineers many decades ago.

DON'T LOOK AT IT AS YOUR OWN MONEY

An important thing to keep in mind is that if you think of the money it takes to carry out your plans as your own personal, hard-earned, bigger-than-life dollars, rather than as resources available to your business based on its rational needs, you may never do anything. Too many small businesspeople don't grow with their businesses because they cannot shake the typical small-business mentality. You have to forget that the dollars are coming out of your own pocket. What are the needs of the practice? If you can get over this hurdle, you will be much better able to run your business objectively, successfully, and, therefore, professionally.

The change in viewpoint from worrying over *your* entrepreneurial dollars and the objective financial needs of the *business* is a critical milestone that few small businesspeople ever reach and a transition that few really make. They simply cannot overcome their innermost fears about the risks of being in business, such as the risk of loss. There is, of course, always the possibility that dollars spent will not earn as high returns as you would like, or that you have planned, but there can be an even greater risk in not doing *anything*. The forgone profits from inaction can be more damaging, ultimately, than even only modest returns on an investment prudently made in good faith. Such an investment is a true "businessman's risk."

Either way, however, there is no better investment than that to be made in your own business. The trick is to have a business that needs capital to grow, and to nurture the business so that it has growth potential. A consulting engineering practice, or any professional practice for that matter, is such a business.

RISK FINANCING

Risk assessment is a natural, vital, ongoing process in which you, as president and chief executive officer of your consulting business, must engage. Even in this real world of almost infinite uncertainties, a rational and professional understanding of risk factors will permit you to at least contemplate, if not design, the array of alternative futures you may encounter. Then, with your eye on what the business needs to survive and progress, rather than how many dollars you are taking out of your pocket, you can lay action plans.

You may well decide to borrow working capital from your banker. If you don't have one, get one. Maybe the bank that holds your mortgage or your auto loan will be suitable. You have a good credit rating with them. If you don't have a credit rating because you don't borrow large sums of money, or because you don't own your own home, you still have credibility as a professional person.

Use local community institutions wherever possible. "Grass roots" contacts can be helpful. People know you, your spouse, or your family. They may know you by reputation. It might make sense to join your local Chamber of Commerce. Many contacts can be made relating to the purely business side of your practice. This point deserves stressing. Being a member of your local Chamber of Commerce and, perhaps, a local service organization such as Kiwanis, Rotary, or Lions can be beneficial in expanding your horizons. There are lots of small businesses that can supply necessary services once you get started. There may even be some potential client contacts and networking possibilities. You will find people eager to help you. In some instances, after all, if they can help your business become successful, you may even become a customer of theirs.

Never lose your entrepreneurial perspective, even though you have to try to forget that the money you spend is coming out of your own pocket. One of the reasons that big business planning fell out of favor in recent years is because it had lost touch with reality. Corporate staff planners—like many corporate executives, in fact—never really learned to think and act entrepreneurially (of course, they couldn't due to the very nature of the staff concept in the corporate environment). Consequently, their "plans" were frequently theoretically neat but technically, economically, managerially, or politically impossible. Notwithstanding, "planning" has not disappeared as a critical business function. It is now a more integral and natural part of management thinking on all levels. It has become a bona fide tool for managing rather than necessarily a separate staff function requiring armies of planners and an organizational straitjacket.

In your own little sphere, you will find that your built-in, conscious, and disciplined concern for the future of your practice will be beneficial. Be rational, realistic, flexible, and alert. Scrutinize your intuition as closely as you can. Marshall facts to support your objective setting, decision making, risk taking, resource allocating, problem defining, and problem solving. Use the commercial intelligence process as adroitly as possible. Don't manage on a makeshift basis if you can avoid it. It is reckless. From your own experience with various employers, no doubt you have seen the sad consequences of this.

WATCH OUT FOR DEFICITS AND CASH FLOW

One of the things you must be particularly alert to is the flow of financial values in your business. Outlays are of two types: those items that can be expensed—such as postage, stationery and supplies, books, gasoline, and so on; and those items that represent capital outlays—such as a desk, typewriter, new automobile, computer, and so on. If you don't understand the difference between the two, you can get into real trouble.

A simple case study will illustrate the kind of trouble you can encounter. The profit position of a small wholesale meat business owner looked great and his gross margin was large, but he was always cash poor and could never pay his bills. He was borrowing money constantly and went to the SBA because he couldn't figure out why he was always broke.

The answer, of course, was simple. He was very busy. He was doing a large and growing volume and needed increasing amounts of meat inventory. So he was continually building inventory and paying bills currently from his bank account.

Analysis showed that he was depleting his bank account due to the increasing quantities of meat that he was buying for resale. The profits were real, of course, but wasn't giving himself time to breathe, financially speaking. Incoming cash was going back into new inventory almost immediately, which is a balance sheet account. The way it turned out, his cash purchases of new meat inventory assets were progressively somewhat larger than the cost of goods sold that was showing up on his P&L for the preceding period, and the cumulative effect was to nibble away at his accumulated cash until he was running an overall deficit that was not being made up by profits, even though robust. That is, the cost of goods sold, a P&L statement account representing withdrawals of meat from inventory that went into sales for the period, was always less than his actual new inventory purchases. He did not realize that there was a gap between the two representing incremental capital outlays for inventory in excess of goods sold as reflected by his income statement.

He was continually spending more than he was taking in. He never stopped to let the cash in his bank account catch up with his purchases of new inventories. So he was always running a cash deficit even though he was making high profits on paper. He hadn't run a "surplus" of cash profits over incremental capital or assets (i.e., inventory) outlays for almost a year. He was looking only at his income statement and not his balance sheet. He was not connecting the two.

So when you make your first $10,000 of profits, don't run right out and spend $12,000 on new office equipment, car down payment, redecorating, and so on. You may need the new equipment, but it may also cause you to run a cash deficit for the period, instead of a surplus despite your profits.

Think about your business on a total financial flow basis. You can need cash even when your income statement keeps saying you are earning profits. The faster your business is growing, and the more incremental assets you need to support incremental sales, the more vulnerable you are to running a deficit. In such a case, you may want to borrow funds from your bank.

Deficits make the world go around. Big businesses borrow money all the time because they are normally running deficits. Even the most successful businesses cannot always generate enough cash from ongoing operations, even in the best of times, to satiate their appetite for new capital. In fact, businesses are most vulnerable to generating deficits precisely when they are thriving and growing the most. They cannot "pay as they go."

Their capital needs are greater than the profits that are generated. If they limited capital expenditures to earnings alone, they could not expand and keep up with the competition, or take advantage of new opportunities. Management would have to "disinvest, decapitalize, and retrench" just when more and more funds are needed merely to stand still in a climate of accelerating change. Later sections will explain this.

The small businessman cited above, and it is a true story from the annals of the SBA, had an unknowledgeable view of financial management. The P&L was all he looked at. He did not have an integrated outlook on his system. He did not understand how the P&L and balance sheet are interrelated and work together. He did not know how to combine them into a single-action concept for purposes of management and business planning, analysis, and control.

Exhibit 7-1 is a formula chart for what can be functionally defined as "financial makeup," or *FMU*. Income after taxes (i.e., income before taxes less federal income taxes) less dividends, plus extraordinary gains (and less extraordinary losses) equals retained earnings. But a business also makes capital acquisitions and disposes of some items of property, plant, and equipment as well. It also makes net acquisitions of current and other assets.

Notice that income before and after taxes also includes the usual deduction for depreciation expense. In fact, income after taxes

EXHIBIT 7-1

RATIO PYRAMID FORMULA FLOW CHART: FINANCIAL MAKEUP

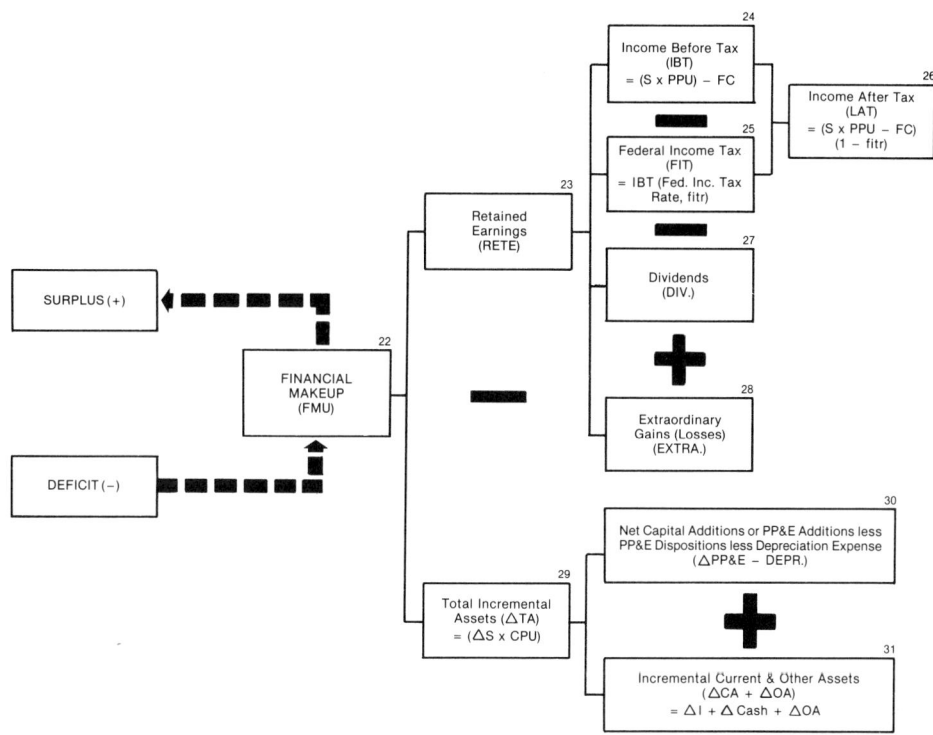

plus depreciation, which is a noncash outlay, is termed "cash flow" by financial and securities analysts, accountants, and executives. But net fixed asset or PP&E additions (i.e., additions less dispositions) are "net" of the depreciation that is deducted from the values of such asset increments and decrements to account for ordinary wear and tear (or deterioration) and obsolescence.

Thus, if we deduct total (actually "net") incremental assets from retained earnings we are essentially cancelling out the noncash depreciation "expense." Through this subtraction, we finally end up with a surplus (if retained earnings are greater than total incremental assets) or deficit (if retained earnings are less than total incremental assets). The result of this subtraction is financial makeup, or *FMU*.

Again, if it is positive, a surplus is generated. If it is negative, financial makeup is actually a deficit that results from ongoing operations.

We will return to surpluses, deficits, and financial makeup a little later. First, we have to lay some additional groundwork in order to be in a position to explore more fully its ramifications for purposes of planning, analysis, and control.

ENTERPRISE STRUCTURAL PROFILES: MANAGEMENT MIND-EXPANDING

A well-run business is no accident. It is a precision instrument because management has made it so. The diverse, antagonistic forces acting upon it may strain it to the limit, but management retains control. Management balances and compensates; management acts today and plans for tomorrow. Management must be entrepreneurial. It must be influential in changing the structural profile of the business to respond to changing needs and new commercial opportunities. It must repair the damage done by adverse economic conditions and/or accommodate to them on a timely basis. The essence of control is doing something about emergent or impending conditions with potentially adverse consequences before their unbridled impact is felt.

A well-run business, like any machine, has an underlying structure—a purposefully designed, rational structure—which responds in an essentially orderly and meaningful manner to the restless and sometimes volatile conditions of its total business environment. And, like a machine, its efficiency of performance can be measured, its responses governed. Or, at least, its underlying structure can be seen to exhibit the pattern implicit in management style or, conversely, the lack of management consistency and methodology underlying an observable out-of-control pattern.

The structural profile of a business is, as a practical matter, clearly discernible from its Profitgraph and Capitalgraph. From these two graphic business aids, the crucial, gross chracteristics of any business may be pinpointed readily for purposes of higher management control and planning, both short- and long-range.

The Profitgraph will undoubtedly be more familiar under its other name, the Break-Even Chart, which portrays the relationship of costs and sales at varying volume levels. Notwithstanding the multiplicity of theoretical, academic objections to the validity of such charts, it can be demonstrated—and consequently categorically stated—that businesses indeed do, when thus charted, actually exhibit rational, average performance patterns—Structural Profiles, if you will.

In other words, when a "least squares" or "linear regression" line (in the language of the statistician) is fitted to, say, the last five years' sales and cost data, the result is an "average" linear or straight line relationship which can be viewed as divided into a fixed and a variable portion. The variable portion may be constructed by dropping a line through the Origin "O" parallel to the "total" trend line. The result is Exhibit 7-2.

In Exhibit 7-2, point BEP is termed the Break-Even Point (i.e., where total costs equal total sales, and income equals zero), distance OF is the amount of the fixed cost component, OV is the line of total variable costs, and line FT is the line of total costs. Note that the least squares line has been fitted to actual sales and costs (represented by the "x" marks) in the years indicated. Notice, further, that for sales levels below break-even sales (i.e., below or to the left of the Break-Even Point), losses are sustained, but for levels above break-even sales (i.e., above or to the right of the Break-Even Point), profits are taken.

EXHIBIT 7-2

PROFITGRAPH

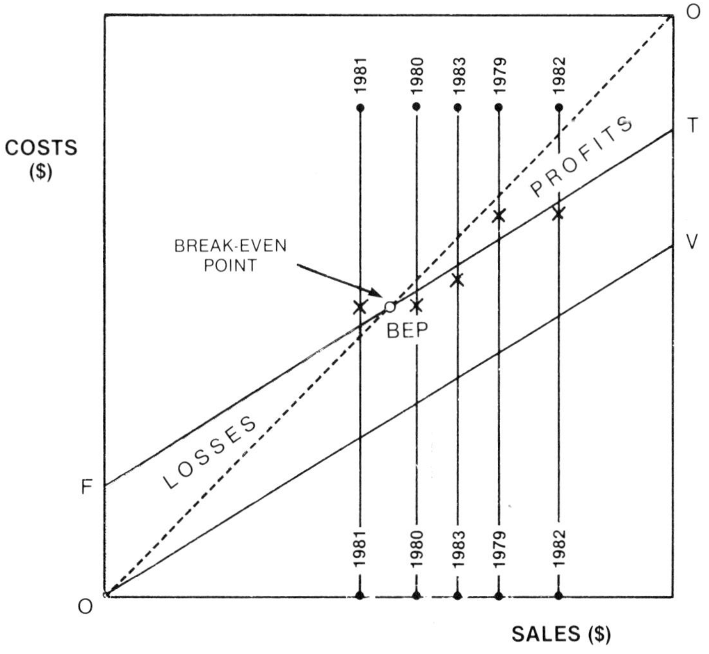

EXHIBIT 7-3

PROFITGRAPH

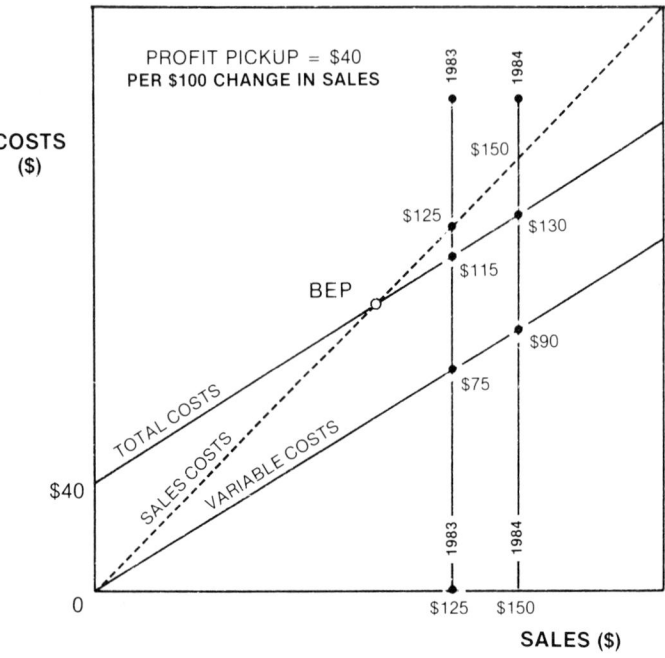

But now let us look at one final characteristic of this Profitgraph—perhaps its most significant from the point of view of understanding the behavior of the business as a whole. First, we'll redraw and simplify Exhibit 7-2, calling it Exhibit 7-3.

Notice that, in Exhibit 7-3, sales volume has increased from $125 in 1983 to $150 in 1984, or by $25, but that total costs have only increased from $115 to $130, or by $15. Consequently, there has been a pickup in profits of $10 (i.e., the sales increase less the cost increase, or $25 less $15).

Observe something else, too, however (which brings up the subject of the marginal income ratio). True enough that the increase in profit of $10 is equal to the increase in sales of $25 less the increase in total costs of $15. However, since by definition "fixed" costs (whatever they may consist of) remain "fixed" or "constant" (over the sales range of interest at any rate), only the "variable" cost component could possibly have contributed to the cost increase.

Note that what we call "marginal income" (i.e., sales less variable costs) for 1983 was $125 less $75 or $50, and for 1984 was $150 less $90 or $60. If we subtract the 1983 marginal income of $50 from the 1984 marginal income of $60, we are left with $10, which is precisely the same as the increase in profits from 1983 to 1984. This is more than coincidence (as you might expect by now), and it is for just this reason that the marginal income decimal ratio (i.e., sales less variable costs all divided by sales), when multiplied by $100, is termed the profit pickup, or *PPU*.

In other words, for every $100 of additional sales made by a business, it can expect on average (since we are here dealing with least squares average trend lines) to earn an additional or incremental profit about equal to the *PPU* (i.e., *PPU* equals the marginal income ratio multiplied by $100) or, in this case, $40 (i.e., the marginal income ratio is

$$\frac{\$150-\$90}{\$150} \text{ in 1984 and } \frac{\$125-\$75}{\$125} \text{ in 1983}$$

each of which equals 0.40, which in turn, when multiplied by $100, equals $40).

The implications of this technique are evident. If the cost structure of the business for the next five years, say, is projected as essentially the same as it is today and has actually been for the past five years (as defined by the least squares trend), then our five-year plan must project, for anticipated volume levels, profit levels that are consistent with the historical Profitgraph structure. But if major profit improvements are desired and forecasted, then any predicted, radical departures from the historical Profitgraph structure must be adequately supported with the essential details of the programs designed to get us there and firm commitments must be made by management now. Needless to say, empty platitudes and mere lip service plus good intentions are just not good enough. A disorganized, incompetent assault on the future will have disorganized, incompetent—and potentially disastrous—consequences. So much for profits and the Profitgraph for present purposes, although there is a great deal more that can be said.

Now let us turn our attention to an equally significant and useful but less familiar graphic tool: the Capitalgraph. Exhibit 7-4 illustrates the companion Capitalgraph to the Profitgraph constructed above.

Exhibit 7-4 has been constructed by plotting the actual values for total capital (assets) and sales for each of the five past years and then computing and constructing their least squares trend line (individual yearly values not shown in this case). That is, capital (or assets) has actually been administered by management so as to produce the structural profile graphically portrayed in the Capitalgraph of Exhibit 7-4. (In passing, it is important to note that the variable

EXHIBIT 7-4

CAPITALGRAPH

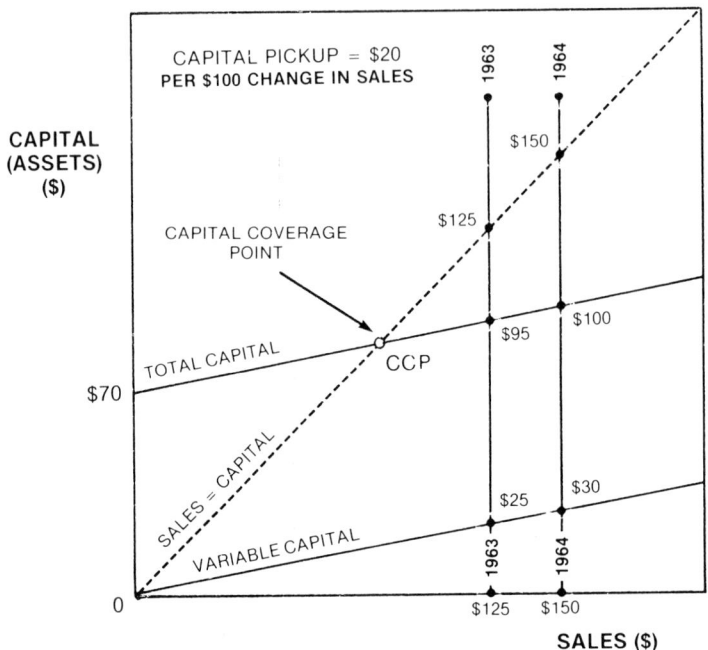

portion does not necessarily consist solely of so-called "Current Asset" items since, for example, inventories include a fixed component; likewise, the fixed portion does not necessarily consist solely of so-called "Fixed" or "Long-Term Assets" since, for example, the Property, Plant, and Equipment account includes Capitalized Developments which depend, to a certain degree, upon sales volume or the level of business activity).

We can develop a set of relationships analogous to the Break-Even Point and the Profit Pickup of the Profitgraph. For example, note that the point marked *CCP* in Exhibit 7-4 (i.e., the Capital Coverage Point) corresponds to the *BEP* of the Profitgraph. *CCP* is the point where total capital equals sales volume or where the capital turnover ratio (i.e., sales divided by total capital) equals unity or one.

The other significant aspect of the Capitalgraph is its underlying Capital Pickup or *CPU*. This may be defined as the added capital required to support an additional $100 of sales volume (again, on average, since we are dealing with least squares trend lines). From the values shown on Exhibit 7-4, CPU may be computed as follows:

$$CPU = \frac{19\_4 \text{ capital less } 19\_3 \text{ capital}}{19\_4 \text{ sales less } 19\_3 \text{ sales}} = \frac{\$100 - \$95}{\$150 - \$125} \times \$100$$
$$= \frac{\$5}{\$25} \times \$100 = 0.20 \times \$100 = \$20$$

Note that it is again only the variable capital component that contributes to the increase in capital required since, by definition, the fixed capital component remains constant over the sales range of interest. Thus:

$$CPU = \frac{\text{Increase in variable capital}}{\text{Increase in sales volume}} \times \$100 = \frac{\$30 - \$25}{\$25} = \$20$$

And, as before, planning the course of the business in the future must take cognizance of our actual past performance even though we may fully intend to improve upon it very substantially.

Thus we have developed two graphic profiles of the business firm. Naturally, the Profitgraph and Capitalgraph must be used in conjunction with all other proven instruments of managerial control. They supplement, rather than supplant, other planning and control tools utilized by management.

The Profit Pickup and Capital Pickup concepts may be summarized conveniently as follows: A business with given structural characteristics (i.e., given Profitgraph and Capitalgraph structures) will require $CPU of added capital to support each $100 of additional sales which, in turn, will require $PPU of added profit. In the example illustrated, $20 of added capital are required to support $100 of added sales which, in turn, will yield $40 of added profit. This sounds pretty good, doesn't it? The return on incremental capital is $40/$20 or 200 percent! But remember, formulas, if conceptually correct, are still merely couriers. As a matter of fact, there are cases where such enormous returns on incremental capital are encountered.

Notice that we have created a ratio, above, of a *PPU* of $40 to the *CPU* of $20, or of *PPU/CPU*. This is, in fact, the mathematical expression for the return on incremental capital employed, or *rice*, which we shall now write, then, as:

$$rice = PPU/CPU$$

In other words, for example, the individual project return on individual new investment project capital is just that kind of ratio. Or, in general (starting from a given sales volume), a sales increase will return $PPU of added profit which, when divided by the added capital (or $CPU required to support the added sales), gives us the return on incremental capital.

Also notice that we are speaking here in terms of return on Incremental Capital. The Actual Total return on capital, rce, is equal to Total Income/Total Capital Employed. For 19_3, going back to our example, total income was $10 and total capital $95, or rce equalled $10/$95 or 10.53 percent. For 19_4, rce was $20/$100 or 20 percent. Thus, although the return on incremental capital was 200 percent, the actual increase in the Total Return on Capital Employed in the business was only 89.93 percent (i.e., 20 percent less 10.53 percent, all divided by 10.53 percent)—a relatively smaller figure, although still an exceptionally high performance level. In this particular case, note the tremendous leverage on rce exerted by a relatively smaller change in sales. That is, sales increased by 20 percent (i.e., from $125 to $150, or $150 less $125, all divided by $125 equals 20 percent) while return on capital employed increased by 89.93 percent!)

Thus, while the 200 percent return on incremental capital was enormously high, it was largely dissipated when viewed from the perspective of the total return on capital for the basic business which was initially substantially beneath it (i.e., 10.53% in 19_3). Obviously, of course, the trick is to embark on lots of larger high-yielding projects so that rce will approach rice. For most companies this is their "impossible dream."

Let us now take a little closer look at rice = PPU/CPU. We must first examine the basic relationship:

RETURN ON CAPITAL (i.e., ASSETS) EMPLOYED = CAPITAL TURNOVER RATIO × INCOME BFIT (i.e., Before Federal Income Taxes) TO SALES RATIO

We shall use the following symbols in the algebraic equations below:

rce = Return on Capital Employed
$ibtr$ = Income BFIT to Sales Ratio
ctr = Capital Turnover Ratio
FC = Fixed Costs
VC = Variable Costs
$FCAP$ = Fixed Capital
$VCAP$ = Variable Capital
IBT = Income BFIT
S = Sales

$$IBT = S - \text{Total Costs} = S - FC - VC \tag{1}$$

$$\text{Total Capital Employed} = FCAP + VCAP \tag{2}$$

so that
$$ctr = S/\text{Total Capital} = S/(FCAP + VCAP) \qquad (3)$$
and
$$ibtr = IBT/S = (S - FC - VC)/S \qquad (4)$$
and so
$$rce = ibtr \times ctr = (S - FC - VC)/S \times S/(FCAP + VCAP) \qquad (5)$$
or
$$rce = (S - FC - VC)/S \div (FCAP + VCAP)/S \qquad (6)$$

Now, if we separate the numerators of the two parenthesized expressions, rce may be expressed as follows:

$$rce = \left(\frac{S-VS}{S} - \frac{FC}{S}\right) \bigg/ \left(\frac{VCAP}{S} + \frac{FCAP}{S}\right) \qquad (7)$$

Studying this expression, we observe that as sales (S) become bigger and bigger, with fixed costs (FC) and fixed capital (FCAP) remaining essentially constant, the two separate ratios (FC/S) and (FCAP/S) each become smaller and smaller until, at the limit (i.e., where sales become so large that Fixed Costs and Fixed Capital are very, very small compared with it) we may say that their value is essentially zero.

If we now actually replace (FC/S) and (FCAP/S) by zero, we have:

$$rce_{\text{limit}} = \left(\frac{S-VC}{S}\right) \bigg/ \left(\frac{VCAP}{S}\right) \qquad (8)$$

But $(S - VC)/S$ equals the marginal income ratio or, when multiplied by \$100, PPU, and VCAP/S equals CPU when multiplied by \$100. Thus we see that for a given set of structural conditions (i.e., a given Profitgraph and a given Capitalgraph), the theoretically maximum return on capital employed that is approached as a limit is:

$$rce_{\text{limit}} = PPU/CPU \qquad (9)$$

which we recognize as the return on incremental capital, rice

In our illustration, this value is 200%. But since, in actuality, we only made 10.53% in 19_3, and 20% in 19_4, the return on capital efficiency was 10.53%/200% or 5.27% in 19_3 and 20%/200% or 10% in 19_4. That is,

$$\text{rce efficiency} = \text{Actual rce} \div (PPU/CPU), \text{ or Actual rce} \div \text{rice} \qquad (10)$$

As a matter of record, as noted before, there are situations in actual practice where the orders of magnitude computed here do actually prevail, both on a corporate-wide scale and for individual investment projects. However, for the most part, even the very large incremental returns illustrated, when encountered in practice, represent such small investments and returns, in absolute dollars, that their transmitted effect on the total return on capital employed of the total basic business is barely perceptible (due to the fixed element in costs and in capital). Most companies have incremental characteristics of much more modest proportions.

Observe that we have been dealing, all this time, with mathematical models. Many subjects may be conveniently developed in their purest, most idealized form. This is true especially where the definition, measurement, and precise separation of component aspects are in practice difficult, if not impossible, tasks. Thus, theoretical models, as oversimplified but meaningful characterizations, possess great utility and flexibility for purposes of analysis, illustration, and mastery of underlying principles.

In practice, the theoretical model must be modified, perhaps even distorted. But normally it retains enough recognizable features so that a considered plan of action based thereon may be attainable. It would be imprudent to strain for perfection in establishing a practical base of operations solely upon a theoretical model, but this is not so much a reflection upon the soundness of the underlying theory as it is upon our own ingenuity and maturity in adapting it to practical situations. It is fully as difficult to develop good practices as it is to create sound theories. Furthermore, misapplication of any analytical tool will yield immaterial, misleading, or meaningless results.

THE WORKING PROFITGRAPH

Now let us redraw Exhibits 7-3 and 7-4, incorporating these ratiorays, as Exhibits 7-5 and 7-6. Studying Exhibit 7-5, the Break-Even Profitgraph, we observe that the marginal income ratio, expressed as Profit Pickup, may be read directly from the percentage scale on the right-hand side of the graph (i.e., the variable cost line coincides with the 40 percent ray). Likewise, the pretax profit margin (in all cases considered here we are speaking of income *before* taxes) may be read directly by measuring the distance from the 0 percent ray to the total cost line. In 1983, IBT % Sales = 8 percent and in 1984, 13.3 percent.

Thus, with the ratio rays drawn directly upon the graph, one

EXHIBIT 7-5

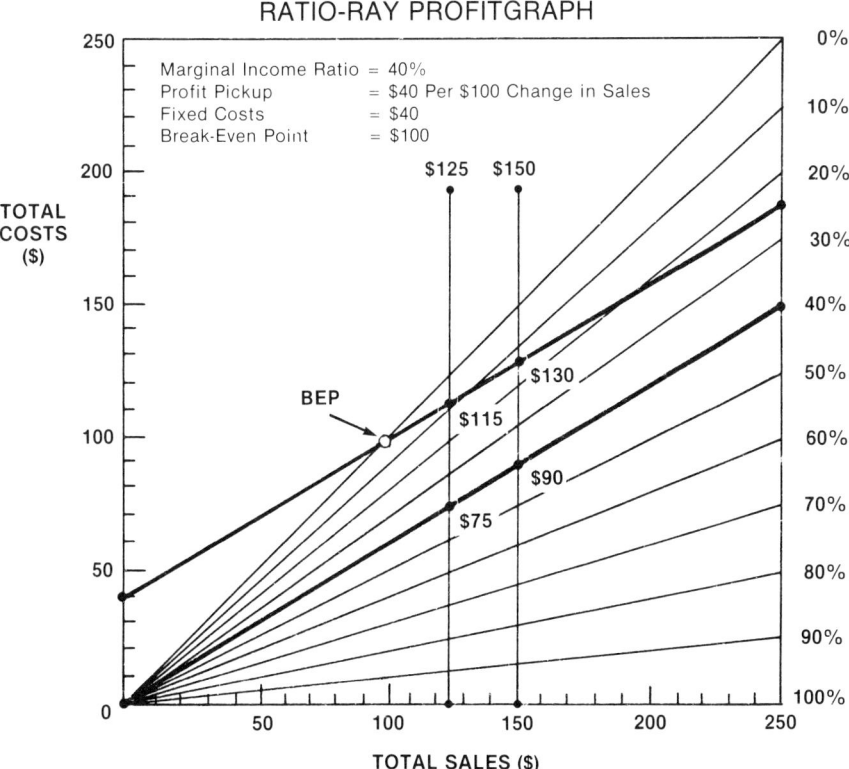

may directly obtain the *PPU* or marginal income ratio and the percentage of pretax margin. And, of course, we also can directly obtain the Break-Even Point, Marginal Income, and Fixed Costs.

A final important observation: Although the line of total costs cuts across profit percentage rays of increasing value as sales increase, note that the percentage of total profit margin does not increase in equal increments even though actual profit itself does increase by equal amounts. To put it another way, profit increases less and less for successive, equal increases in sales. Successive percentage point increases in total profit margin come harder and harder; conversely, it takes a larger and larger increase in sales to generate equal, successive increments in the total profit margin percent. Table 7-1 illustrates this diminishing or "decelerating" incremental profitability of the total business as volume increases (particularly take note of Column 5 in Table 7-1).

EXHIBIT 7-6

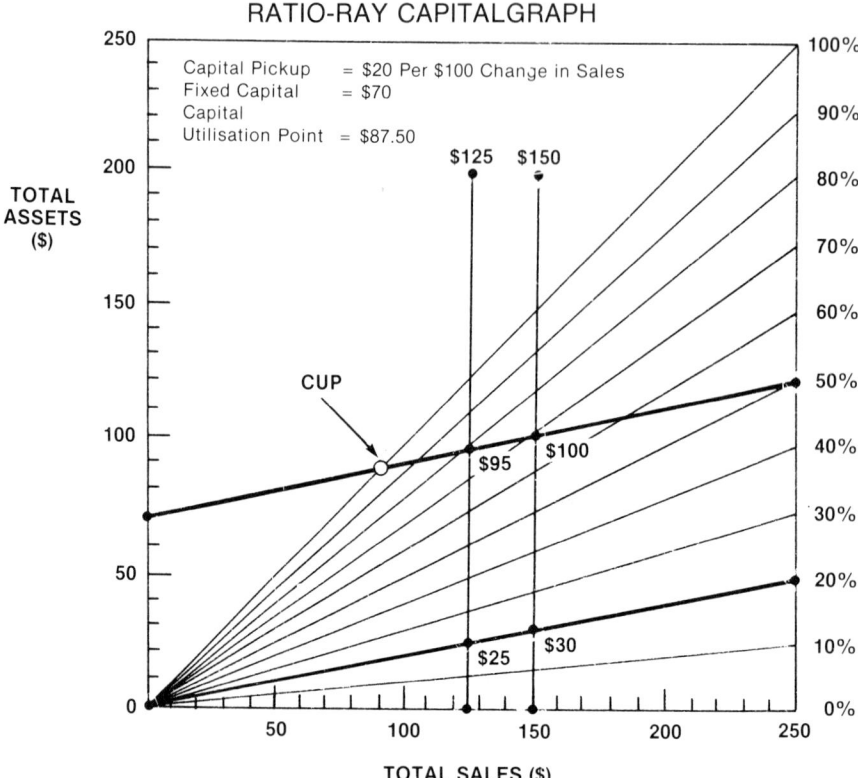

THE WORKING CAPITALGRAPH

Turning to the Capitalgraph of Exhibit 7-6, we can directly read the Capital Pickup, fixed capital, and CUP (i.e., Capital Utilization Point) in a manner analogous to that described above for the Profitgraph. If one is more familiar with capital turnover as opposed to total assets as a percentage of sales, Exhibit 7-7 shows capital turnover in place of the CPU values. In this case note, however, the adjustment necessitated in the construction of the rays and the greater difficulty in interpolating turnover values between successive rays since the scale is *not* linear.

Finally, note that the line of total capital on Exhibit 7-6 intersects ratio rays of successively decreasing values as sales increase; thus, for a given average capital structure, the amount of total capital

TABLE 7-1

ILLUSTRATION OF DIMINISHING INCREMENTAL PROFITABILITY WITH INCREASING SALES VOLUME

1 Line	2 Sales	3 Income	4 Income % of sales	5 % points of incremental income for equal sales increments	
1	$100	$ 0	0.00%	—	▼
2	110	4	3.64	3.64 % Points	▼
3	120	8	6.67	3.03	Decreasing Rate of Increase or Decelerating Incremental Profitability
4	130	12	9.23	2.56	
5	140	16	11.43	2.20	
6	150	20	13.33	1.90	▼

EXHIBIT 7-7

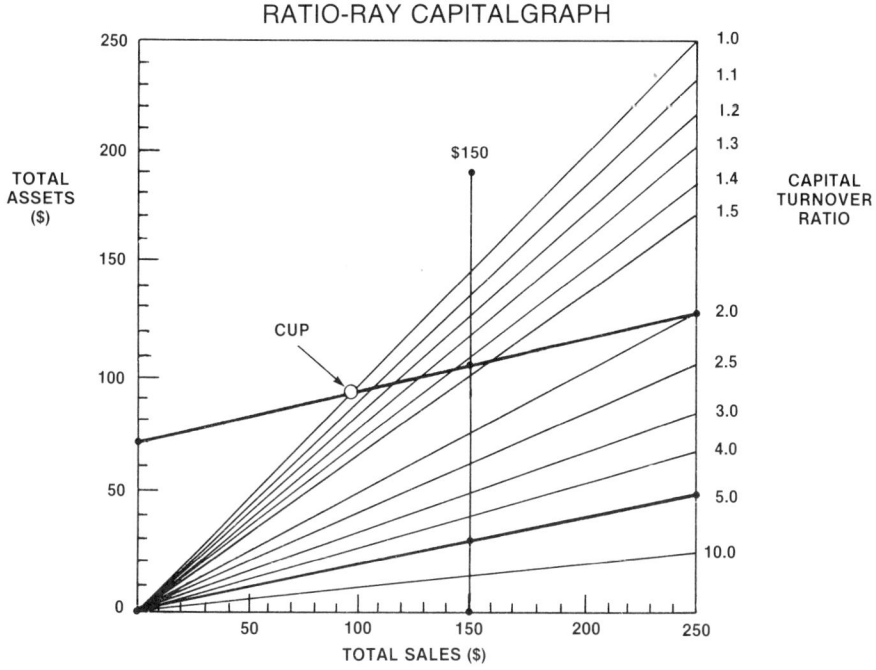

RATIO-RAY CAPITALGRAPH

required to support successively larger total sales volume does not increase as rapidly as sales. But, again, analogous to the case of the Profitgraph, in terms of capital turnover ratio the number of capital turns does not increase in equal increments even though the amount of capital does increase by equal amounts.

That is, capital turnover increases less and less for equal, successive increments in volume. Table 7-1 illustrates this diminishing or "decelerating" incremental capital utilization efficiency of the total business as volume increases.

EFFICIENCY OF PERFORMANCE: A STILL CLOSER LOOK

To borrow some terms from the language of engineering, the Profitgraph and Capitalgraph might be termed "equations of state" or "condition curves" of the business. That is, if the business exhibits a certain set of structural characteristics (i.e., a specific Profitgraph and Capitalgraph), then it is not managerial incompetence that is responsible for diminishing incremental profitability and diminishing incremental capital utilization efficiency (provided, of course, that it is simply not possible or not feasible to substantially alter the Profitgraph or Capitalgraph characteristics of the particular enterprise without impairing its longer-term economic well-being).

Thus, in the sense that these graphs represent "built-in" relationships, they are "normal" curves for the business as a whole and "state points" lying along them represent those levels of costs and capital that will accompany various sales volumes (so long as management does not attempt to alter its methods of conducting the business, but continues to run it in essentially the same way as during the period covered by the Capitalgraph and Profitgraph). Therefore, since these "characteristic curves" of the business assume one level of capital and costs at a given activity level, it is simply irrelevant to say the business is less efficient here than at a lower activity level where the increase in profits and the increase in turnover were greater for the same increase in absolute dollars of sales volume. Tables 7-1 and 7-2 illustrate this reasoning. In Table 7-1 note that, as before, although profit margins increase with sales, the incremental income does not increase as fast. Clearly, although it is quite true that the incremental efficiency of the business is not as great at higher and higher levels of sales, management cannot be meaningfully accused of poor performance.

Given a specific set of characteristics or characteristic curves for the business (reflecting its underlying operating and financial

TABLE 7-2

ILLUSTRATION OF DIMINISHING CAPITAL UTILIZATION EFFICIENCY WITH INCREASING SALES VOLUME

1 Line	2 Sales	3 Assets	4 Assets % of sales	5 % Points of assets decrease for equal sales increments	6 Capital turns	7 Number of incremental turns for equal sales increments
1	$100	$ 90	90.000%	—	1.111	—
2	110	92	83.636	6.364 % Points	1.196	0.085 Turns
3	120	94	78.333	5.303	1.277	0.081
4	130	96	73.846	4.487	1.354	0.077
5	140	98	70.000	3.846	1.429	0.075
6	150	100	66.667	3.333	1.500	0.071

Column 5: ▶▶ Decreasing Rate of Decrease or Decelerating Capital Utilization Efficiency ▶▶

Column 7: ▶▶ Decreasing Rate of Increase or Decelerating Incremental Turnover ▶▶

structure), higher and higher activity levels will necessarily be accompanied by lower and lower incremental profitability and capital utilization efficiency. According to the "condition curves," the business can never attain higher incremental efficiencies except at the lower activity levels (again, we must remind ourselves, barring drastic structural changes in the conduct of the business).

Notice throughout our discussion the emphasis on incremental profitability and incremental capital utilization efficiency. Also note, in passing, that we will not attempt to evaluate the view that, notwithstanding this propensity of the business relating to diminishing incremental efficiencies, it always pays to make that extra dollar!

There are some interesting ramifications here. If you do not continue to support the business, it may wither away and die. On the other hand, at some point your "incremental capital" may be capable of being applied to better advantage elsewhere. What to do? The answer may not be easy.

Exhibits 7-8 and 7-9 are blank chart masters for your use in plotting financial data for your own firm or other illustrative example. Exhibit 7-10 is a data collection sheet. Collect, chart, and ana-

EXHIBIT 7-8

PROFITGRAPH

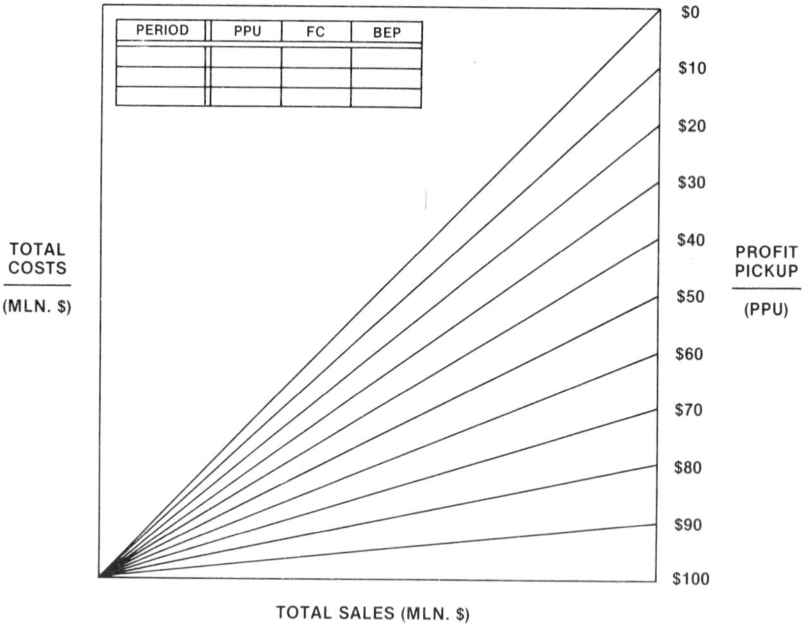

PERIOD	PPU	FC	BEP

TOTAL COSTS (MLN. $)

PROFIT PICKUP (PPU)

TOTAL SALES (MLN. $)

EXHIBIT 7-9

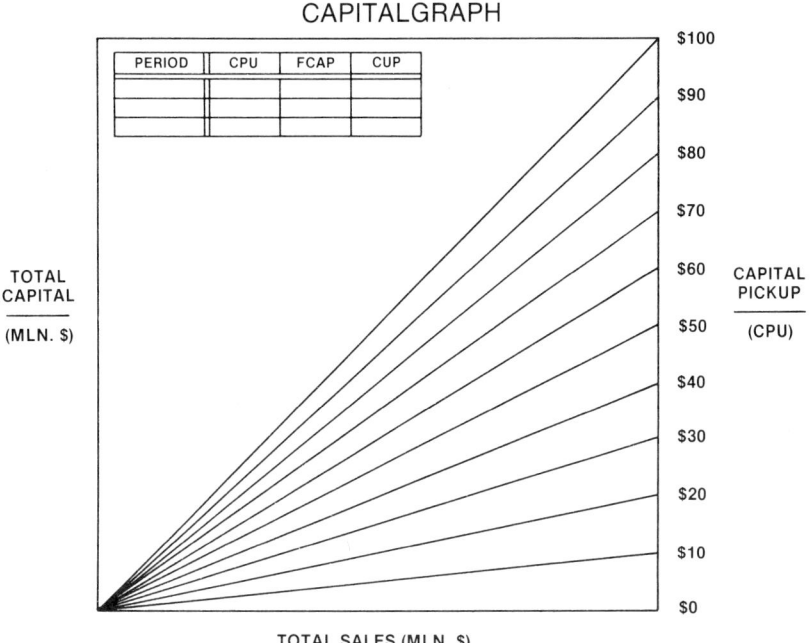

lyze data for your own example for the last five years, the five years prior to that, and for the last ten years. You may be very surprised at the results. Is your employer under control? If so, what are its "enterprise" or "structural" profiles and enterprise or structural "constants" (i.e., PPU, CPU, FMU, $rice$, and rce_{limit}, etc.)?

Thus, the concept of efficiency of operation does not apply to a single condition curve by itself, but only to a set of alternative condition curves for the same business, or condition curves for different businesses when compared. Observe, incidentally, that this statement is true even considering our use of the efficiency concept before in computing return on investment efficiency, since we there measured actual condition curves against theoretically "maximum" curves. Thus, a comparison was implicit.

In such cases of interfirm and alternative structure analyses, for the same level of operation, the business (or alternative condition curve for the same business) that yields the highest profitability and turnover is the more "efficient."

The only sense in which an increase or decrease in activity level along a single condition curve may be said to be indicative of a more or less efficient operation is that at higher and higher activity

EXHIBIT 7-10

FINANCIAL FACT SHEET*

Line	Symbol	Item/Ratio Description	19	19	19	19	19	19	19	19	19	19
1	S	Sales	$	$	$	$	$	$	$	$	$	$
2	TC	Total Costs										
3	IBT	Income Before Taxes										
4	FIT	Federal Income Taxes										
5	IAT	Net Income										
6	DIV	Dividends										
7	RE	Retained Earnings										
8	NNCO	Depreciation & Amortization										
9	CF	Cash Flow										
10	RDE	Research/Development/Eng. Expense	$	$	$	$	$	$	$	$	$	$
11	rder	RD&E to Sales Ratio	%	%	%	%	%	%	%	%	%	%
12	TA	Total Assets @ 12/31	$	$	$	$	$	$	$	$	$	$
13	ATA	Average Total Assets for Yr.	$	$	$	$	$	$	$	$	$	$
14	ΔTA	Incremental Assets										
15	EQ	Shareholders' Equity										

16	FMU	Financial Makeup	$	$	$	$	$	$	$	$	—
17	CFMU	Cumulative Financial Makeup									—
18	PPU	Profit Pickup ($/$100)									
19	CPU	Capital Pickup ($/$100)									
20	rice	Return on Capital Efficiency									
21	FC	Fixed Costs ($)									
22	FCAP	Fixed Capital (Assets) ($)									
23	BEP	Break-Even Point (Sales) ($)									
24	CUP	Capital Utilization Point ($)									
25	ibtr	Profit Margin (BFIT)	%	%	%	%	%	%	%	%	—
26	ctr	Capital Turnover Ratio	X	X	X	X	X	X	X	X	—
27	rce	Return on Capital (BFIT)	%	%	%	%	%	%	%	%	—
28	flr	Financial Leverage Ratio	X	X	X	X	X	X	X	X	—
29	req	Return on Equity (BFIT)	%	%	%	%	%	%	%	%	—
30	rcee	Return on Capital Efficiency	%	%	%	%	%	%	%	%	—
31	U	No. Physical Product Units Sold	U	U	U	U	U	U	U	U	—
32	U	Total No. Units in Field Service	U	U	U	U	U	U	U	U	—
		COMPANY NAME & PRODUCT LINE DESCRIPTION									

(*) Dollar amounts in ($000,000) unless otherwise specified.

levels the total profit margin as a percentage of sales and the total capital turnover ratio are higher and higher. This view, however, completely ignores the incremental effects of activity-level changes. If the incremental approach is used, the effect on efficiency is just the opposite, as we have already seen (i.e., as sales increase incremental efficiency decreases and, conversely, as sales decrease, incremental efficiency increases—from a percentage efficiency point of view). Of course, the behavior of both total profitability and turnover, and incremental profitability and turnover, are of significance to management. We are here making the distinction in order to lay bare the business anatomy.

UNCERTAINTY

The reader who is grounded in the field of financial analysis or engineering economy will, no doubt, have already pounced upon a statement made earlier regarding efficiency of performance; namely, that points lying along the "normal" curves (i.e., Profitgraph and Capitalgraph total cost and total capital lines, respectively) represent those levels of costs and capital that will accompany various sales volumes. One would immediately recognize that the real-world operation of a business does *not* truly lie along a straight line (among others things!). He or she will remark, and correctly so, that if we insist upon characterizing business performance as a straight line, we should at least draw three lines of total costs and three lines of total capital. That is, we should draw an upper level of total costs defining a maximum or "optimistic" estimate or limit, a lower line defining a minimum or "pessimistic" estimate or limit, and a line somewhere between the two defining a "best" estimate or "most probable" average performance—or something like that—and the same for capital. "Control chart" thinking can be applied here, as elsewhere, complete with all the statistical ramifications that go along with it, including considerations of standard deviation, confidence intervals, upper and lower control limits, and so on.

THE CASE FOR COMPARATIVE ANALYSIS

"Interfirm Comparison" and "Comparative Analysis," or planning and control by the comparative method, are such important management tools that it is appropriate to spend a few moments on it. Much of the benefit of Profitgraph and Capitalgraph analysis lies in management's willingness to compare itself with the competition in addition to making comparisons among various segments of the business itself.

To cite an old European Economic Community study: "All firms try to be as efficient and as rational as possible. It is therefore natural that they should want a measure to assess in what proportion their efforts at scientific management yield any results. They also want to be able to answer the question: Is the firm more scientifically managed and more efficient than previously?

". . . A measure of the degree of accomplishment—a measure of efficiency—should rather bring out the difference between the effective amount of profit and the ideal amount that could have been attained under given external conditions. This measure, however, can hardly ever be assessed practically, and it remains very questionable whether it is at all possible to define theoretically how it should be calculated under ideal conditions.

". . . It is, therefore, necessary to reduce somewhat such an ambitious project and to content ourselves with a measure of efficiency which provides incomplete information. If it is not possible to deduce how far the firm falls short of the minimum cost possible today, there are two related questions of almost as much interest which it is possible to answer, to some extent at least, namely:

1. How efficient is the firm under present conditions in relation to some earlier period? and
2. How efficient is the firm under prevailing conditions in relation to other firms in the same field?

". . . The advantage in introducing this basis of comparison—the same firm at another period or a similar firm—is that it is no longer necessary to measure the absolute distance to the minimum, but simply to compare two similar distances. This is often considerably easier."

This, then, is the case for interfirm comparison which is universally recognized as a significant means of raising the performance of individual businesses. It follows that management ratios established through interfirm comparison are significant tools in helping managers improve the efficiency of their businesses. Ratios have three main uses:

1. When compared from one period to another, they throw light upon changes in the health of a business and possible causes thereof;

2. When compared among similar firms they help the managements concerned to establish whether, and if so for what reason, their business is lagging behind that of their competition; and
3. Whether they realize it or not, managers are bound to base future plans on expected relationships among such items as sales, costs, profits, assets, liabilities, manpower utilization, and so on (i.e., generally, relationships between resources applied and results achieved, and between assets, liabilities, and their components).

The strong case for interoperation comparisons notwithstanding, one occasionally encounters objections to the use of comparative analysis (i.e., comparative productivity and management ratio analysis). When candid objections are raised as to the validity of comparisons per se, it generally develops that they are based upon a half-truth. We must reject that defensive and unprofessional viewpoint where the motive for objecting is that unless the comparison shows us in a favorable light it isn't any good. This is merely the corporate "Id" ("*I'm different*") speaking—or, rather, complaining.

Obviously, if comparisons are made on the basis of yardsticks having no strong or plausible causal relationship with the parameters to be measured, the results will be useless. Such misapplications of perfectly good analytical tools by inexperienced or incompetent practitioners often have the unfortunate effect of undermining confidence in their true worth. Past experience with just such misapplications of comparative analysis, therefore, may tempt many otherwise extremely able managers to turn their backs upon it and take the intractable and disquieting position that "We're different from everybody else!"

About "being different," it has been said that the first basic principle is that there is a virtue in firms *not* being identical (insofar as systems, procedures, methods, and styles are concerned). One virtue in *not* being identical in methods, systems, and ways of doing things is that it provides a differential or discriminant among operations and companies that can be isolated and compared financially and economically. This facilitates a determination of what methods and management techniques provide the most for the least.

CHARTING TOTAL RETURN ON CAPITAL

To this point our discussion of graphs has involved, on the one hand, costs versus sales (i.e., the Profitgraph) and, on the other,

EXHIBIT 7-11

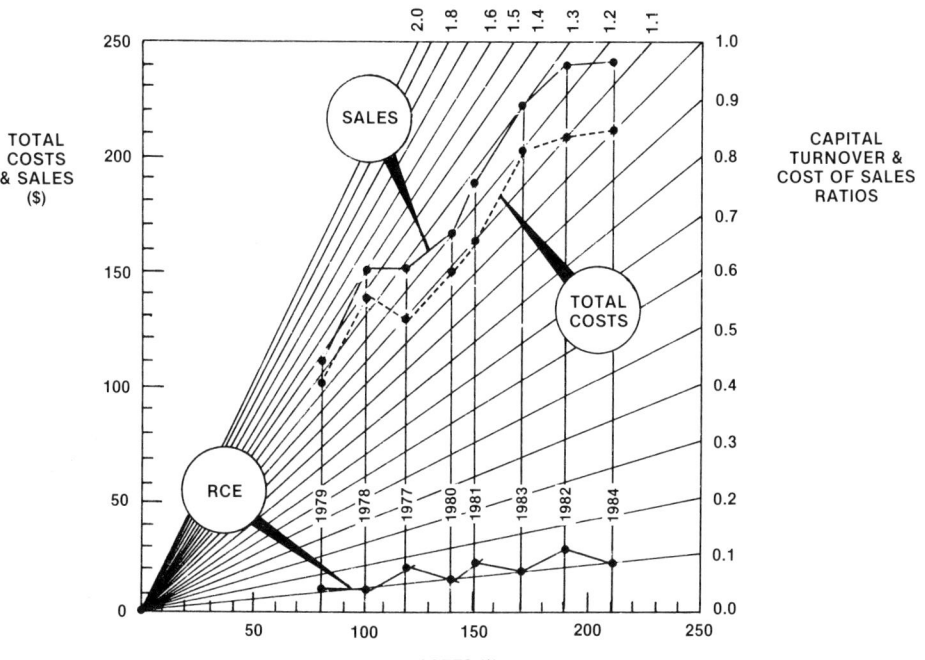

capital or assets versus sales (i.e., the Capitalgraph). However, to comprehend more fully the interplay of cost and capital behavior patterns portrayed thereon, it would be helpful to have a graphical representation of return on capital.

Return on capital is ordinarily viewed as the product of the capital turnover ratio and the income-to-sales ratio (i.e., Sales ÷ total capital or capital turnover ratio, *ctr*, all multiplied by income ÷ sales or the income-to-sales ratio, *ibtr*). But, if we rewrite the formula so that return on capital is equated to the difference between the capital turnover ratio and the cost of sales-to-capital employed ratio, then we can plot these two ratios for successive years as shown in Exhibit 7-11. The vertical distance between the sales trend line and the total costs trend line directly reads (by measured difference) the actual income (profit or loss), and the difference between the two ratios corresponding to any given set of sales and total cost values (also directly read using the ratio-ray principle) is the return on capital employed, which is separately plotted at the base of the chart. Algebraically speaking:

$$\text{Return on Capital, } rce = \frac{\text{Income Before Taxes}}{\text{Total Capital}} = \frac{IBT}{TCAP} \quad (11)$$

However,

$$IBT = \text{Sales} - \text{Total Costs} = S - TC \quad (12)$$

$$rce = (S - TC)/TCAP = (S/TCAP) - (TC/TCAP). \quad (13)$$

In other words, by plotting (S/TCAP) and (TC/TCAP) as separate trend lines on a ratio-ray chart having Sales and Total Costs as its vertical axis and Total Capital as its horizontal axis we obtain, by subtraction, a useful plot of return on capital employed. As indicated, we can directly read the capital turnover ratio and the cost of sales ratio as well as the return on capital (when separately plotted as shown).

The ratio-ray return on capital chart proves useful where a large amount of information is desired from a single chart form, since both ratios and absolute dollar amounts are shown. Also, this type of chart is more convenient for tracking an individual operation from period to period or year to year than the rather unconventional geometry of the perhaps more familiar representation of return on capital shown as Exhibit 7-12. This latter chart is excellent, however, for comparing industries, companies, divisions, product lines, and so on for the same period.

Note that on Exhibit 7-12, the vertical scale is total capital or assets turnover and the horizontal, percentage profit margin. On this type of graph each individual hyperbolic curve shown represents a single value of return on investment. That is, whereas on the ratio-ray type of graph each ray represented A (i.e., a given value on the vertical axis) divided by B (i.e., a given value on the horizontal axis), on this hyperbolic chart form, each hyperbolic line or curve represents A multiplied by B (i.e., capital turnover ratio multiplied by the income-to-sales ratio). Several values are shown on the graph to illustrate its principle of operation. Exhibit 7-13 is a blank master form for your own use.

But $(S - VC)/S$ equals the marginal income ratio or, when multiplied by $100, PPU, and $VCAP/S$ equals CPU when multiplied by $100. Thus, we see that for a given set of structural conditions (i.e., a given Profitgraph and a given Capitalgraph), the theoretically maximum return on capital employed that is approached as a limit is:

$$rce_{lim} = PPU/CPU$$

which we recognize as the return on incremental capital employed, $rice$.

In our illustration, this value is 200%. But since, in actuality,

Financial Engineering for Effective Business Management

EXHIBIT 7-12

RETURN ON CAPITAL GRAPH: HYPERBOLIC TYPE

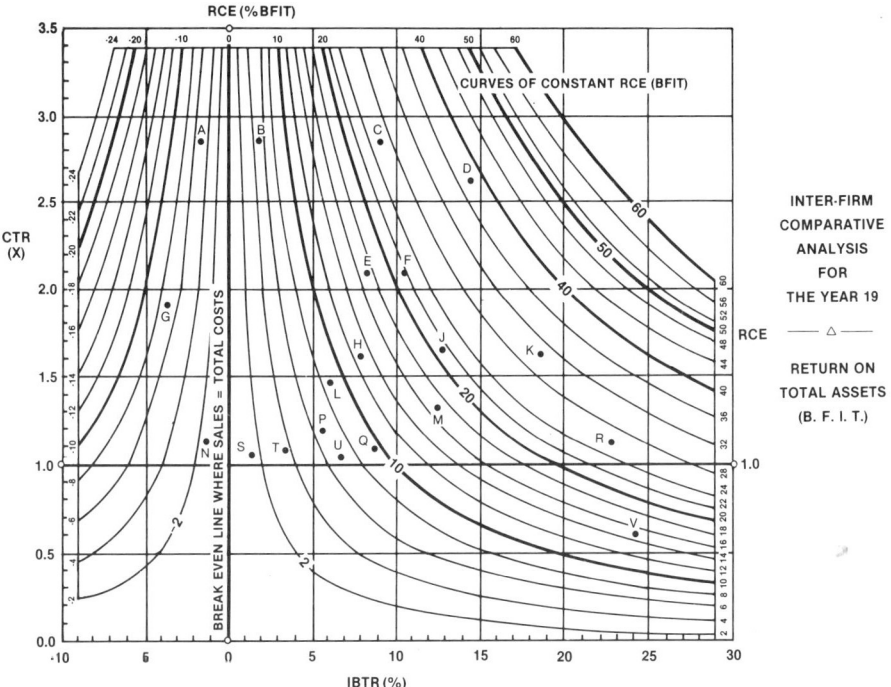

we only made 10.53% in 19_3 and 20% in 19_4, the return on capital efficiency was 10.53%/200% or 5.27% in 19_3 and 20%/200% or 10% in 19_4.
That is

$$\text{rce efficiency} = \text{Actual rce} \div \text{rice} \qquad (10)$$

As a matter of record, as noted before, there are situations in actual practice where the orders of magnitude computed here do actually prevail, both on a corporate-wide scale and for individual investment projects. However, for the most part, even the very large incremental returns illustrated, when encountered in practice, represent such small investments and returns, in absolute dollars, that their transmitted effect on the total return on capital employed of the

EXHIBIT 7-13

INTERFIRM COMPARATIVE ANALYSIS FOR THE YEAR 19■■

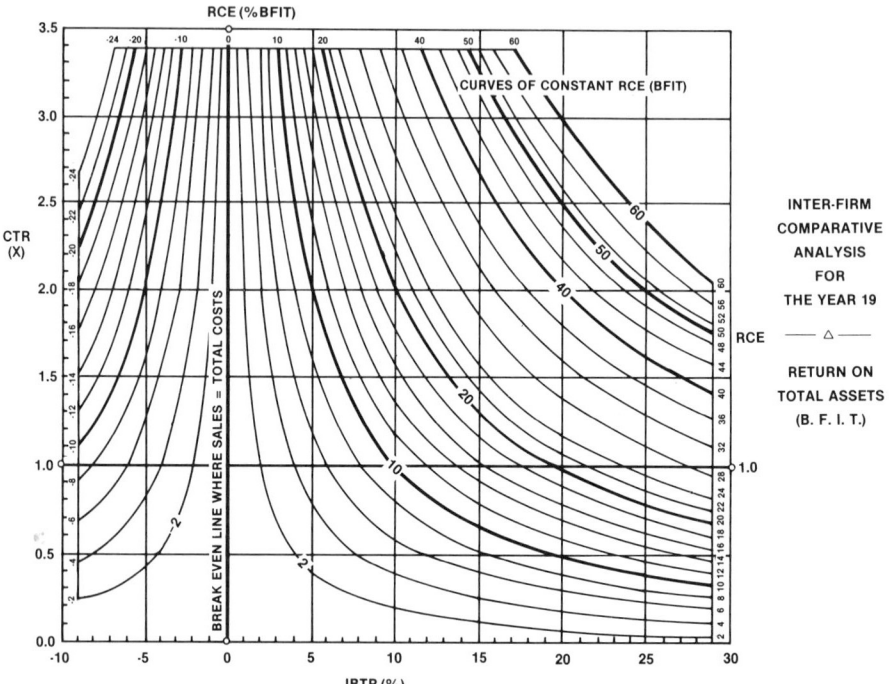

total basic business is barely perceptible (due to the fixed element in costs and in capital). Most companies have incremental characteristics of much more modest proportions.

Observe that we have been dealing, all this time, with mathematical models. Many subjects may be conveniently developed in their purest, most idealized form. This is true especially where the definition, measurement, and precise separation of component aspects are in practice difficult, if not impossible, tasks. Thus, theoretical models, as oversimplified but meaningful characterizations, possess great utility and flexibility for purposes of analysis, illustration, and mastery of underlying principles.

In practice, the theoretical model must be modified, perhaps even distorted. But normally it retains enough recognizable features so that a considered plan of action based thereon may be attainable. It would be imprudent to strain for perfection in establishing a practical base of operations solely upon a theoretical model. But this is not so much a reflection upon the soundness of the underlying theory as it is upon our own ingenuity and maturity in adapting it to practical situations. It is fully as difficult to develop good practices as it is to create sound theories. Furthermore, misapplication of any analytical tool will yield immaterial, misleading, or meaningless results. And, as noted before, there is nothing so practical as a sound theory.

CHAPTER 8

FINANCIAL ENGINEERING AND THE PROFITOGRAPHY SYSTEM

THE NEED FOR OUTSIDE CAPITAL

We will now return to another interpretation of the interplay between profit pickup and capital pickup that alerts management to a very significant tendency of the business which may be due to external economic factors, internal management policies and practices, or a combination of both. It is an all too familiar tendency which, however, has heretofore not been developed in the financial literature.

All businesses need new capital to grow and prosper, whether such incremental capital be needed for expansion of manufacturing or research facilities or to finance needed working capital for expanded current operations. On the surface it would certainly seem to be desirable—indeed ideal—if a business could completely finance its capital requirements from internal sources alone. But this is, in practice, rarely possible. Furthermore, as we shall see shortly, such self-financing capability of a business may even be symptomatic of a severely deteriorated or deteriorating condition.

The basic relationship to which we referred above is the difference between profit pickup and capital pickup (i.e., $PPU - CPU$). Although this is an oversimplification that we will explore more fully later, let us see what it begins to tell us. We will return to the concept of financial makeup and cash flow in our travels.

If the pretax profit pickup is $30 of added profit for each additional $100 of sales volume which, in turn, must be supported by an additional amount of capital, or a capital pickup, of $20, the business can theoretically finance its increased capital requirement entirely out of earnings. On the other hand, if the capital pickup is $60 instead of $20, the business cannot finance the added capital demand out of earnings and will have to go outside.

Again, roughly speaking (i.e., neglecting, for the moment, considerations relating to dividends, income taxes, etc.), if next year's anticipated volume is $100 greater than last year's and if $PPU = \$20$ and $CPU = \$60$, then next year's incremental income of $20 will *not* cover the incremental capital of $60 required to finance the added $100 of sales. This deficiency must be supplied from the outside. Note another interesting fact, however. If the new level of sales is maintained for, say, the next two years, the incremental capital of $40 (i.e., after deducting from the capital pickup of $60 the first year's $20 profit pickup contribution) will be recovered through the cumulative, internal profit increment. Table 8-1 illustrates these points. Column 5 shows the outstanding debt after partial debt recovery (i.e., through application of each year's incremental profit contribution).

The $60 of added capital need be obtained only once. It remains in the business as a kind of "float" or "reservoir" of value which need not be replenished each year. That is, the $460 of capital does not have to be obtained every year. The same $460 may be reused over and over so long as sales remain at the $600 level. However, in each year (for years 1 through 3) there will be sales revenues of $500 which will yield $70 of profit, also each year. Thus, due to the incremental capital injection of $60 in year 1, there will be $20 of added profit for each year of years 1 through 3, or a cumulative profit increment of $20 + \$20 + \$20 = \$60$ over the three-year period. This cumulative profit increment equals the $60 of incremental capital required. After the initial $60 has been recaptured, the business may again expand without having any residual debt carryover that would result if management had expanded as shown in Table 8-2.

Notice that one might thus consider profit pickup (or, more

TABLE 8-1

ILLUSTRATION OF SELF-FINANCING CAPACITY AND DEBT RECOVERY: CASE II

Year	Sales	Income	Capital	Residual debt carryover
0	$500	$ 50	$400	$–
1	600	70	460	40 or $60 – $20
2	600	70	460	20 or $40 – $20
3	900	130	640	120 or $20 – $20 + $180 – $60
4	900	130	640	60 or $120 – $60
5	900	130	640	0 or $60 – $60

properly, retained earnings, as we shall presently see) to be a kind of internal "growth" factor to the extent that each $100 of sales this year can finance a sales increase next year equal to the profit pickup plus some correction factor.

In Table 8-2 for year 3, instead of being free and clear of debt, the company has incurred an added debt burden (i.e., the $20 – $20 item is the recovery carryover from the first increment of capital of $60; the $180 – $60 is the added debt burden due to the second capital increment of $180 required to boost sales from $600 to $900, or ($900 – $600) × $60 capital pickup per $100 of sales increase). Note that in year 5, however, the company will have recouped this second capital outlay also.

But we must recognize that the entire amount of incremental, pretax profits generated (by virtue of the profit pickup characteristic of the Profitgraph) is not available for reinvestment. This incremen-

TABLE 8-2

ILLUSTRATION OF SELF-FINANCING CAPACITY AND DEBT RECOVERY: CASE I

Year	Sales	Income	Capital	Residual debt carryover
0	$500	$50	$400	$–
1	600	70	460	40 or $60 – $20
2	600	70	460	20 or $40 – $20
3	600	70	460	0 or $20 – $20

tal profit must also cover income taxes and dividends. This leaves us with retained earnings (i.e., net income or income before taxes less income taxes, less dividends), which must thus cover the incremental capital demanded by the Capitalgraph.

Let us set down these relationships as algebraic formulas and see where they lead us. The following abbreviations shall be used, with equation numbering continuing from the previous chapter:

IBT = Income before taxes
FIT = Federal income taxes
DIV = Dividends
ΔRE = Incremental retained earnings from the year's operations
$\Delta TCAP$ = Incremental (Δ) capital
IAT = Income after taxes
$fitr$ = Federal income tax ratio (i.e., FIT/IBT)
$dpor$ = Dividend payout ratio i.e., DIV/IAT)

S = Sales
ΔS = Incremental (Δ) sales
VC = Variable costs
FC = Fixed costs
PPU = Profit pickup (to be applied in decimal form hereafter; viz. PPU = $30 or 0.30)
CPU = Capital pickup (to be applied in decimal form hereafter; viz. CPU = $60 or 0.60)

$$\Delta RE = IBT - FIT - DIV = IAT - DIV \qquad (14)$$

and, for 100% self-financing,

$$\Delta TCAP = \Delta RE = IBT - FIT - DIV \qquad (15)$$

But, by definition, we know that:

$$IBT = S - VC - FC = S \times PPU - FC \qquad (16)$$

$$\Delta TCAP = \Delta S \times CPU \qquad (17)$$

where

$$\Delta S = S - S_o \qquad (18)$$

where, further, S_o is this year's sales and S is next year's estimated volume. Thus, substituting in Equation (15), we have:

$$(S - S_o)CPU = S \times PPU - (FC - FIT - DIV) \qquad (19)$$

Now, if we let:

$$r = S/S_o \qquad (20)$$

$$R = PPU/CPU, \qquad (21)$$

then $\quad (S - S/r) \times (PPU/R) = S \times PPU - FC - FIT - DIV \qquad (22)$

Collecting terms, we finally obtain:

$$S \times PPU(r - 1)/Rr - S \times PPU = -FC - FIT - DIV \qquad (23)$$

or, finally,

$$S \times PPU \left(\frac{r-1}{Rr} - 1\right) = -FC - FIT - DIV, \quad (24)$$

if self-financing is possible (i.e., for $\Delta TCAP = \Delta RE$ or $\Delta RE - \Delta TCAP = 0$). If self-financing is not possible, $\Delta RE - \Delta TCAP$ does not equal zero but is as follows:

$$\Delta RE - \Delta TCAP = FMU = \text{Financial Make-Up} \quad (25)$$

(i.e., amount of outside funds required) or, from the above,

$$FMU = S \times PPU \left(1 + \frac{1-r}{Rr}\right) - FC - FIT - DIV \quad (26)$$

Studying Equations (20) and (21) more closely, we will recognize that the ratio r represents the growth of next year's sales, S, over this year's sales S_o. Also the ratio R is the return on incremental capital employed (rice) developed previously (refer back to the section on incremental return on incremental investment).

Consequently, we may rewrite Equations (20) and (21) thus:

$$r = S/S_o = 1 + i, \quad (27)$$

where i is the annual sales growth factor expressed as an interest rate (i.e., $i = S/S_o - 1 = (S - S_o)/S_o$), and

$$R = PPU/CPU = \text{rice} \quad (28)$$

Substituting Equations (27) and (28) in Equation (27), we have:

$$FMU = S \times PPU \left(1 + \frac{1 - i - 1}{(1 + i)\text{rice}}\right) - FC - FIT - DIV \quad (29)$$

or

$$FMU = S \times PPU \left(\frac{1 - i}{(1 + i)\text{rice}}\right) - FC - FIT - DIV \quad (30)$$

From Equation (30), if $FMU = 0$, the business can internally finance next year's anticipated sales growth, i; if FMU is positive the business will not require all of its retained earnings to finance the sales growth anticipated; but if FMU is negative or less than zero, then the business will require $FMU of incremental capital (or assets) to compensate for the deficit generated.

Thus, the year's retained or reinvested earnings determine the maximum growth potential for next year without outside financing. That is, since each $100 of added sales must be supported by $CPU of new investment (i.e., ΔS requires $\Delta S \times CPU$ of new investment, $\Delta TCAP$), then the year's retained earnings will support $\Delta RE/CPU$ dollars of added sales *without the need for external financing*. This may be converted to an internally generated growth "rate" defined by $(100\% \times \Delta RE/CPU)/S_o$, where S_o is last year's sales level and $(\Delta RE/CPU)$ now defines the added sales which may just be supported by last year's retained earnings.

For example, if last year's sales were $10,000,000, income after taxes was $500,000, dividends were $200,000, and capital pickup is $60 (i.e., $60 per $100 of added sales or 60% or, again, in decimal form, 0.60), then retained earnings are $500,000 − $200,000 = $300,000 and the incremental sales that may be supported by retained earnings alone are

$$\Delta RE/CPU = \$500,000/0.60 = \$833,333$$

Thus, in this case, the growth "rate" is:

$$(100\% \times \Delta RE/CPU)/S_o = (100\% \times \$833,333)/\$10,000,000 = 8.33\%$$

That is, the $1,200,000 of added debt (i.e., the product of the capital pickup of $60 or 0.60 and the $2,000,000 of incremental sales, or $0.60 \times \$2,000,000 = \$1,200,000$) will be paid off after

$$\$1,200,000/\$150,000 = 8 \text{ years}$$

To determine the number of years, N (as suggested previously) from the equation $N = \Delta TCAP/\Delta RE$

$$N = (\Delta S \times CPU)/(IBT - FIT - DIV) = (\$2,000,000 \times 0.60)/(\$600,000 - \$3,000,000 - \$150,000)$$

$N = \$1,200,000/\$150,000 = 8$ years, as before

Table 8-3 shows this computation in tabular form.

TABLE 8-3

ILLUSTRATION OF SELF-FINANCING CAPACITY AND DEBT RECOVERY: CASE III

Year	+IBT	−FIT	−DIV	−ΔTCAP	FMU
1	$ 600,000	$ 300,000	$ 150,000	$1,200,000	−$1,050,000
2	600,000	300,000	150,000	—	+ 150,000
3	600,000	300,000	150,000	—	+ 150,000
4	600,000	300,000	150,000	—	+ 150,000
5	600,000	300,000	150,000	—	+ 150,000
6	600,000	300,000	150,000	—	+ 150,000
7	600,000	300,000	150,000	—	+ 150,000
8	600,000	300,000	150,000	—	+ 150,000
TOTAL	$4,800,000	$2,400,000	$1,200,000	$1,200,000	0

One additional observation should be made at this point. Although the foregoing analysis of surpluses and deficits can assist management in measuring the propensity of the business for outside capital, this analytical approach is meant to *supplement but not supplant* the very important, conventional process of cash budgeting.

CASH FLOW

Speaking of surpluses, deficits, and cash, the reader who has come to grips with the myriad aspects of financial analysis has probably already noticed some suspicious parallels in the above discussion to an analysis of the "cash flow" concept. This is entirely intentional.

A complete analysis of cash flow lies outside the scope of our discussion. However, we are interested in those aspects of the concept that bear upon profit pickup and capital pickup—that is, which bear upon the structural characteristics of the business.

Without appraising its advantages and disadvantages, then, cash flow is frequently defined as net income plus noncash outlays (such as depreciation). However, in Equation (25) we have carried the basic idea still further and converted the cash flow concept into a working relationship measuring the self-financing capability of the business.

Although the familiar noncash item of depreciation generally encountered in discussions of cash flow is not explicitly incorporated in Equation (25), we can show that it is, in fact, automatically included. Specifically, the following noncash transactions may typically occur in a large corporation not all of which, however, reflect net changes in book values or are of sufficient magnitude to warrant inclusion by the security analyst in his computation of cash flow:

1. Depreciation allowances
2. Depletion allowances
3. Amortization (of leaseholds, bond discounts and premiums, goodwill, organization costs, research and development costs, etc.)
4. Foreign exchange devaluations (in countries A, C, E, and G)
5. Foreign exchange revaluations (in countries B, D, F, and H)
6. Undistributed earnings of foreign subsidiaries
7. Deferred income taxes
8. Deferred executive compensation

9. Stock dividends
10. Write-off of undepreciated balances of abandoned properties, etc.
11. Prepayment chargeoffs, unearned rents, unrealized profits, etc.
12. Unfunded pension costs, etc.

The net effect of these noncash transactions we shall term Net Non-Cash Outlays, or NNCO. Now, if we examine Equation (25) more closely, we can satisfy ourselves that depreciation and all other items are, in fact, implicitly included. Equation (25) or $\Delta RE - \Delta TCAP = FMU$ may be rewritten as:

$$(IAT + NNCO) - DIV - (\Delta TCAP + NNCO) = FMU \quad (31)$$

Note that we have not changed the equation at all since we first added and then subtracted NNCO. But notice, too, that we now have the use of the term $(IAT + NNCO)$, which is conventionally the definition of "cash flow."

The problem remains of interpreting the expression $(\Delta TCAP + NNCO)$. We know that $\Delta TCAP$ already includes depreciation for the period as a deduction from the plant and equipment asset account. Analogously, all the other noncash items can be shown to be incorporated in $\Delta TCAP$, but of the reverse sign from that assumed by $(+NNCO)$, with the exception of such items as deferred income taxes and stock dividends, which cancel out on the liabilities side of the ledger. We may thus interpret $(\Delta TCAP + NNCO)$ to be the net change in total assets exclusive of noncash additions (i.e., foreign exchange revaluations on foreign investments and undistributed earnings of foreign subsidiaries) and deletions (i.e., depreciation, depletion, amortization, and foreign exchange devaluations).

Hence, the security analyst and investor may continue to utilize their familiar cash flow calculations without destroying the significance and utility of Equation (25) and its derivative, Equation (29), which are first cousins, if you will, of the conventional cash flow concept.

Observe that if:

ΔTL = Change in total liabilities (including long-term debt)
ΔCS = Change in capital stock
ΔRE = Change in retained earnings

then, since the change in total assets, $\Delta TCAP$, equals the total change in the sum of liabilities, capital stock, and retained earnings, we can write:

Financial Engineering and the Profitography System

$$\Delta TCAP = \Delta TL + \Delta CS + \Delta RE \tag{32}$$

But $\Delta RE - \Delta TCAP = FMU$ from Equation (25), and therefore we see that if we know $\Delta TCAP$, ΔRE, and ΔCS, we can estimate the approximate change in liabilities that will accompany incremental sales, or again,

$$FMU = \Delta RE - \Delta TCAP - \Delta TL - \Delta CS \tag{33}$$

Note, finally, that FMU = Cash Flow $-$ DIV $-$ ($\Delta TCAP$ + $NNCO$) since, by definition, Cash Flow = $IAT + NNCO$.

SURPLUSES AND DEFICITS: A CLOSER LOOK

Cash flow surpluses and deficits, like the weather, are neither good nor bad in and of themselves. For example, a surplus may be interpreted to mean:

1. The company is exceedingly profitable and has ample funds generated by profits, and so on to cover the financing of sales expansion with its attendant increased demands for capital.
2. The company is stagnating and cannot find profitable new investment opportunities to exploit to raise sales. Thus, profits exceed incremental capital requirements. Or, again, new investment projects yield incremental returns less than the company's present return on total investment so that it does not pay to reinvest in sales expansion since the returns from added sales will tend to erode the present profit position.
3. The company is stagnating and, in effect, is cannibalizing itself by liquidating its assets and reducing sales and costs. A company phasing itself out of business falls in this category.
4. The company is desirous of increasing capital for future investment opportunities and is not reinvesting profits in sales expansion for the present. Rather, it is hoarding cash.
5. The company is, perhaps, adjusting to a downturn in sales or is striving for greater efficiency and profitability and is liquidating inventories and other asset items to a substantial degree while also paring costs.
6. The company may have closed down an unprofitable plant and abandoned a product line with substantial capital losses upon liquidation due to the forced sale of equipment, inventories, receivables, and so on. However, significant cash proceeds have been realized from the liquidation.

By the same token, the generation of cash flow deficits could well mean that:

1. The company is very unprofitable and cannot obtain enough funds from operations to finance sales increases that are available to it.
2. The company is investing heavily in new investment opportunities and expects exceptional returns on its incremental investments sometime in the future, which will more than justify any deficits generated now.
3. The company is obtaining increased sales at a rate that demands working capital additions beyond the incremental funds available from operations alone. Liquidity of the business may be adversely affected and insolvency could result in spite of the fact that the higher activity level superficially appears most desirable.
4. The company may be liquidating liabilities and retiring debt at a very great rate or perhaps is heavily engaged in buying its own stock (for treasury) on the open market, using large-scale ongoing outside sourcing. (If existing cash on hand is used, decapitalization and even surpluses can be generated, however.)

Every one of these selected situations is familiar to us. In certain cases the company would have no problem in borrowing the additional funds needed from its bankers; in other cases the company is definitely a bad risk.

PICKUP CHARACTERISTICS OF INDIVIDUAL COST AND CAPITAL COMPONENTS

In the same way that the Profitgraph and Capitalgraph are constructed from total cost, total asset, and total sales figures, scattergraphs may be constructed showing the "pickup" characteristics (per $100 of sales change) for cash, receivables, inventories, property, plant and equipment, liabilities, direct labor costs, direct material costs, overhead costs, selling expense, general and administrative expense, and so on.

At this point the reader will be able to visualize these relationships in his or her mind's eye. Thus, each component of cost and capital may be viewed as consisting of a fixed and a variable portion that can be graphically described. The familiar break-even chart has, of course, long been dissected into its several components. But up to now only "working capital" (i.e., current assets less current liabilities), rather than total assets, has been related to sales diagrammatically in the historical literature of finance, engineering economics, and managerial accounting.

THE PROFITOGRAPHY SYSTEM OF FINANCIAL ANALYSIS AND CAPITAL BUDGETING

The above section heading refers to determination and graphic representation of the firm's operating and financial structures or "profiles" for purposes of analysis, planning, and control.

Our purpose has been well served if you recognize (and, hopefully, agree) that from the detailed Profitgraph and Capitalgraph characteristics we can reasonably and readily prepare the following pro forma accounting documents for alternate forecast sales volumes (for total company, individual divisions, individual product lines, etc.) based upon historical data:

1. Statement of Financial Condition (note that both asset and liability items may be constructed from the Capitalgraph and the "explosion" of the basic balance sheet equation: Total Assets = Total Liabilities + Capital)
2. Statement of Income
3. Statement of Funds Flow

Similar statements may be based upon alternate management policies and detailed programs specifically designed to modify the historical profit pickup and capital pickup properties in the direction of greater profitability and return on investment.

Profit planning should not be treated or used merely as a survival kit to avert disaster or as a "bag of tricks" to impress shareholders and securities analysts. It must be a disciplined, scientific approach to optimizing business results; an approach worthy of truly professional management to be applied on a continual, strategic basis.

Notice that we use the term *strategic* rather than *tactical*. This is because profit planning is administrative and long-range in nature, while the tactical aspect of managerial control is short-run and executed on an individual "mission," "milestone," or "task" basis. Budgetary control represents the tactical aspect of business planning and control. The budget is a firm commitment specifically designed to support and implement the broader objectives of the long-range profit plan.

While ratio-ray graphs can be employed to portray flexible budget relationships for the various operating and financial factors, Profitgraphs and Capitalgraphs are *not* flexible or variable budgets as such. Their greatest utility lies in the analysis and portrayal of the gross business structure and alternate long-range profit plans. Limited adaptations of the Profitgraph and Capitalgraph idea may assist management in controlling *current* operations but, in general, the economic and accounting characteristics of Profitgraphs and Capitalgraphs hinder their successful tactical application.

The gross characteristics of a profit plan can be illustrated graphically. Exhibit 8-1, which we shall call the Cash Flow Profit Planner, or CFPP chart, is a diagram of a five-year plan based upon Profitgraph, Capitalgraph, and cash flow characteristics of the business.

CFPP data can readily be constructed from relationships previously developed. Assuming that the Five-Year Profit Plan projects sales levels S_1 through S_5 and that $TCAP_0$ = this year's total capital or assets, S = this year's sales volume, IBT_0 = this year's pretax income, $(D + T)$ = this year's dividends and income taxes, and FMU_0 = this year's financial makeup, if any, all other items may be computed from the following relationships:

$$CFPP = 1: \quad \Delta TCAP = \Delta S \times CPU$$

$$CFPP = 2: \quad \Delta IBT = \Delta S \times PPU$$

$$CFPP = 3: \quad \Delta RE = \Delta IBT - FIT - DIV$$

$$CFPP = 4: \quad FMU = \Delta RE - \Delta TCAP$$

Observe that the analysis assumes that profit pickup, capital pickup, fixed costs, and fixed assets are known. That is, these latter four characteristics of the business may be derived from:

1. Historical Profitgraph and Capitalgraph analyses
2. Targeted Profitgraph and Capitalgraph structures
3. A combination of (1) and (2).

An example illustrates the use of the CFPP. Exhibit 8-2 shows a situation where:

$$TCAP_0 = \$\ 8 \text{ (all data in millions of dollars)}$$
$$S_0 = \$10$$
$$IBT_0 = \$\ 2$$
$$S_1 = \$15$$

Financial Engineering and the Profitography System

EXHIBIT 8-1

CASH FLOW PROFIT PLANNER

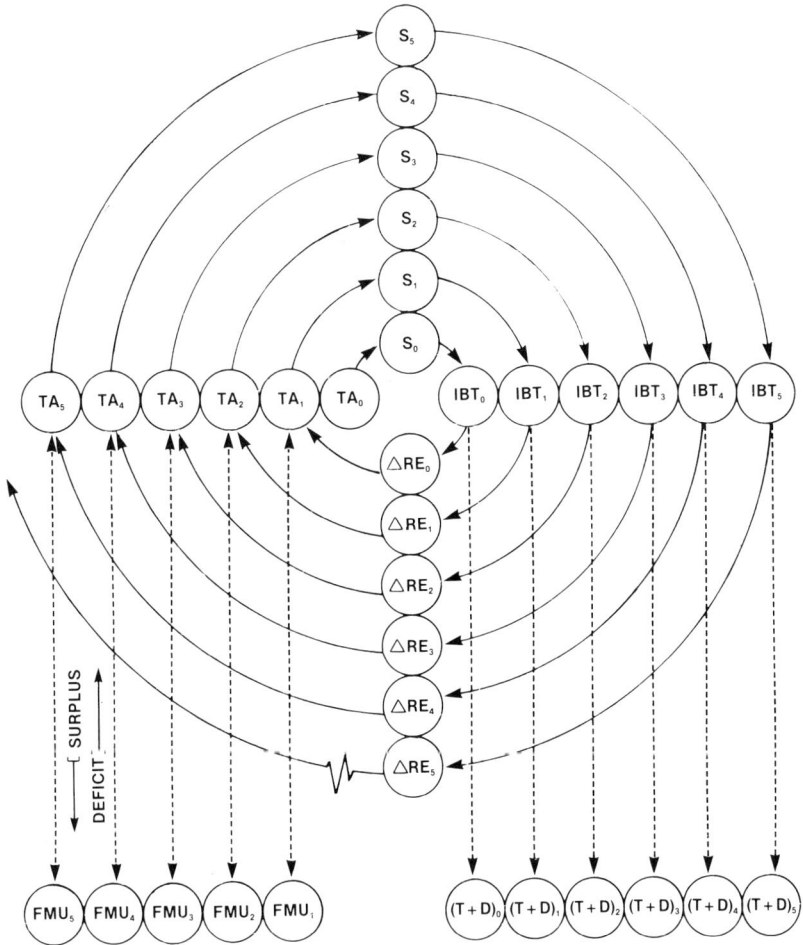

$$S_2 = \$15$$
$$S_3 = \$15$$
$$S_4 = \$12$$
$$S_5 = \$15$$

and *PPU*, *CPU*, *DIV*, and *FIT* are as shown by Table 8-4. Table 8-5, in turn, shows the calculation of *TCAP*, *IBT*, $(D + T)$, and *FMU* for years 1 through 5 of the profit plan. Note that *PPU*, *CPU*, *DIV*, and

EXHIBIT 8-2

CASH FLOW PROFIT PLANNER: CASE STUDY

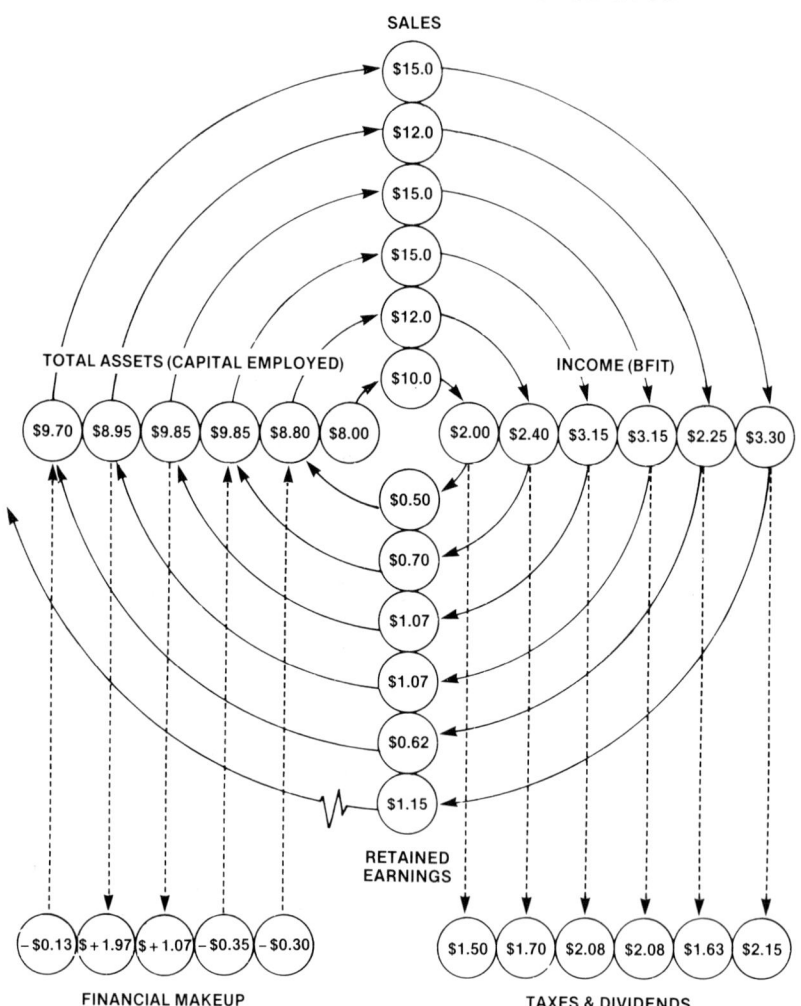

FIT are taken from Table 8-4. Exhibit 8-3 is a blank master CCPP chart for your use in working through your own examples.

THE CONCEPT OF LEVERAGE

We will recall from our previous discussion of incremental return on incremental investment that a twenty percent increase in sales in-

TABLE 8-4

CASH FLOW PROFIT PLANNER: ASSUMED DATA
FOR EXHIBIT 8-2

Year	PPU	CPU	DIV	FIT
0	$20	$40	$500,000	50% of IBT
1	20	40	500,000	50% of IBT
2	25	35	500,000	50% of IBT
3	25	35	500,000	50% of IBT
4	30	30	500,000	50% of IBT
5	35	25	500,000	50% of IBT

duced an increase in return on capital of almost 90 percent. At that time, we termed this relationship one of tremendous "leverage" by sales upon investment return.

The ratio-ray graphs presented in our discussions likewise portrayed the patterns of ratio change for various operating and financial relationships. In particular, two significant, common characteristics of ratio-ray graphs stood out: (1) ascending trend lines whose Y-intercepts lie above the origin intersected ratio rays of *decreasing* $A:B$ value (refer back to Exhibit 7-6) and (2) for equal increases in B, the $A:B$ ratio did *not* decrease in equal decrements but actually decreased less and less with each succeeding increase in B. Tables 7-1 and 7-2, we will recall, demonstrated this interesting behavior. Thus, a given change in B induced a certain change in A or in some function of A and B (i.e., if A = Costs and B = Sales, an important function of A and B is income, which equals the difference between B and A or $B - A$). As we observed above, a small change in one factor—as, for example, sales—can induce a disproportionate change in others—as, for example, in income or return on investment.

Thus, by the very descriptive term "leverage" we mean that a relatively small change in one factor may induce a much greater change in another (or vice versa). Expressed slightly differently, the former factor has increased power of action, amplification, or magnification—or a great "leverage"—upon the latter. "Leverage" lies at the heart of what is alternately called "differential" or "incremental" cost and profit analysis by different writers. And since the Profitgraph and Capitalgraph, and so on, are basically graphical representations of leverage relationships, we will explore the subject a little.

Broadly speaking, every mathematical equation might be said

EXHIBIT 8-3

FINANCIAL MAKEUP ANALYZER

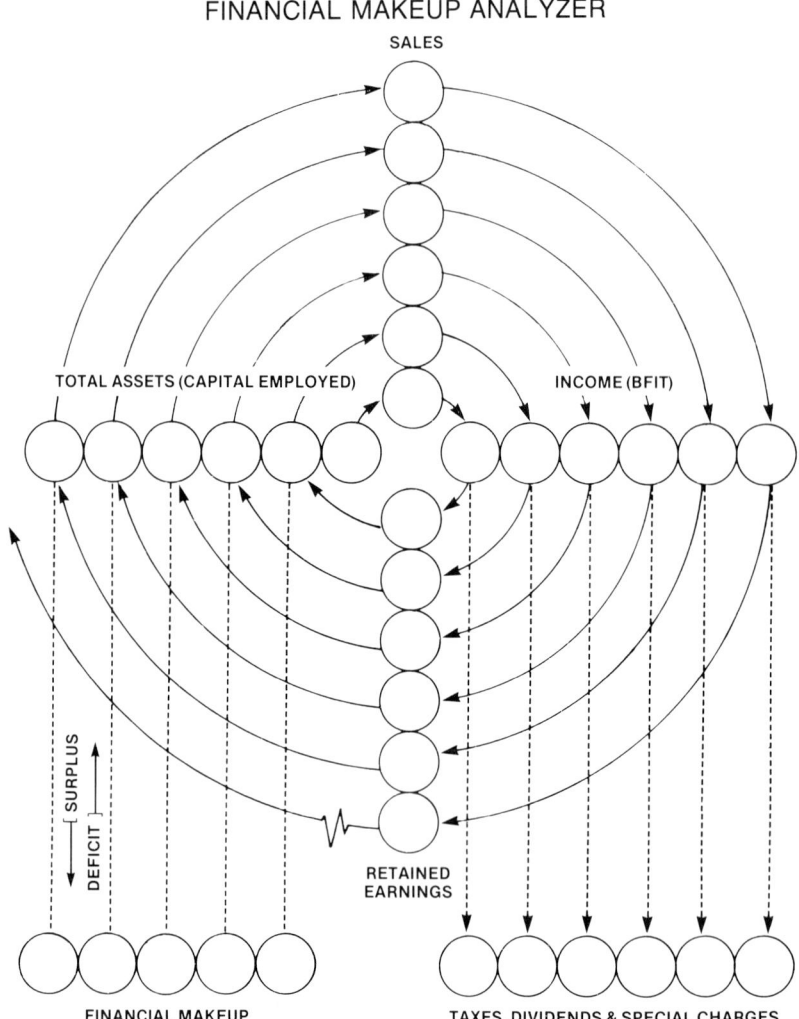

to express a "leverage" relationship. However, we will limit our use of the term to those relationships expressing the resultant "change" in one factor due to a given "change" in another. Thus, $y = f(x)$ is *not* a leverage equation, but $\Delta y = f(\Delta x)$ *is* a leverage equation (where, again, Δ signifies a "change" in x or a "change" in y).

Let us construct the leverage expressions for some of the more common operating and financial ratios and work out some illustra-

TABLE 8-5

CASH FLOW PROFIT PLANNER: COMPUTED DATA FOR EXHIBIT 8-2

Year	ΔS	$\Delta TCAP = \Delta S \times CPU$	$TCAP$	$\Delta IBT = \Delta S \times PPU$	IBT	$\Delta RE = IBT - FIT - DIV$	$FMU = \Delta RE^* - \Delta TCAP$
0	—	—	$8.00	—	$2.00	$2.00 − $1.00 − $.50 = $0.50	—
1	$2.0	$2.0 × 0.40 = 0.80	8.80	$2.0 × 0.20 = $0.40	2.40	$2.40 − $1.20 − $.50 = $0.70	−$0.30 (DEFICIT)
2	3.0	3.0 × 0.35 = 1.05	9.85	3.0 × 0.25 = 0.75	3.15	$3.15 − $1.58 − $0.50 = $1.07	−$0.35 (DEFICIT)
3	0.0	0.0 × 0.35 = 0.00	9.85	0.0 × 0.25 = 0.00	3.15	$3.15 − $1.58 − $0.50 = $1.07	+$1.07 (SURPLUS)
4	−3.0	−3.0 × 0.30 = −0.90	8.95	−3.0 × 0.30 = −0.90	2.25	$2.25 − $1.13 − $0.50 = $0.62	+$1.97 (SURPLUS)
5	3.0	3.0 × 0.25 = 0.75	9.70	3.0 × 0.35 = 1.05	3.30	$3.30 − $1.65 − $0.50 = $1.15	−$0.13 (DEFICIT)

* In this computation, ΔRE is that for the prior year.

tive examples. In our derivations, we will use the following abbreviations:

S = Sales
TC = Total Costs
TVC = Total Variable Costs
TFC = Total Fixed Costs
$TCAP$ = Total Capital or Assets
$VCAP$ = Variable Capital
$FCAP$ = Fixed Capital
IBT = Income Before Taxes
BEP = Break-Even Point
EQ = Shareholders' Equity
MI = Marginal Income

The following ratios will be of interest:

$$ibtr = \text{Income Before Tax Ratio} = (S - TC)/S$$
$$= (S - TVC - TFC)/S = IBT/S \tag{34}$$

$$ctr = \text{Capital Turnover Ratio} = S/TCAP$$
$$= S/(VCAP + FCAP) \tag{35}$$

$$rce = \text{Return on Capital} = IBT/TCAP = ibtr \times ctr \tag{36}$$

$$flr = \text{Financial Leverage Ratio} = TCAP/EQ \tag{37}$$

$$req = \text{Return on Equity} = IBT/EQ = rce \times flr \tag{38}$$

$$mir = \text{Marginal Income Ratio} = (S - TVC)/S \tag{39}$$

$$msr = \text{Marginal Safety Ratio (also termed the ``Margin of Safety'' and ``Profit Elasticity'' Ratio)} = (S - BEP)/S = IBT/MI \tag{40}$$

where

$$BEP = \text{Break-Even Point (Sales)} = TFC/mir \tag{41}$$

and

$$bepr = \text{Break-Even Point Ratio} = BEP/S \tag{42}$$

From the above basic relationships, the following are derived readily and may be easily verified if the reader will take a few moments to make the indicated substitutions:

$$ibtr = mir \times msr \tag{43}$$

$$rce = ctr \times mir \times msr \tag{44}$$

$$req = flr \times ctr \times mir \times msr \tag{45}$$

$$msr = 1 - bepr \tag{46}$$

This last equality yields the following round of relationships, in turn:

$$ibtr = mir(1\text{-}bepr) \tag{47}$$

$$rce = ctr \times mir(1\text{-}bepr) \tag{48}$$

$$req = flr \times ctr \times mir(1\text{-}bepr) \tag{49}$$

Thus, it is seen that many of the familiar concepts of modern financial management can be incorporated into a few key expressions. Notice that although most of the above formulas relate to the Profitgraph, analogous expressions may be derived for the Capitalgraph.

Now, to derive the leverage equations for each of the above it is necessary to utilize the "Delta" method of differential calculus. In the following equations, Δ again denotes the change in value of the term following, or the difference between the second (or terminal) value and the first (or initial) value. Thus, $\Delta S = S_2 - S_1$, $\Delta ibtr = ibtr_2 - ibtr_1$, $\Delta BEP = BEP_2 - BEP_1$, and so on, where the subscripts 2 and 1 denote the second and first values of the term, respectively:

$$\Delta mir = (\Delta S(1-mir_1) - \Delta TVC)/S_2 \tag{50}$$

$$\Delta msr = 1 - \Delta bepr = 1 - (\Delta BEP - \Delta S \times bepr_1)/S_2 \tag{51}$$

$$\Delta ibtr = (\Delta mir \times msr_1) + (\Delta msr \times mir_2) \tag{52}$$

$$\Delta ctr = (\Delta S - \Delta TCAP \times ctr_1)/TCAP_2 \tag{53}$$

$$\Delta rce = (\Delta ibtr \times ctr_1) + (\Delta ctr \times ibtr_2) \tag{54}$$

$$\Delta req = (\Delta rce \times flr_1) + (\Delta flr \times rce_2) \tag{55}$$

$$\Delta BEP = -(\Delta mir \times BEP_1)/mir_2 \tag{56}$$

But in equation (53), above, TCAP consists of two components: variable capital, VCAP (i.e., cash, inventories, and accounts receivable or at least the variable portions thereof) and fixed capital, FCAP (i.e., the fixed portions of inventories, property and plant and equipment, etc.). Now, since by definition fixed capital is fixed, $\Delta TCAP$ can only be a function of $\Delta VCAP$, and so, rewriting Equation (53), we have:

$$\Delta ctr = (\Delta S - \Delta VCAP \times ctr_1)/TCAP_2 \tag{57}$$

In a similar manner, although

$$\Delta flr = (\Delta TCAP - \Delta EQ \times flr_1)/EQ_2 \tag{58}$$

since the change in $\Delta TCAP$ is only in the $VCAP$ component, the expression can be rewritten as:

$$\Delta flr = (\Delta VCAP - \Delta EQ \times flr_1)/EQ_2 \tag{59}$$

In all of the above equations, if any of the factors is considered to be constant for a particular application, then the Δ term is simply considered to be zero, namely, ctr = constant, then $ctr_1 = ctr_2$ and $ctr_2 - ctr_1 = \Delta ctr = 0$.

Frequently, in the public press, articles can be found stating that XYZ Corporation increased sales by 10 percent and profits by 20 percent, or sales dipped 10 percent but profits *plunged* 20 percent compared with a previous period. Such relationships reflect the operating leverage characteristic of the business, which can be developed as follows:

$$IBT = S - TVC - TFC = MI - TFC$$

where

$$MI = S - TVC = \text{Marginal Income} \tag{60}$$

Dividing through by S, we have:

$$IBT/S = MI/S - TFC/S = mir - TFC/S \tag{61}$$

or

$$IBT = mir \times S - TFC \tag{62}$$

Taking the derivative to obtain ΔIBT, we find that:

$$\Delta IBT = (\Delta mir \times S_1) + (\Delta S \times mir_2) - \Delta TFC \tag{63}$$

But, by definition, TFC is constant and so $\Delta TFC = 0$, or

$$\Delta IBT = (\Delta mir \times S_1) + (\Delta S \times mir_2). \tag{64}$$

Dividing through by IBT_1 and S_1, we now obtain

$$(\Delta IBT/IBT_1)/S_1 = (\Delta mir \times 1/IBT_1) + (\Delta S/S_1) \times (mir_2/IBT_1). \tag{65}$$

But, ordinarily, it can be assumed that the marginal income ratio has not drastically been changed from period to period and so we can approximate $mir_2 = mir_1$, and therefore:

$$(\Delta IBT/IBT_1)/S_1 = (\Delta mir/IBT_1) + (\Delta S/S_1) \times (mir_1/IBT_1) \tag{66}$$

Now, multiplying through by S_1, and noting that $IBT_1/S_1 = ibtr_1$, we have

$$\Delta IBT/IBT_1 = (\Delta mir/ibtr_1) + (\Delta S/S_1) \times (mir_1/ibtr_1) \quad (67)$$

Finally, if we let $(\Delta IBT/IBT_1) \times 100\%$ be the percentage change in income, and $(\Delta S/S_1) \times 100\%$ the percentage change in sales, and further note that since mir_2 is taken as essentially equal to mir_1 so that $\Delta mir = 0$, then:

$$\% \text{ change in } IBT = \% \text{ change in Sales} \times mir_1/ibtr_1. \quad (68)$$

Note also that, in Equation (72), if $\Delta mir = 0$ or $mir_2 = mir_1 = mir$, then $\Delta IBT = \Delta S \times mir$ or, since $mir = PPU/\$100$, then $\Delta IBT = \Delta S \times PPU/\$100$.

Therefore, knowing mir_1 and $ibtr_1$, plus the percentage change in sales, we can compute the resultant percentage change in IBT. In the equation, the term $(mir_1/ibtr_1)$ is the "operating leverage factor."

Another result of even more interest is extractable. Let us recall that $ibtr = mir \times msr$, so that $mir/ibtr = 1/msr$. Also, $msr = 1 - bepr = 1 - BEP/S$. Consequently, we can rewrite Equation (68), above, as

$$\% \text{ change in } IBT = \% \text{ change in Sales} \times 1/mrs_1 \quad (69)$$

$$= \% \text{ change in Sales} \times 1/(1 - bepr_1)$$

$$= \% \text{ change in Sales} \times 1/(1 - BEP_1/S_1)$$

$$= \% \text{ change in Sales} \times S_1/(S_1 - BEP_1)$$

or

$$(S_1 - BEP_1) \times (\% \text{ change in } IBT) = S_1 \times \% \text{ change in Sales} \quad (70)$$

or

$$S_1 - BEP_1 = (S_1 \times \% \text{ change in Sales})/\% \text{ change in } IBT \quad (71)$$

and, finally,

$$BEP_1 = S_1 - S_1 \times \% \text{ change in Sales}/\% \text{ change in } IBT \quad (72)$$

or, factoring out S_1,

$$BEP_1 = S_1 \times (1 - \% \text{ change in Sales}/\% \text{ change in } IBT) \quad (73)$$

Thus, if quarterly P&L statements are issued by company ABC, it can be assumed that the break-even characteristics have not been altered dramatically from quarter to quarter and thus one can intelligently "guesstimate" one of the most closely guarded secrets of modern corporate managements (note, again, that we have reasonably assumed the marginal income ratio to be a constant for the periods under discussion) since, from these reports, we can obtain the percentage change in sales, the percentage change in income, and the initial sales volume, S_1.

It is an amusing sidelight, perhaps, to observe that this highly confidential business statistic is laid bare, using a little algebra, from numbers supplied by management itself. However, this is only another instance where "overt" commercial intelligence techniques prove themselves just as potent as the cloak-and-dagger varieties one reads about from time to time. It just takes a little "applied intelligence" assisted by elementary engineering thinking.

We can also illustrate the leverage principle applied to return on capital, taking into account both cost and capital variability. In this example, the derivation is omitted but the technique of derivation is similar to the above analyses. Equation (74) states that a resultant percentage change in return on capital equals the percentage change in sales multiplied by the "Return on Capital Leverage Factor" which equals $(mir/ibtr_1) - (ctr_1 \times vcapr)$, where $vcapr$ = variable capital ratio = $VCAP/S$ = $CPU/\$100$. That is:

$$\% \text{ change in } rce = \% \text{ change in Sales} \times (mir/ibtr_1 - ctr_1 \times vcapr)$$
$$= \% \text{ change in Sales} \times (PPU/ibtr_1 - ctr_1 \times CPU) \quad (74)$$

where % change in $rce = \Delta rce/rce_1$,
  % change in sales = $\Delta S/S_1$, and
  mir and $vcapr$ are fixed in value for purposes of this
  formulation and PPU and CPU are expressed in decimal form.

Another even more useful relationship yielding the percentage change in return on capital can be derived as follows:

$$rce = IBT/ACE \quad (75)$$

or

$$rce \times ACE = IBT \quad (76)$$

Now, differentiating, Equation (76) becomes:

$$\Delta rce = (\Delta IBT - rce_2 \times \Delta ACE)/ACE_1$$
$$= \Delta IBT/ACE_1 - (rce_2 \times \Delta ACE)/ACE_1 \quad (77)$$

Dividing through by rce_1, we can write:

$$\Delta rce/rce_1 = (\Delta IBT/rce_1 \times ACE_1) - (rce_2/rce_1) \times (\Delta ACE/ACE_1) \quad (78)$$
$$= \Delta IBT/IBT_1 - (rce_1 + \Delta rce)/rce_1 \times \Delta ACE/ACE_1$$

or

$$\Delta rce/rce_1 = \Delta IBT/IBT_1 - (1 + \Delta rce/rce_1) \times (\Delta ACE/ACE_1) \quad (79)$$

or

$$\Delta IBT/IBT_1 = \Delta rce/rce_1 + \Delta ACE/ACE_1 + (\Delta ACE/ACE_1) \times (\Delta rce/rec_1)$$
$$= \Delta ACE/ACE_1 + (\Delta rce/rce_1) \times (1 + \Delta ACE/ACE_1) \quad (80)$$

or

$$\Delta rce/rce_1 + (\Delta IBT/IBT_1 - \Delta ACE/ACE_1)/(1 + \Delta ACE/ACE_1) \quad (81)$$

Finally, since $(\Delta rce/rce_1) \times 100\%$ = % change in rce, $(\Delta IBT/IBT_1) \times 100\%$ = % change in IBT, etc., we can write Equation (81) as follows:

% change in rce = (% change in IBT
   − % change in ACE)/(100% + % change in ACE)
$$(82)$$

To illustrate the interplay of price changes, physical output changes, and variable cost changes as they affect income, another leverage relationship may be constructed. Again, the detailed derivations are omitted but the technique of arriving at the formulas is the same as before. In Equation (83), price, output, and cost changes are so arranged that their net effect is to keep income at the same level as before. Thus, for $IBT = 0$:

% price change + % physical volume change × (% price change
+ % mir_1) = % variable unit cost change
   × (100% + % physical volume change)
   × (100% − % mir),   (83)

where % price change = $(\Delta p/p_1) \times 100\%$ and p = unit price,
   % physical volume change = $(\Delta n/n_1) \times 100\%$, and n = number of physical output units, and
   % variable unit cost change = $(\Delta VUC/VUC_1) \times 100\%$ and VUC = variable unit cost.
   In Equation (84), both ΔIBT and VUC = 0. Thus,

% price change = − (% phys.vol.chg.) × (% price chg. + % mir_1)
$$(84)$$

In the general case where all the factors vary, and where $S_1 = n_1 \times p_1 =$ total initial sales volume,

$$\Delta IBT = \%\text{ price change} \times S_1 + \%\text{ phy. vol. chg.} \\ \times S_1 \times (\%\text{ price chg.} + \%\ mir_1) \\ - \%\text{ var. unit cost chg.} \times S_1 \\ \times (100\% + \%\text{ phys. vol. chg.}) \quad (85) \\ \times (100\% - \%\ mir_1)$$

Also, the new sales volume, S_2, as a percentage of the old, S_1, for any of the above cases, is

$$(S_2/S_1) \times 100\% = (n_2 p_2 / n_1 p_1) \times 100\% \\ = (100\% + \%\text{ price chg.})(100\% + \%\text{ phys. vol. chg.})/100\% \quad (86)$$

We have not developed leverage relationships for Capitalgraph factors. However, these are highly significant since it can be shown that:

$$rce = \frac{PPU \times per}{1 + caper(CPU - 1)} \quad (87)$$

where $caper =$ capital elasticity ratio $= (S - CUP)/S$ and $per =$ profit elasticity ratio $= (S - BEP)/S$. Note that per is the same as the marginal safety ratio, but is preferred to the latter for the sake of uniformity in terminology (of course, we might have used some such term as "capital safety ratio" instead).

It can also be proven that:

$$caper = 1 - (CUP/BEP)(1 - per) \quad (88)$$

and that:

$$per/\Delta caper = BEP/CUP \quad (89)$$

or, in percentage form:

$$\%\text{ chg. in } per/\%\text{ chg. in } caper = (BEP/CUP)(caper_1/per_1) \quad (90)$$

Furthermore,

$$\Delta rce = \Delta per \left[\frac{PPU - rce_1(CPU - 1)(CUP/BEP)}{1 + caper_2(CPU - 1)} \right] \quad (91)$$

or, in percentage form:

$$\%\text{ chg. in } rce = \Delta per \left[\frac{PPU/rce_1 - (CPU - 1)(CUP/BEP)}{1 + caper_2(CPU - 1)} \right] \times 100\% \quad (92)$$

Financial Engineering and the Profitography System

In the above equations PPU and CPU are assumed to be constants of the business expressed in decimal form and BEP and CUP equal $FC/mir = FC/PPU$, and $FCAP/(1 - CPU)$, respectively.

EXAMPLES

The following ten examples are presented to illustrate the application of leverage relationships.

Example 1:

The XYZ Co. has a break-even point of $20,000,000 and a marginal income ratio of 70% or 0.70. At a sales volume of $50,000,000, total capital employed in the business is at a level of $30,000,000. What are the ratios of income ($BFIT$) to sales and to total capital employed? Also, what are the inherent fixed costs of the business?

Solution: From Equation (47), $ibtr = mir\,(1\text{-}bepr)$; also, $mir = 0.70$ (given), and $bepr = BEP/S = \$20,000,000/\$50,000,000 = 0.40$ and $(1\text{-}bepr) = 0.60$. Therefore, $ibtr = 0.70 \times 0.60 = 0.42$ or 42%, and $rce = ctr \times ibtr = S/TCAP \times ibtr = \$50,000,000/\$30,000,000 \times 0.42 = 0.70$ or 70%, and $TFC = BEP \times mir = \$20,000,000 \times 0.70 = \$14,000,000$.

Example 2:

If XYZ Co. management was able to increase the marginal income ratio of 0.70 by 10 points to 0.80, what would be the new break-even point?

Solution: From equation (56), $\Delta BEP = -(\Delta mir \times BEP_1)/mir_2$. Thus, $\Delta mir = mir_2 - mir_1 = 0.80 - 0.70 = 0.10$, and $\Delta BEP = -(0.10 \times \$20,000,000)/0.80 = -\$2,500,000$ and the new BEP is: $BEP_2 = BEP_1 + \Delta BEP = \$20,000,000 - \$2,500,000 = \$17,500,000$.

Example 3:

The XYZ Co.'s arch competitor, the IOU Corp., releases only the barest minimum of data. However, it does issue quarterly reports to its shareholders and the public press. The *National Business Daily*, in commenting upon the IOU Corp.'s performance for the last quarter, noted that with an increase in sales of 10% over the previous quarter sales of $40,000,000, profits rose 60% from the previous quarter mark of 10% $BFIT$. What are the approximate levels of the IOU Corp.'s marginal income ratio and break-even point, information which IOU management guards jealously and considers proprietary?

Solution: From Equation (68), % change in IBT = % change in sales $\times$ $mir_1/ibtr_1$. Therefore, $0.60 = 0.10 \times mir_1/0.10$, and $mir_1 = 0.60 \times 0.10/0.10 = 0.60$ or 60% (estimated). Now, from Equation (73), $BEP_1 = S_1 (1 - $ % change in sales/% change in $IBT)$ and, therefore, $BEP_1 = \$40,000,000 (1 - 0.10/0.60) = \$40,000,000 \times 0.833 = \$33,320,000$ (estimated).

Example 4:

What will be the percentage change in the return on capital employed if XYZ Co. raises its sales volume by 20% from the $50,000,000 level? Use XYZ Co. characteristics already given in Example 1, above.

Solution: From Equation (74), the return on capital leverage factor must be computed and multiplied by the 20% volume change given. From Example 1, above, $mir = 70\% = 0.70$, and $ibtr_1 = 42\% = 0.42$. Also, from Example 1, $ctr_1 = \$50,000,000/\$30,000,000 = 1.67$, and from Example 10, $vcapr = CPU/\$100 = \$50/\$100 = 0.50$. Now, substituting these values in the formula for the return on capital leverage factor,

rce leverage factor = $(0.70/0.42) - (1.67 \times 0.50) = 0.835$.

Finally, multiplying the above by the 20% sales volume increment, we obtain our answer:

% change in rce = $0.20 \times 0.835 = 0.167$ or 16.7%

Example 5:

If the XYZ Co. is obliged by competition to lower its prices on Turbotrons by 10% and, at the same time (due to negotiating a new labor contract) increase its variable costs by 10%, what must be the physical output level of Turbotrons in order to retain a profit ($BFIT$) of $2,000,000? What must be the new sales volume? Turbotrons are produced by the Turbotronics Division, whose annual sales are $10,000,000; marginal income ratio is 40%; total fixed costs are $2,000,000; volume is 100,000 Turbotrons annually; and price is $100 per Turbotron unit.

Solution: We must use Equation (83). Substituting the given volume, we have the following:

% price change = $-10\% = -0.10$

% physical volume change = ?

% variable unit cost change = $10\% = 0.10$

and

$$mir = 40\% = 0.40$$

so that

$$-10\% + \% \text{ phys. vol. chg. } (-10\% + 40\%)$$
$$= 10\%(100\% + \% \text{ phys. vol. chg.})(60\%),$$

and working through this equation, we find:

$$\% \text{ phys. vol. chg.} = 66.667\% \text{ increase}$$
$$n_2 = 166.67\% \times 100{,}000 = 166{,}667 \text{ Turbotrons}$$
$$S_2 = 166{,}667 \times \$90 = \$15{,}000{,}000$$

As a check, let us see what the new profit actually is. If our calculations are correct (as well as the basic formula), it should be $2,000,000, as before.

$$S_2 = \$15{,}000{,}000$$
$$FC_2 = FC_1 = \text{constant @ } \$2{,}000{,}000$$
$$VUC_1 = (1 - mir) \times S_1/100{,}000 = \frac{(1 - 0.40) \times \$10{,}000{,}000}{100{,}000}$$
$$= \$60/\text{Turbotron unit}$$
$$VUC_2 = 110\% \times \$60 = \$66/\text{Turbotron unit}$$
$$IBT_2 = S_2 - FC - TVC_2 = \$15{,}000{,}000 - \$2{,}000{,}000$$
$$- \$66 \times 166{,}667 \text{ units} = \$15{,}000{,}000 - \$2{,}000{,}000$$
$$- \$11{,}000{,}000 = \$2{,}000{,}000.$$

This checks out and illustrates how large a leverage effect upon volume relatively smaller price and variable cost changes can exert. Normally, in a case such as this, it is unlikely that a company could increase its volume enough in the near-term to offset price and cost changes of the magnitude indicated.

Example 6:

For XYZ Co.'s Turbotronics Division, what was the original profit pickup? The new *PPU*?

Solution: For the original conditions where $mir = 40\%$, $PPU = mir \times \$100 = \$40$ per $100 of increased sales. For the new condition, where variable costs increased 10% (i.e., $TVC_2 = 110\%\ TVC_1$), from Example 5, above, $TVC_2 = \$11,000,000$ and $S_2 = \$15,000,000$, so:

vcr_2 = variable cost ratio = $TVC_2/S_2 = \$11,000,000/\$15,000,000 = 73.333\%$

$mir_2 = (1 - vcr_2) = 1 - 0.733 = 0.267$ or 26.7%

or, from Equation (75),

$$\Delta mir = (\Delta S(1 - mir) - \Delta TVC)/S_2$$

$$= (\$5,000,000(1 - 0.40) - \$5,000,000)/\$15,000,000$$

$$= -0.133 \text{ or } -13.3\%,$$

and therefore,

$$mir_2 = mir_1 + \Delta mir = 40\% - 13.3\% = 26.7\%$$

as above.

Thus, the new $PPU = \$26.70$ per $\$100$ of added sales. Consequently, a 10% increase in variable costs and a 10% decrease in prices have the compound effect of reducing PPU by 33.2%.

Example 7:

Using the profit pickup concept, compute IOU Corp.'s break-even point given the following data:

$S_1 = \$40,000,000 \qquad IBT_1 = \$4,000,000$

$S_2 = \$44,000,000 \qquad IBT_2 = \$6,400,000$

Solution: From our discussion of Exhibit 7-2 at the very outset, $PPU = (\Delta IBT/\Delta S) \times \$100$. This relationship is proven below. Thus, $(\$6.4M - \$4.0M/\$44M - \$40M) \times \$100 = \$60$ per $\$100$. Also, it can be shown that $BEP = S - (IBT/PPU) \times \$100$, which is also proven below. Now using subscript 1 values of S and IBT, $BEP = \$40M - (\$4M/\$60) \times \$100$ and $BEP = \$33,330,000$. This value should and does check with the solution found in Example 3, above, since the Sales and Income values given in Example 7 underlie the data given in Example 3.

Now, we can prove that $PPU = (\Delta IBT/\Delta S) \times \$100$ as follows:

$$mir = (S - VC)/S \qquad (93)$$

or

$$S \times mir = S - VC \qquad (94)$$

Financial Engineering and the Profitography System 169

Differentiating Equation (94) we have:

$$S_1 \times \Delta mir + mir_2 \times \Delta S = \Delta S - \Delta VC \quad (95)$$

But, assuming that mir is a constant so that $\Delta mir = 0$ or $mir_2 = mir_1 = mir$,

$$mir_2 \times \Delta S = \Delta S - \Delta VC = mir \times \Delta S \quad (96)$$

Also, $mir = PPU$ (in decimal form) and so:

$$PPU \times \Delta S = \Delta S - \Delta VC \quad (97)$$

or

$$PPU = (\Delta S - \Delta VC)/\Delta S \quad (98)$$

Now, since

$$IBT = S - FC - VC \quad (99)$$

then, by differentiation of Equation (99),

$$\Delta IBT = \Delta S - \Delta FC - \Delta VC \quad (100)$$

Also, by definition, FC is a constant or $\Delta FC = FC_2 - FC_1 = 0$, and so

$$\Delta IBT = \Delta S - \Delta VC \quad (101)$$

Finally, substituting Equation (101) in Equation (98), we can write:

$$PPU = (\Delta IBT/\Delta S) \times \$100 \quad (102)$$

which was to be proved.

To prove that $BEP = S - (IBT/PPU) \times \$100$, we proceed as follows. The marginal income ratio, expressed in terms of profit pickup, facilitates computation of the break-even point even more readily than Equation (73), above, as follows:

$PPU = mir \times \$100$ per $\$100$ added sales volume (by definition)
$$\quad (103)$$

$$PPU = (\Delta IBT/\Delta S) \times \$100 \text{ (Equation (102), repeated)} \quad (104)$$

or

$$mir = (\Delta IBT/\Delta S) = PPU/\$100 \quad (105)$$

Now, from Equation (41), $BEP = TFC/mir$. Also,

$$TFC = S - TVC - IBT = S - (S - S \times mir) - IBT \quad (106)$$

or

$$TFC = S - (S - S \times \Delta IBT/\Delta S) - IBT \quad (107)$$

and

$$TFC = (S \times \Delta IBT/\Delta S) - IBT \tag{108}$$

Substituting in Equation (41) for TFC and mir, we have:

$$BEP = (S \times \Delta IBT/\Delta S - IBT)/(\Delta IBT/\Delta S) \tag{109}$$

Dividing through by $(\Delta IBT/\Delta S)$,

$$BEP = S - (IBT/\Delta IBT) \times \Delta S \tag{110}$$

But, from Equation (105), $mir = (\Delta IBT/\Delta S) = PPU/\$100$ so that

$$BEP = S - (IBT/PPU) \times \$100 \tag{111}$$

which was to be proved.

Note that since $S - BEP = MS$ or the margin of safety, by definition, we can rewrite Equation (111) as:

$$S - BEP = (IBT/PPU) \times \$100 = MS \tag{112}$$

or

$$MS = (IBT/PPU) \times \$100 \tag{113}$$

Thus, the break-even point may also be computed as sales less the ratio of income BFIT to profit pickup multiplied by $100 or, in terms of marginal income ratio, as the difference between sales and the ratio of income before taxes to marginal income ratio. Also, the margin of safety (MS) equals $100 multiplied by the ratio of income before taxes to profit pickup or, in terms of marginal income ratio, MS equals the ratio of income before taxes to the marginal income ratio.

Let us now continue with our examples:

Example 8:

If XYZ Co. wishes to increase its return on capital by 20% next year and it anticipates an increase in profits of about 50% due to intensive profit improvement programs, the elimination of several marginal product lines, warehousing operations, and factories, what must be the change in assets even if this will entail a massive change in capital management policy?

Solution: From Equation (82),

$$\% \text{ change in rce} = \frac{\% \text{ chg. in } IBT - \% \text{ chg. in } ACE}{100\% + \% \text{ chg. in } ACE},$$

where

$$\% \text{ change in } rce = 20\%$$
$$\% \text{ change in } IBT = 50\%$$
$$\% \text{ change in } ACE = ?$$

To solve for % change in ACE we must simplify Equation (82), as follows:

$$(100\% + \% \text{ chg. in } ACE)(\% \text{ chg. in } rce)$$
$$= \% \text{ chg. in } IBT - \% \text{ chg. in } ACE \qquad (114)$$

or

$$\% \text{ chg. in } rce + (\% \text{ chg. in } ACE \times \% \text{ chg. in } rce)$$
$$= \% \text{ chg. in } IBT - \% \text{ chg. in } ACE \qquad (115)$$

or

$$\% \text{ chg. in } rce - \% \text{ chg. in } IBT = -\% \text{ chg. in } ACE$$
$$\times (100\% + \% \text{ chg. in } rce) \qquad (116)$$

and so:

$$20\% - 50\% = -\% \text{ chg. in } ACE \times (120\%)$$

or

$$-30\% = -120\% \times \% \text{ chg. in } ACE$$

and, finally,

$$\% \text{ change in } ACE = 30\%/120\% = 25\%$$

Example 9:

From the data given in Example 7, how great a decrease in sales can IOU Corp. sustain before profits disappear? Assume sales are at a level of $44,000,000.

Solution: From Equation (113), and the PPU computation of Example 7, we have: $MS = (IBT/PPU) \times \$100$ and $MS = (\$6,400,000/\$60) \times \$100 = \$10,667,000$ (rounded). That is, XYZ sales can decline by over $10.5 million before a loss is sustained. This can be checked by adding $10,667,000 to the computed BEP of $33,330,000 of Example 7. This sum should and does equal $44,000,000.

Example 10:

At the new sales volume of $60,000,000 XYZ Corp.'s rce increases by 16.7% (from Example 4). If the original financial leverage ratio was 2.0, the capital pickup is $50 or 0.50 (in decimal form), and the original total assets were $30,000,000, what is the new return on stockholders' equity?

Solution: From Equation (55), $\Delta req = (\Delta rce \times flr_1) + (\Delta flr \times rce_2)$. Now, from Example 1, XYZ's rce at $50,000,000 sales is 70% and so $\Delta rce = 16.7\% \times 70\%$ or 11.69 percentage points. At $60,000,000, with $\Delta rce = 11.69\%$, $rce_2 = 70\% + 11.69\% = 81.69\%$. Also, we are given $flr_1 = 2.0 = TCAP_1/EQ_1 = \$30,000,000/EQ_1$ and $EQ_1 = EQ_2 = \$15,000,000$. Thus, since $CPU = 50\%$ and $\Delta S = \$10,000,000$ (i.e., sales increase 20% from $50,000,000 to $60,000,000), $\Delta TCAP = \Delta S \times CPU/\$100 = \$10,000,000 \times 50\% = \$5,000,000$ and $TCAP_2 = \$35,000,000$, or finally, $flr_2 = \$35,000,000/\$15,000,000 = 2.33$.

Thus, substituting in Equation (55), where we now have:

$$\Delta rce = 11.69\%$$
$$flr_1 = 2.00$$
$$\Delta flr = 2.33 - 2.00 = 0.33$$

and

$$rce_2 = 81.69\%$$
$$\Delta req = (0.1169 \times 2.00) + (0.33 \times 0.8169)$$
$$= 0.2338 + 0.2723$$
$$= 0.5061 \text{ or } 50.61\%$$

We thus see that the leverage upon return on equity of a 20% sales increase is quite large.

The foregoing algebraic gymnastics provide important insights into some of the major interrelationships among financial ratios, of the manner in which changes in ratio components affect total ratios, and of how incremental percentages can be used to advantage in typical applications, including interpretation of common information reported in the daily financial press. If the reader will take the time to apply the same reasoning processes to other familiar characteristics of the business which can be mathematically expressed, including many of those previously developed in our discussion, he or she will extract additional useful relationships that are not at all obvious to the casual analyst (or even the seasoned "financial analyst"). This is the essence of "financial engineering."

Interestingly, the concept of the Profitgraph dates back to pre-World War I, when its ramifications for managerial planning, analysis, and control were originally formulated and termed "profit engineering" (this was actually the title of a book that appeared around 1933). The subject blossomed later into reformulation of the subject matter where the Profitgraph became the Break-Even Chart. And the "profitless point" of the original Profitgraph became the "break-even point" of later literature in the field.

However, there has been no parallel development of the Capitalgraph concept, or "Enterprise" or "Structural" constants in the literature, except for a Capitalgraph charting "Working Capital," as noted before, But all of these important management tools were developed by industrial engineers. It certainly appears that the engineering profession has done its fair share to advance the art, science, and cause of financial management.

SUMMARY OF SOME IMPORTANT RELATIONSHIPS

Exhibit 8-4 is a handy summary of some basic financial sensitivity or leverage relationships, sometimes also called *incremental* or *differential relationships*. For the more intrepid reader, the subject of Profitography can keep you occupied for endless days and weeks. The important thing is that the microeconomics of the firm have been laid bare before your eyes. In fact, the preceding sections supply you with more analytical know-how than is typically possessed by even the most sophisticated financial analyst. This is because they represent an engineering approach to financial analysis based upon proven, if elemental, mathematical modeling approaches widely utilized in the engineering sciences and, to some extent, in engineering economics methodology.

BASIC RATIO ANALYSIS

We will essentially conclude our journey through some of the more interesting terrain of Financial Engineering with a chart showing a few basic financial relationships, as used by managers and analysts. Exhibit 8-5 illustrates the ratio pyramid or ratio relationships lying behind return on assets and return on equity.

The ratios illustrated on this chart are all computed *before* federal income taxes and extraordinary items, and so on. This way they offer a more operations-oriented view of business performance than after-tax computations. Of course, securities and financial analysts are interested in both before- and after-tax ratios since stockholder and management interests include both operating performance (most usefully expressed in before-tax terms), and "net" management performance (typically expressed in after-tax computations).

EXHIBIT 8-4

THE BASIC SENSITIVITY, INCREMENTAL, DIFFERENTIAL, OR FINANCIAL LEVERAGE RELATIONSHIPS: I

If: $A = B \times C$
and: $B = A/C$,

then: $\Delta A = C_1 \Delta B + B_2 \Delta C$
and: $\Delta B = \dfrac{\Delta A - B_2 \Delta C}{C_1}$

Also: $\%\Delta A = \%\Delta B + \%\Delta C + [\%\Delta B \times \%\Delta C]$
and: $\%\Delta A = \dfrac{\%\Delta A - \%\Delta C}{1 + \%\Delta C}$

where Δ = Change or Increment (or Decrement) in A, etc.
and $\%\Delta A$ = Percentage Change in A, etc. and *not* Percentage "Points"

The above formulas can be applied to any of the basic financial performance ratios such as the following, where:

S = Sales
TC = Total Costs
TVC = Total Variable Costs
TFC = Total Fixed Costs
$TCAP$ = Total Capital (Assets)
$VCAP$ = Variable Capital (Assets)
$FCAP$ = Fixed Capital (Assets)
$\Delta TCAP$ = Incremental Capital (Assets)

$NNCO$ = Net Non-Cash Outlays (i.e., Depreciation)
IBT = Income Before Federal Income Taxes
IAT = Income After Federal Income Taxes
DIV = Dividends
EQ = Shareholders' Equity
MI = Marginal Income
PPU = Profit Pickup (Least Squares Fit)
CPU = Capital Pickup (Least Squares Fit)

$ibtr$ = Income Before Tax Ratio = $(S - TC)/S$
$= (S - TVC - TFC)/S = IBT/S$ (1)

ctr = Capital Turnover Ratio = $S/TCAP = S/(VCAP + FCAP)$ (2)

rce = Return on Capital (Assets) Employed = $IBT/TCAP = ibtr \times ctr$ (3)

Financial Engineering and the Profitography System

$$flr = \text{Financial Leverage Ratio} = TCAP/EQ \tag{4}$$

$$req = \text{Return on Equity} = IBT/EQ = rce \times flr \tag{5}$$

$$vcr = \text{Variable Cost Ratio} = TVC/S \tag{6}$$

$$mir = \text{Marginal Income Ratio} = (S - TVC)/S = 1 - (TVC/S) \text{ where } PPU = mir \times \$100 \tag{7}$$

$$msr = \text{Marginal Safety Ratio (also termed the Margin of Safety and Profit Elasticity Ratio)} = (S - BEP)/S = IBT/MI, \text{ where} \tag{8}$$

$$BEP = \text{Break-Even Point (Sales)} = TFC/mir, \text{ and} \tag{9}$$

$$bepr = \text{Break-Even Point Ratio} = BEP/S \tag{10}$$

$$rice = \text{Incremental Return on Incremental Capital} = rce_{\text{limit}} = PPU/CPU \tag{11}$$

$$CF = \text{Cash Flow} = IAT + NNCO \text{ (where } NNCO \text{ includes Depreciation, etc.)} \tag{12}$$

$$FMU = \text{Financial Makeup} = CF - DIV - (\Delta TCAP + NNCO) \text{ where, if } FMU \text{ is } (+), \text{ the business has generated a Surplus or, if } (-), \text{ a Deficit} \tag{13}$$

$$ibtr = mir \times msr \tag{14}$$

$$rce = mir \times msr \times ctr \tag{15}$$

$$req = mir \times msr \times ctr \times flr \tag{16}$$

$$msr = 1 - bepr \tag{17}$$

Also:

$$ibtr = mir\,(1 - bepr) \tag{18}$$

$$rce = mir\,(1 - bepr) \times ctr \tag{19}$$

$$req = mir\,(1 - bepr) \times ctr \times flr \tag{20}$$

Now, applying the leverage relationships to the return on capital employed or return on assets ratio, rce, we have the following as an illustration:

$$rce = ibtr \times ctr \text{ (this is formula (3) repeated), and} \tag{21}$$

$$\Delta rce = ctr_1 \Delta ibtr + ibtr_2 \Delta ctr, \text{ or} \tag{22}$$

$$\%\Delta rce = \%\Delta ibtr + \%\Delta ctr + [\%\Delta ibtr \times \%\Delta ctr] \tag{23}$$

(continued)

EXHIBIT 8-4 (Continued)

To give a specific simple example, suppose that profits on sales (*ibtr*) go up from 10% to 12% from one year to the next, and that the capital turnover ratio (*ctr*) increases from 1.25X to 1.50X. From Formula (22) we have:

$$\Delta rce = ctr_1 \Delta ibtr + ibtr_2 \Delta ctr$$
$$= (1.25 \times 0.02) + (0.12 \times 0.25)$$
$$= 0.025 + 0.030$$
$$= 0.055 = 5.5 \text{ percentage points (as distinguished from 5.5\%)}$$
$$\text{or 5.5 pp}$$

Since the original return on assets (*rce*) was 10% × 1.25X = 12.5%, the new value for return on assets would be 12.5% + 5.5% (points) = 18%.

Also, using Formula (23), for an increase of 20% in *ibtr* (i.e., 2% on 10%) as well as a 20% increase in *rce* (i.e., 0.25 turns on 1.25 turns):

$$\%\Delta rce = \%\Delta ibtr + \%\Delta ctr + [\%\Delta ibtr \times \%\Delta ctr] =$$
$$20\% + 20\% + [20\% \times 20\%] = 44\%$$

Or, since $rce_1 = 12.5\%$, $rce_2 = 12.5\% + 12.5\% \times 44\% = 12.5\% \times 144\% = 12.5\% \times 1.44 = 18\%$, as before.

Exhibit 8-6 is a variation on 8-5. This time we see how the marginal income ratio, *mir*, the marginal safety ratio, *msr*, capital pickup, profit pickup, and variable cost and capital are integrated into the more classical or conventional formulations of Exhibit 8-5. Other variations are possible, and the Profitography System approach can be further developed to yield greater and greater detail for purposes of ultimately getting to return on assets or return on equity. Notice, incidentally, that Exhibit 8-6 is titled: "Ratio Pyramid Formula Flow Chart: Return on Investment," and that the final computation (on the left) is Before Tax Return on Equity instead of Return on Assets or Total Capital Employed. We are here simply deferring to one use of the term "Return on Investment" that is common in the financial community. Our point is that you should be careful to use terms consistently, and to understand what is being defined by terms that you come across. Define terms for your own clients. Otherwise, they may misinterpret them due to the sometimes conflicting applications of the same term.

EXHIBIT 8-5

RATIO PYRAMID ILLUSTRATING COMPONENTS OF RETURNS ON CAPITAL AND EQUITY

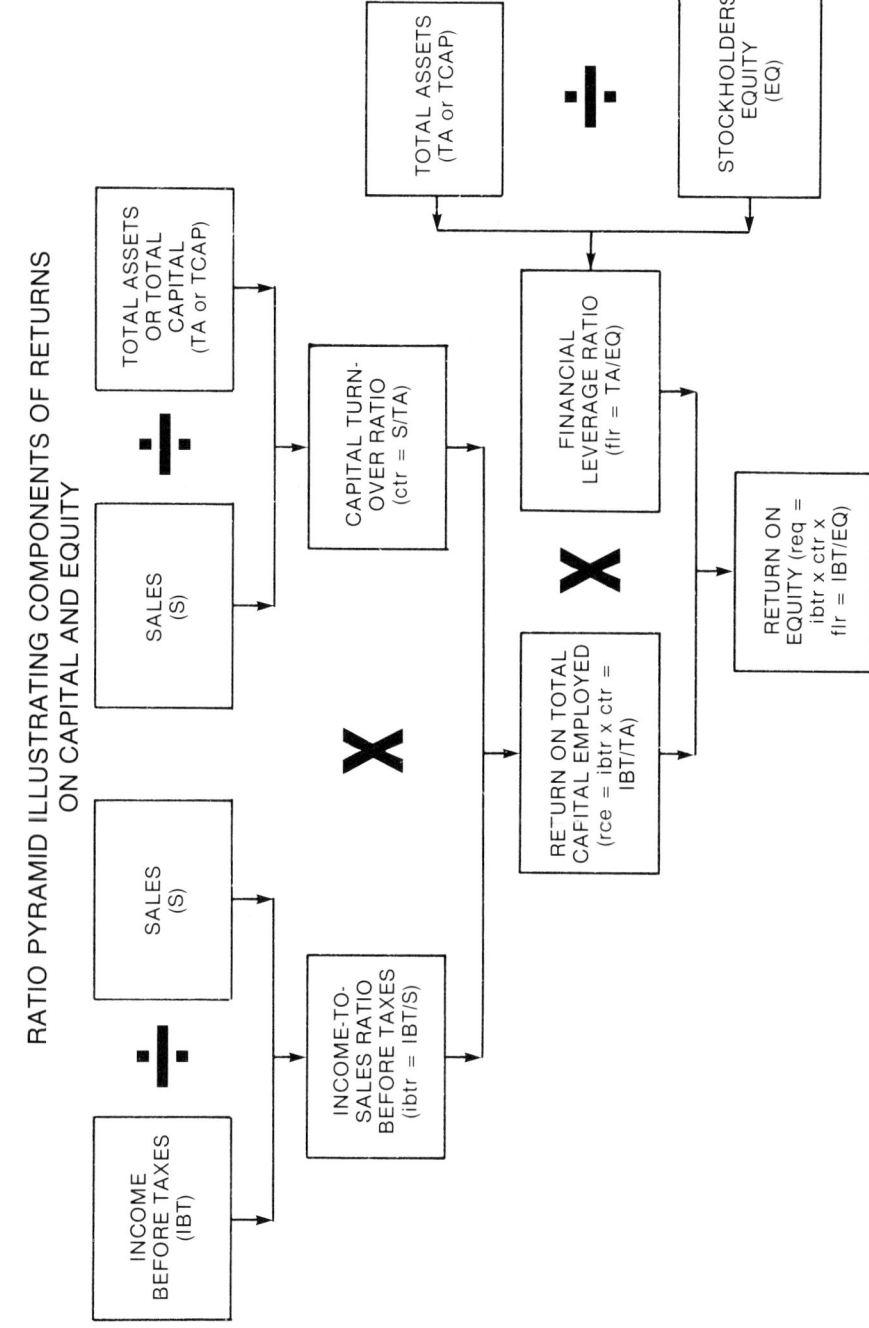

EXHIBIT 8-6
RATIO PYRAMID FORMULA FLOW CHART: RETURN ON INVESTMENT

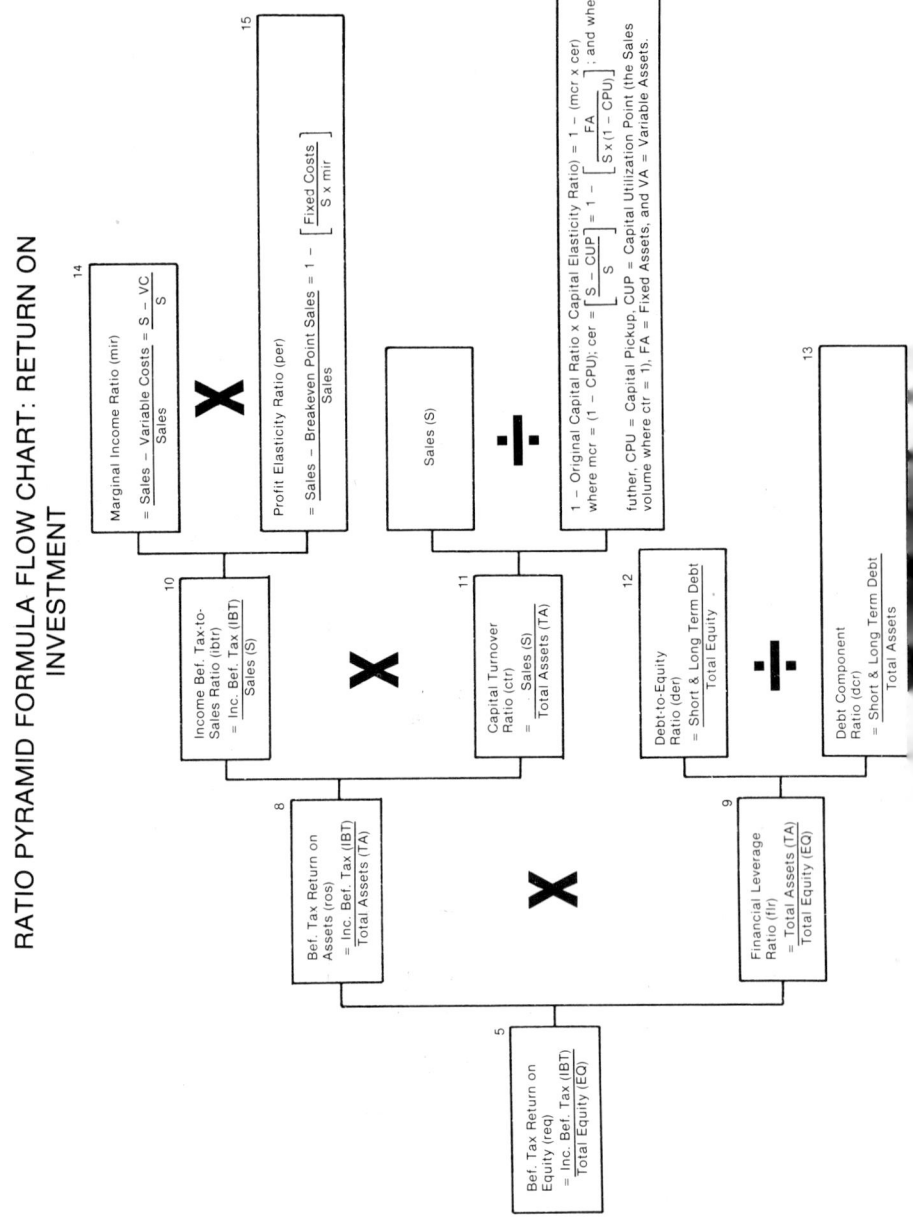

"Return on investment," "return on equity," "return on capital," and some other terms can mean completely different things, depending upon who is using them and how they are being defined. Be certain you understand the definition of "Earnings per share." Financial analysts usually define it in terms of after-tax income, whatever else it may also exclude. A formula chart can be created for this computation also. Try it.

Also remember that for operations-focused managerial control, again, before-tax analysis can be more useful than after-tax computations. But after-tax data are important for overall business management purposes. Keep the two separated in your mind and when you are working through financial analyses of projects. After-tax data are more accounting-oriented. Profitography and leverage-type analysis is more economics-focused. Both are important in their place. Accounting "modifies" the economics and brings it down to earth; but you must use both methodologies for purposes of financial and managerial planning, analysis, and control, and for purposes of comprehensive profit planning, including longer-range strategic planning.

SUMMARY OF FINANCIAL ENGINEERING BENEFITS

Profitgraph, Capitalgraph, and Financial Makeup analyses are strategic (rather than tactical) management tools which facilitate the determination of cost and asset trends and patterns in relation to sales and changes therein.

Not only may actual historical trends and patterns be readily portrayed, but comparisons may be made such as the following:

1. Historical or past performance versus projected or planned performance (i.e., as set out in the five-year longer-range plan)
2. Year-by-year actual performance versus planned performance for those years (based upon any given longer-range plan)
3. Successive longer-range plans
4. Company versus competitor firms or versus different companies producing identical or similar products/services
5. Budgeted performance for the year (or budgeted, annualized performance for the year-to-date) versus annualized, actual performances for year-to-date, and so on

Many important judgments may be made on the basis of the foregoing lineup of possible applications, including:

1. The quality of past, present, and planned performance of the business and its components (i.e., cost and asset components)
2. The feasibility of attaining planned performance in light of past and present performance without massive changes in cost structure, capital or assets structure, and management policies
3. Whether the business and its component parts are under control or out of control (i.e., direct labor, direct material, factory burden, and other cost components; receivables, inventories, plant and equipment, and other asset components, etc.)
4. Whether the business, even if very successful by "rate of return" standards, is generating a net *surplus* or a net *deficit* (i.e., whether it has been, is, and will be able to sustain sales growth from internally generated "funds" or whether management will have to go to the money markets and, if so, for how much)
5. Determination of the magnitudes of massive structural changes required to attain targeted, strategic objectives

Other applications are possible, too, such as the analysis of prospective acquisition candidates to determine if their managements have been beefing them up in anticipation of sale, or whether prospects can support themselves. Different screening levels are feasible, depending upon the depth of available financial data.

These graphical presentations are intended to supplement—*not supplant*—conventional statistical tabulations and other analytical tools that financial and management people have been using successfully for many years. The bottom line is that managements can gainfully employ just as many perspectives on the intricate problems confronting them as they can get their corporate hands on.

Naturally, everything in these past two chapters can be programmed for computer application, including the graphics.

WE'RE NOT QUITE FINISHED . . .

This chapter has presented some very important concepts and tools. However, there are also major omissions, as we alluded to at the outset. It has not been our intention to rehash standard materials commonly found in almost any good text on managerial finance, management accounting, engineering economics, controllership, capital budgeting, budgetary control, and so on.

Any engineering consultant should have a shelf of a select few books on the foregoing subjects. Resurrect your old text on engineering economics or, better yet, get an up-to-date model. It will probably cover such topics as decision making under uncertainty, time value of money in all of its ramifications, project and equipment selection, retirement and replacement, lease versus buy analysis, conventional break-even analysis (which, incidentally, is almost never treated the way you will find it in this and the last chapter), the cost of capital, depreciation and depletion, linear programming, the mathematics of probability, and other topics. And there will probably be a full complement of compound interest tables, logarithms, probability data, and so on.

In addition, quality assurance-type control charts have come into wide use in a variety of applications. Therefore, it would be well to purchase a good book on quality control and bone up on the use of control charts in its diverse aspects. The concepts of upper and lower control limits; standard deviations; optimistic, pessimistic, and most likely values; ranges; confidence intervals; normal curves and 95% limits; percentiles; and so on can be of potential importance to you.

As a consultant, sooner or later you will need a knowledge of engineering economics and Financial Engineering in product and systems design, project planning and control, and even forensic practice. The economics of safety are becoming more and more important for purposes of objective setting, decision making, risk taking, resource allocating, and general problem solving. The "she-devil" of modern technology (i.e., safety, health, and environment concerns) is more and more a critical factor on both micro- and macro-economic levels. Companies, entire industries, and even nations are more and more cognizant of and vulnerable to the economic, social, and political consequences, both short- and long-term, of their actions—and inactions.

Furthermore, the upper limit to project scale continues to expand. Larger and larger systems—sometimes termed macroengineering—demand deeper and deeper analysis, including financial and economic. Adverse events potentially affect larger and larger numbers of people. Even national borders are no guarantee of hazardous effects containment.

However, even smaller-scale technological problem solving, such as that more typically encountered by consulting engineers, may well be part and parcel of larger-scale management interests. Engineers can no longer function in a vacuum and be content to address only the small piece of the total picture that seems, on the surface, to be involved.

In fact, as time marches on it may well be that only the engineering profession is capable of addressing larger-scale problem solving, decision making, and risk taking where technology is concerned.

From the foregoing viewpoints, adequate insight in relation to many types of management problem solving must be based not only upon competent and responsible technical engineering but, in addition, competent and broad-gauged Financial Engineering.

CHAPTER 9

BUSINESS-BUILDING REPORTS

THE CRAFT OF WRITING

Actionable reports are the stuff of which the successful consulting practice is made. There is magic in words that are correctly and creatively used. Writing is an art, a craft—it is also a discipline. Before we write, we must think. We must say what we mean and mean what we say, especially in the exacting and unforgiving environment of the engineering sciences.

Few people are born with the talent to mold and shape words the way the artist creates a painting. However, even the person with ordinary writing skills can learn to be effectively expressive both in speaking and in writing. Public speaking should be preceded by private thinking. This is also true in writing.

One secret to business and technical writing is organization and structure. If we have properly analyzed our problem, and if we then create a detailed framework or outline for our report, then we can easily fill in individual sections with meaningful, succinct content. Inexperienced writers usually encounter a stumbling block because they try to write everything at once, without a disciplined structure. The writing, in turn, stumbles, meanders, goes in and out of focus, and may miss its mark. It tends to be more of a stream of consciousness, and does not hang together. On the other hand, each section, if it is bite-sized, will be easier to prepare. It is when the inexperienced writer tries to sustain a lengthy discourse that problems typically arise.

THE ENGINEERING AND SCIENTIFIC ADVANTAGE

Skill in writing, as in everything else, comes only with practice. As an engineer or scientist, you certainly know how to think problems through. You are not sloppy in your thinking about technical problems. You have been trained to be disciplined, rigorous, methodical, and careful. That is half the battle won right there.

The other half is that your report should be "craftsmanlike." It is capable of being a most potent emissary. If effectively done, it can be a valuable sales and marketing tool.

Don't merely write as you speak. Think about what you want to put on paper in terms of its phraseology. When we speak we take many liberties that look terrible in print. You will also find that our spoken vocabulary is far smaller than our written vocabulary. Clarity, precision, naturalness, and simplicity are important guiding principles in writing. Grammar and syntax are also important, but with a good structure, if you can avoid the temptation to be fancy, you will have it made even if your grammar and syntax are not perfect. Grammar is merely the system of rules used to construct sentences. Syntax is that part of grammar dealing with sentence structure. Since rules and conventions are fundamental to grammar, rhetoric, and composition, as an engineer you are almost home free. Your thoughts and thinking are good. The next task—putting them down correctly and creatively—is, in a sense, really the easy part.

Grammar deals with the parts of speech, word-form variations and their different applications or constructions, and the relationships among words in sentences, phrases, paragraphs, and so on. Rhetoric is basically the craft or art of writing good prose. It concerns diction, the choice of words, and style, the effective arrangement of words. Composition is simply the art of putting it all together so that it is literate and, it is hoped, even interesting as well as relevant.

Many good study aids in grammar, rhetoric, and composition

Business-Building Reports 185

are readily available. They are very inexpensive. Spend time scanning them. Write something—almost anything—about anything. English that is effective in writing and speaking comes only with practice, but it can be the best investment of time and effort you have ever made. Buy a style manual or two. Get a good book on report writing and one on technical writing.

Don't leave anything to the imagination. This should be easy, as engineers are accustomed to learning difficult subject matter from books. Improving your writing ability can be duck soup through use of the same problem-solving techniques.

SOME HINTS

Letter writing can be very good practice, as letters are usually short and to the point. You can teach yourself how to write better by being more attentive to the letters you write. They, too, can be potent sales and marketing tools.

In writing, we must let our thoughts rather than our egos show through. We professionals frequently have egos that are bigger than life, a fact that is not friendly, certainly not practical, and most probably not warranted.

Don't let your next project interfere with writing that report on the last assignment you had. Give every project its full due, and don't try to rush through a report-writing job. On some occasions you can spend days agonizing over a single, simple, short letter—or even a paragraph or two—to be certain that you have said what you intended to say, that you have really thought things through, and that you have said it all the best way you know how. A letter can mean big dollars.

Writing is really a pretty high-powered activity. It uses up far more intellectual energy than one may realize. It is highly charged, and can be exhilarating. Don't insult yourself—or your reader or client—by trying to slip or squeeze the writing in at odd times. Concentrate; be at your best. Write when you are refreshed and intense. Writing can be a hobby, too, but good writing is *not* a relaxing endeavor. It requires high energy levels. It can even be explosive. After all, meaningful reading takes intense concentration and high mental energy. Why should we expect effective writing to take less?

Since you have already set up an office, you have the makings of a reasonably good writing environment. Avoid distractions. Take breaks when you start going numb from the effort, which will happen if the project is intricate or otherwise difficult. Don't necessarily try to complete a report in a single sitting. You can grow stale. When you reach some key milestone, stop. Go back to the project later, even the next day. But never, *never* stop a writing project short of a true milestone.

Format and appearance are important marketing aspects of reports. Even if grammar and syntax are not masterful, we can go a long way toward creating a favorable impression if a report looks good. Physically, it should look as though it was done by a real professional. This means good-quality paper and binding, attractive type face; adequate "white space," and so on. As for grammar, chances are that the client is not a professional writer or an English professor either. Therefore, although you should make an ongoing effort to improve the technical aspects of your writing (i.e., grammar, rhetoric, style, etc.) you really needn't be frightened of writing. Nobody is going to try to make a fool out of you. No one can. The client knows less about your subject than you do. That is why you are the consultant.

If you are the type who freezes up or gets writer's block every time you take a pen to paper, take heart. Remember that what you are about to write is of extreme, if not urgent, importance at least to your client. He is paying you at your usual professional-level consulting rates to help him make money. All you want out of the engagement is a satisfied client who will come back to you for more professional consulting in your specialty and who trusts your judgment, impartiality, and ethics. You're not out to win a Pulitzer prize.

But strive to write well. A really literate report is a great marketing tool. It can enhance your reputation and be important for that extra measure of success. It will give you an added competitive edge.

THIRTEEN RULES FOR EFFECTIVE WRITING

The following rules will give you a good start on effective writing:

1. Use short words, short sentences, and short thoughts.
2. Avoid run-on and intricate sentences.
3. Avoid long paragraphs.
4. Avoid jargon, buzzwords, slang, too many acronyms, triteness, and euphemisms that obscure your otherwise sharp and clear thinking.
5. Vary sentence rhythm structure to make your reading more interesting and less monotonous.
6. Use subheadings, indentations, lists, underlinings, bullets, capital letters, and other devices to make the report visually inter-

esting and attractive and to give different parts or points emphasis or high impact. You will want some parts of your report to jump out from the page and hit the reader on the nose.
7. Write *to* your client (or other audience) rather than *at* him or her.
8. Use visual aids wherever appropriate.
9. Use active and dynamic words and descriptions rather than passive ones.
10. Don't be afraid to write in the first person (i.e., "I", "we"), but not to a fault.
11. Make your reports practical, complete, concise, clear, readable, intellectually honest, and "self-teaching" to the extent feasible.
12. Your report should be as long as is necessary to accomplish your objectives.
13. Rewrite, reread, rewrite, reread, and then rewrite and reread again.

Remember, too, that the longer your report or, at least, the longer individual narrative sections are, the more opportunities there will be for your reader-client to see less-than-perfect writing. Without doubt, a carefully thought out report structure with subheads can go a long way toward masking your technical and stylistic writing deficiencies.

Although good writing is, again, primarily an acquired skill, even if your craftsmanship in writing is not perfect it will still have some style because you wrote it. You are unique. Each of us is unique. Even if you are not an experienced or polished writer, you possess all the necessary elements to become one. You are a specialist in an important field. No doubt you have something to say on what you have learned in the field or from books. You know how to learn rules, whether they apply to scientific problem solving or grammar. You have read good writing in many forms, both technical and nontechnical. You certainly know where to get the information you need to study writing-related subjects. But most important, you now have a totally different motive for wanting to write well—perhaps one you never had before. You're going to make money at it. This can make all the difference in the world. It does wonders for motivation.

Becoming an experienced and polished writer is not difficult in either its level of complexity or knowledge content. It is only difficult in that it demands that you practice, practice, and then practice some more. Almost anyone can become an excellent writer. We each start out with a blank slate. But once we write down that very first sentence we are on our way.

THE TECHNICAL REPORT

There are a number of generalized report types: preliminary reports, interim or progress reports, final reports, comprehensive narratives, short-form reports, summary reports, and executive highlights. They can be nonnarrative and consist primarily of mathematical derivations or specific, quantitative problem solving. They can be design studies, plant or site layouts, cost estimates, feasibility or other engineering economy studies, accident reconstruction reports, product safety engineering investigation reports, in-depth case studies, market analyses, appraisals, plans, comparative analyses, and so on. They can be classy-looking reports bound like books, or they can be memorandum reports of one page or less.

But in each instance the engineering consultant is writing for a specific client on a specific issue or project and for a specific purpose. The report writer must have a full comprehension of the objectives of the report and engagement. All reasonably accessible information sources must be identified. A decision has to be made as to what specific sources are to be drawn upon. Data must be gathered with explanatory text, then sifted and selected, classified and otherwise assimilated or synthesized, abstracted, analyzed and interpreted, and finally reported and summarized in the most meaningful and actionable form. At every stage of the report-making process the consultant has to have his audience clearly in mind, as well as the time, staff, financial, and data constraints that will invariably be present.

You should also take into account temperament and limitations of your reader-client. Some clients will not be engineers, believe it or not. They may be totally uncomfortable with graphs, formulas, tables, and charts. Others may want maximum detail. Still others will cringe at the very thought of anything that smacks of being statistical or mathematical. Most people are not quantitative; they are qualitative. Yet every single client has a bona fide need that only you can fill. You have to find a way to communicate effectively and efficiently. The client must be able to use your report for problem identification, problem solving, objective setting, decision making, risk taking, and resource allocating. Even the format of the report should take into account the type of client and the specific application to which the report will be put. In fact, there is a good chance that the application will dictate format, content, and style, if we are sensitive to this factor.

So know your audience. Define your objectives, determine report form, design the outline, search and sources, analyze, and then sit down and write. If you approach the writing project like any other engineering project you will not get writer's block. There's no reason to fear that blank piece of paper. Just think of its potential in the right hands.

Once you get into the swing of things you'll see how easy the process of writing is. Your thoughts will probably run way ahead of your ability to put things down on paper. Writing is only difficult to someone who hasn't tried it.

REPORT APPEARANCE AND VISUAL AIDS

As engineers, we feel very much at home with charts, graphs, flow diagrams, tables, statistical summaries, and the like. We may not all be artists, but we know how to read engineering drawings and understand perspectives, exploded views, transition diagrams, circuit diagrams, layered transparencies, slide charts, nomographs, and so on. We also know how to use the various kinds of visual presentation equipment that are available.

A transparency maker can be very useful to prepare high-impact color-transparency-type visual aids for use with an overhead projector. You can even do the charts and diagrams yourself. You have a wide choice of extremely attractive and eye-catching visuals of very high impact, and they can be prepared on preframed plastic film.

An overhead projector will do wonders for presentation of your report results to clients, but professional-level transparencies will really make your credibility skyrocket. A 3M transparency-making machine runs about $700. If you are going to have to make a lot of presentations, perhaps including proposals, look into it—it can be a good investment. Also look into a Kroy Lettering Machine and GBC Binding Machine. The binding machine puts nineteen slots down the side of an eleven-inch-long piece of paper and then permits binding of a report booklet with a nice black or other color spiral-type spine. Other types of bindings and report covers are also available, but the GBC-type binding looks very professional and says to your client, "I respect you; therefore, I am giving you the courtesy of providing you with the best job I can." This type of binding permits the pages to be laid out flat, just like a book. The GBC Binder costs around $500 or $600, but I think it is well worth it. You may also consider getting a laminating machine at some point.

Naturally, you will need a copier. Get one that gives you crisp, high-contrast copies, preferably one made by a company with a reputation for reliability. Buy it locally so that the servicing company can provide good response when you have a problem. You can do all kinds of tricks with a copier that has reduction and enlargement features.

Along the way you'll probably have occasion to see reports produced by your competition. Study them. Make copies of them. Collect them. What are their strengths and weaknesses in content, format, and style? Can they help you improve your own reports? How serious a threat are your competitors based on what you now see in their actual reports? What do their reports imply as to competitive resources, capabilities, capacity, and potential? What kinds of equipment does your competition use to produce reports? How much has your competition spent on report-presentation-related equipment? Apply the full battery of commercial intelligence techniques to analysis of your competitors through their reports. After all, a consulting engineering report is the end product—the bottom line—of the consulting engagement in its purely professional aspects.

In short, which competitors are "class acts"? Which are unprofessional and unworkmanlike? Make a full comparative analysis of your reports versus those of your competition. List assets and deficits. Lay plans to build upon your strengths and eliminate your weaknesses.

THE REPORT AS PRELUDE

A report represents the results of engagement casework. However, you should view it as only the beginning of still another important process: the follow-up relationship. This can be as important as the initial engagement. Now the chips are down. You have to attempt to ensure that the client studies your report, understands it, and, most important, applies it. Stay in close touch with the client. If you have not heard from him or her within thirty days, telephone. The client is now part of your "family," with a vested interest in you. You must show the client how to maximize his or her investment in you. There may be follow-up consulting casework, it is true. But whether there is or not, you must make every effort to help the client use to advantage the work for which he or she has paid cold, hard cash. If you follow through properly, you will be amply rewarded much of the time. The client will come to you for more advice. If your reports are useful and professionally done he or she will look upon you, likewise, as useful and professional. It is as simple as that.

Your reports should exude a willingness to share your expertise. Don't intimidate the client by writing in a secretive manner. Don't insult him or her by being obscure or using highly technical terms that merely feed your own ego. Write reports so that they are self-teaching. Make a client feel like a full partner in the consulting process, rather than some outsider who is forced to use you against his will. Make the client feel that you and he or she are a team and that you are trying to help him or her understand the problem and its solution in such a way that the team effort comes alive. Make the consultant-client relationship effective, comfortable, and natural.

A good report will be shown around. Count on it. You want to create reports so that prospective clients will say, "I saw one of your reports and I thought it was excellent, so I thought I'd call. Can you help me?" You may even obtain totally unsolicited engagements where clients want you to redo analyses ineptly done by other consultants. I know consultants who have benefited in both of these ways due to their own care in report preparation versus lack of attention by their competitors.

PROBLEMS AS OPPORTUNITIES

Never be afraid to suggest modifications, additions, or deletions to your reports. After you have submitted a report to a client, if new information comes to light, or even if you have some occasion to review your casework and find it wanting, don't ever hesitate to call your client. Merely explain that you feel a change is in order that will strengthen it. Remember that engineering is a process of cumulative progress over time. The more time that elapses, the more likely it is that you will have or will acquire additional insights.

If you discover an error, don't charge to rectify it. Just hope it hasn't done any harm. If you think follow-up work is in order, discuss the possibilities with your client. Focus on the client's needs rather than yours. Remain action-oriented. Has the client implemented your recommendations? What, exactly, has he or she done with your findings and opinions? Has the client followed up? If your proposal has not been implemented, why not? If so, what has been the result? What can be learned? Are results as you predicted? If not, why not? Has the problem changed? How? Can report recommendations be salvaged? What needs to be done to optimize the investment your client made in the consulting engagement in order to adapt it to new conditions? Is the client mad? Why?

But good, bad, or indifferent, *stay in touch*. That is the only way to resolve problems and make progress. You can avoid a lot of misunderstandings if you stay in touch with your client. Confront problems directly and swiftly; don't shrink from them. This is particularly applicable to the human aspects of the client-consultant relationship.

Reports can take other, less well-known forms. For example, a report can be prepared in response or rebuttal to another report. This is particularly common in the field of forensic consulting. A report by one engineering expert will be followed by a report from another. There may be as many as six, seven, or more reports by various experts involved in a single case. If you put them all together you sometimes see a collection of very interesting viewpoints, some valid, some not. Some are smokescreens; some are even unethical. We will talk about this later on. The ethical content of your report is particularly visible in this type of consulting work. Be sure you understand what safety engineering and product safety engineering are really all about, instead of what some professor may have told you very peripherally, offhandedly, or even incompetently way back when. These subjects go well beyond safety factors and published safety codes and standards. A product, workplace, building, or other facility will not necessarily be safe merely because its design and construction comply with applicable codes and standards. There are superseding and controlling rules of practice. More on this later, too.

EVALUATE YOUR REPORT

To evaluate your report, the following six standards may be applied:

1. How well is it planned out and organized overall?
2. How logical and persuasive are the relationships between your supporting data and conclusions or opinions?
3. Are concreteness and specificity sufficient to support your arguments?
4. Are the breadth and thoroughness of development of your arguments adequate (and even ethical)?
5. Does it reflect maturity of thought and judgment in relation to both the problem and the client's real needs?

6. Is its "language"—style, grammar, and so on—acceptable as standard English?

These factors reflect the need for competency, effectiveness, efficiency, and style. In short, your report should reflect true professionalism from start to finish.

BE RESPONSIVE TO CLIENT NEEDS

As your practice expands, continually develop your reporting formats. Make any report a model in its class, from a purely professional standpoint. Design it to be ultra-responsive to client needs at every turn. Serve real rather than perceived but unsubstantiated needs of clients. These can be vastly different.

What a client theoretically needs, or even what he says he wants, may be quite different from what you perceive or imagine his requirements to be. One of the challenges in consulting practice is to determine the differences among stated, real, and perceived needs. Ultimately, your services must provide a client with results that satisfy his stated needs. But you must also give him the flexibility to accommodate his real needs, which, it is hoped, will be the same as your perception of them.

There may not be standards for the preparation of reports in your own specialty. Continue to experiment over the years and develop useful reporting techniques and materials. Remain alert to client needs. Watch for signs of overburdening clients with results of your investigations. Try to get feedback from clients as to the utility of your reporting efforts and presentation style. The whole point is to be of as much help as possible. Clients will work with you because they will appreciate that you are trying to work with them. Standardize report formats to the extent feasible.

CHAPTER 10

IDENTIFYING, UNDERSTANDING, AND SURPASSING YOUR COMPETITION

GETTING A FIX ON YOUR COMPETITION

When you first start up a consulting engineering practice you have to assume that no matter how superior or novel you believe your services or skills are, there are lots of professionals out there who are potential competitors with roughly the same education, experience, and resources in your areas of interest. This should not be a discouraging factor, but merely a safe and "surprise-free" viewpoint. You must assume that your competition is just as smart and just as savvy as you are.

But once you start getting a few clients, especially if you do a good job for them, you will soon be told gratuitously how good or how bad your competition really is. Welcome to the real world.

This kind of "field intelligence" will snowball. In fact, start thinking both intensively and extensively about the concept of "commercial intelligence." Soon you will learn who and where the clients are—aside from such information you develop without benefit of being told by clients themselves—and you will also learn who and where all your competitors are. Clients love to ask if you know this one or that one.

Most consultants advertise. One important bit of intelligence is to find out where your competitors advertise. To find out, you should keep up with all your trade journals, newspapers, and so on. Ask your clients where a good place might be to advertise your services to the market you are targeting, or ask them where other engineers like you advertise. They may even have a file they keep on consultants.

If you are lucky, you might be able to get a client to show you his consultants file. It may be very informal and very compact. The client's selfish interest in giving you information about established consultants in your field is that, if you continue to develop your own practice in a way that benefits him and saves him the trouble of dealing remotely, or inefficiently, with consultants or with a number of them, he may be making it easier and perhaps even more economical for himself in the long run. For example, when it comes to my oldest and most dependable clients, I provide a lot of nonbillable counsel because I know that they are not going to simply use the information I give them and go to someone else. A client has a great deal to gain in helping you get started and perhaps develop your practice in a way that can benefit his business. Never forget this "enlightened selfishness" of old and dependable clients. When you raise your consulting rates, raise theirs last; when you come upon something that is important and relevant to their business, expose them to it first.

In any event, you will get more information from a client's consultants file than you know what to do with. Try not to ask potential or actual competitors too directly about the marketplace or consulting practice. They may smell the rat and quite possibly try to mislead you. This is one reason why, again, you should try to find a mentor who will be truthful—even if brutally so—with you. But don't let anybody dissuade you, not even a mentor. You are capable of overcoming almost any obstacle to successful independent consulting practice except your own negative thinking. This includes overcoming somebody else's negative thinking that yours feeds on if you don't recognize the problem at the outset.

YOU VERSUS THEM

Make a list of your own strengths and weaknesses. When you start learning about your competitors, make a comparison chart of "Yours" versus "Theirs." Such a chart should include factors you can identify as real and/or potential business assets versus liabilities, business builders versus builder detractors, and so on. For example, what is the minimum educational credential needed to do the work in your field? How important are advanced degrees? Who has them? Do you need one? Is a professional engineering license critical? Necessary? Helpful? Is it illegal to practice in your specialty and consult for your potential targeted client universe without engineering registration? What certifications are there? Who has nontraditional or external degrees? What good are they?

Is your field crowded? If so, how many of your fellow professionals actually do consulting? Do they do it in your particular areas of interest? To what extent? Where are the gaps and voids? Where are the clients?

How do clients purchase services in your field? Can you identify a competitive advantage you enjoy? What are the advantages of your competitors? Where do your competitors have their offices relative to your own location? Are you near or far from the market? Is age a factor? Who has what kind of office facilities? Laboratories? What kinds of ties or "old boy" networks are there in your field? Are competitors relatively remote? If so, is such remoteness an asset or a liability? Why? If you have reason to believe that a remote competitor is doing a lot of business in your territory, can you parlay this into an advantage for yourself? Just how local is your prospective target market? Is there a logical or built-in territory for your specialty or is the whole world your stage? Would relatively more local prospects prefer to do business with a local consultant? If so, why? If not, why not? Don't give up on this point. It could be very important to you.

Never stop learning about your competitors. Continue to make comparative analyses as your own activity increases and as you circulate more and more in your field. Comparative analysis has great profit potential. It can also save you headaches if you get to know the styles of your competitors and how your competitors interact with clients. Does your worthy competitor in subspecialty X spend more time consulting, working at a full-time job, or playing a violin in his local symphony orchestra? How responsive to client requirements, schedules, and deadlines is your competitor in specialty Y?

A useful little device for analyzing yourself and your competitors from any one of a number of standpoints is something known as "The Ten Little Men." It goes like this:

1. Who I Am
2. Who I Think I Am
3. Who I Am Trying to Be
4. Who I Think You Are
5. Who I Think You Are Trying to Be
6. Who You Are
7. Who You Think You Are
8. Who You Are Trying to Be
9. Who You Think I Am
10. Who You Think I Am Trying to Be

PART-TIME VERSUS FULL-TIME CONSULTING

If you are looking forward to the probability of full-time consulting, you will be planning and doing things differently from the way your colleagues do things who are only looking to do part-time consulting. Sure, you have to start off small on a part-time basis, but you are seriously keeping your eyes and ears open for opportunities capable of liberating you from that well-paying but professionally unrewarding full-time job. You are looking two, three, maybe five years down the road. You will soon see how differently you think about things as a prospective entrepreneur, compared with the "employee mentality."

But keep these dreams and plans to yourself. Few, if any, employers care to think that a star professional is dissatisfied and eventually looking to leave. In this respect, employers are myopic, since they truly believe in many instances that if they do all of the things an employer "should" do, they can "own" you—or at least hang on to you.

STAYING POWER AND "FIREPOWER"

In consulting, superior staying power does not necessarily mean having lots of money behind you. It can mean merely having the courage of your convictions. You know you're good. You know you have something to offer and, therefore, to sell. There is a market for your knowledge and experience or some variation thereof. Don't give up. Keep at it. Keep doing client development, not until it hurts, but until it works. That's real staying power. Some of your fellow engineers who also tried consulting will not make it because they won't continue to do what it takes to make it work.

You also have an open field to surpass competition. Be certain that you acquire as much knowledge as you can in your chosen field or specialty. Do all of the right things: education, contacts, library,

PR, advertising, client development, equipment, and so on. Don't let anyone be in a position to say, "I can do it better." *You* do it better. Problems with which you are confronted by clients may differ in detail, but they are probably 85 percent the same in substance.

Gather colleagues around you to give you depth. They are the start of a network of consulting associates. Technology in depth is real "firepower." Team operation can also give you a multidisciplinary capability. Clients know one man cannot possibly know everything about everything. Don't try to convince them that *you* do. This is the best way to lose a prospective client and engagement. Sweat out the details of a multidisciplinary style at the beginning. You have a general idea of the types of problems you are likely to see and the types of specialties that will be required. Predict. Play it by ear but then fine-tune as many times as necessary. And don't be afraid to change course. You may find yourself two or three years down the pike doing something just a little different, or even vastly different, from what you started out focusing upon.

If you have a competitive advantage or edge of any type you have a head start, of course. But you have to keep ahead because your competitors will be trying constantly to catch up once they are on to you. Try to develop your own novel style, proprietary formats, and so on that will tell clients you are different. The data base may be the same, the body of knowledge may be the same, but look for ways to utilize and present them differently so your work is more compatible and in tune with the true needs of the client.

Engineers are accustomed to uniformity and standardization. These techniques are put to good use in technical reports and other client interactions. Naturally, uniformity and standardization are beneficial and necessary, but sometimes the standard approach can be a hindrance. Be creative not only in actual problem solving but also in how you present your capabilities and solutions to clients before and after you get the engagement.

Remember that men, machines, materials, money, and the other factors of production are available on the open market. But what makes one organization—or, in this case, consultant—different from another is management.

CONTAIN THE COMPETITION BY INCREASING YOUR MARKET SHARE

Professional counterparts in your specialty who are also doing consulting are trying to increase their practice, just like you. If, as is likely, the business is spread all over the place from one end of the country or state to the other, there is plenty of room for everybody to move around. But in many consulting specialties the business is relatively local. It may be limited to a few nearby cities or industrial centers, or perhaps to a two- or three-state contiguous area, with some exceptions where consultants do a little bit of more remote work on a catch-as-catch-can basis, depending upon opportunities that present themselves or specific efforts made to penetrate a specialized field or industry that is spread out.

Containing the competition is probably not a realistic objective if the concept means setting unattainable goals that are not significantly under your control. You have little control over competitor competence, aggressiveness, style, and reputation. However, under expanding market conditions, we can define "containment" as an effort to minimize market share growth—which is what it is all about. A business may continue to grow in terms of absolute dollars of revenue, yet be stagnant in relation to market share by percentage. It can even fall behind in percent of market, yet still be increasing in sales. Market share is defined as the percentage of market captured. As market size grows, you must run just to keep up, let alone move ahead.

You must expect that, in an expanding marketplace, everyone will shoot for at least their fair share based upon past performance and market share, at minimum. But if you can increase your market share by offering new or expanded services, or by developing the market more aggressively such as through new advertising efforts, greater exposure where the clients are, and so on, you will be in a position to contain competitors or even cause them to lose market share. In a solo consulting practice, of course, there is a limit to your personal growth potential in terms of the number of hours you have to spend working. Thus, when you are bumping against your ceiling you will have to consider raising your fees or else figure out how to give yourself more leverage. More about this later on.

MOVING INTO THE TERRITORY

If you hear that a particular prospect was dissatisfied with the work of one of your competitors, don't wait for such a potential client to call you; "develop" his interest tactfully without overtly attacking your competitor. In professional practice it is unethical to do this.

Competitors are also trying to increase their practices. It is a never-ending effort. But competition is healthy and ultimately benefits competitors and clients alike. It tends to raise the standards of practice and serves somewhat as a self-monitoring control within the entire field. It is not totally efficient, of course. It cannot be. Therefore, if you help it along by continually offering new and improved services in an aggressive and professional manner, you can increase your own credibility and posture, and contain the competition.

On balance, the best way to contain your competitors is to pull out in front of them. To the creative, demanding professional, this should not be too much to ask. Don't rest on your laurels. Some of your competitors will. Your target should be your most worthy competitors. Stay up there with them and run the race for its own sake, if nothing else. Be possessed by the drive to be among the best in your field. And it's not the kind of a race where somebody has to win and somebody has to lose; it's the kind of race where everybody wins—you, your company, your clients, and the public at large. That is what being a true professional is all about. Your handsome consulting engineering income is only one way to keep score. On the other hand, you should strive to attain such a high degree of professionalism that, for all practical purposes, you don't have any competition, at least in your own mind and heart.

SERVICE IS THE KEY

Your sincere effort to be of greater and greater service to clients, striving to be responsible to their needs even if it periodically means modifying, adjusting, or fine-tuning your approach, will prove to be an important factor in your success. This, too, is a method of containing the competition. At minimum, you get to keep your market share.

It all boils down to a good attitude toward your work, your clients, and yourself. If you have a solid enough opinion of yourself, your capabilities, and your value, it will be easier for you to function on the highest professional plane. You are not trying to prove anything to yourself or your clients. You are simply responding to needs and treating problems as opportunities to grow even better at what you do. Thereby, you will become of even greater value to clients, old and new. Your consulting practice will continue to be challenging and rewarding. It will keep you thinking and feeling young.

DON'T COMPETE ON FEES

The one way you don't want to compete is in the area of fees. You might want to go in at a few dollars less per hour than your competition. But remember that either you are presenting yourself and treating yourself like a seasoned professional or you are not. If your expertise is valuable, as engineering expertise most assuredly is, then don't undermine your credibility—or that of the profession—by working cheap. In the long run you will only be sorry and much the worse for wear.

SURPASS YOUR COMPETITION BY SURPASSING YOURSELF

It should come as no surprise that the advanced degree is a mark of excellence—and commitment. The Bachelor of Science degree in engineering is just a point of entry. Many men and women stop there. Even many engineers who also hold their P.E. licenses stop there.

While formal engineering education, at any degree level, must be combined with experience in order for the individual to optimize his or her potential, the advanced degree that is acquired at age twenty-five, thirty-five, forty-five, or fifty-five is a mark of lifelong commitment to the profession.

The potential for professional growth and achievement through formal education is available to all engineers. Relatively few elect to undergo the rigors of advanced study, mainly because the pay is so good even at the bachelor's level, and people get diverted from the more extended courses of professional study. Somehow, they never go back to school.

If you contemplate going into consulting practice, whether earlier or later in your career, you should make the effort and the investment in yourself to get an advanced degree. Your current employer will also be a beneficiary, so he has a stake in encouraging you to move ahead. Most companies have tuition refund plans, which also tells us something about the value of advanced degree programs and even individual postgraduate courses on a nonmatriculated continuing education basis.

Your objective should be to practice at the top of your field. Even most P.E.s do not have advanced degrees. Therefore, you will be one step ahead of them whether you go into private practice or stay in industry. If you have a B.S. degree, work toward your master's. If you have an M.S., work toward your Ph.D. even if it takes five to eight years. Who said life was easy? However, we have it within our power to make it interesting, exciting, and fruitful.

In the engineering field it is feasible to obtain advanced degrees on a part-time basis in evening curricula at accredited engineering schools all around the country. This is a marvelous opportunity that does not appear to be appreciated by many engineers. Alternatively, you may wish to take an advanced degree in an allied field that better reflects your area of specialization, or even a degree in a second engineering discipline to complement your primary field of study.

For example, some engineers take civil or electrical engineering if they are mechanical engineers. Or a civil engineer will study for an architectural degree, a degree in city planning, environmental engineering, or even real estate. Chemical engineers take courses of study in industrial hygiene in order to qualify for various certifications in that field. Other engineers take advanced degrees in occupational safety and health, human factors, psychology, bioengineering, or even public administration. And, of course, some engineers go on to study law, with the thought of practicing in the patent field.

Advanced degree programs will broaden you immensely. There will be an identifiable and very large synergistic gain. It is becoming more and more recognized that the future is in the hands of the multidisciplinary professional who can look at problems competently from several different perspectives. On the other hand, there is an important place for the professional who looks more and more deeply into his primary field of study, and, through advanced degree programs, postures him or herself to make significant contributions to the field.

In fact, broadening your professional horizons through advanced degree programs may be the easiest way of all to leapfrog over the competition. You are still in a position to specialize, of course. But just think of what you can accomplish in your specialty if you can look at the problems from a number of bona fide, recognized specialty viewpoints. Just think of the value you can have to a client or employer and, therefore, to yourself. Naturally, you also increase your worth to society.

Nowadays we hear a lot about people changing careers in midlife. The engineer does not have to stray far afield, however. The progress of science and technology and the emergence of more and more specialties and entire fields even over the last ten to fifteen years provide ample proof that the engineer need only turn to a second or even third area of interest within the broad universe of engineering technology. There is plenty of room to move around while remaining dedicated to a lifelong career in engineering or allied fields. And we are just at the threshold of what lies ahead.

The elapsed time between a bachelor of science degree in engineering and a master's degree can be two years or twenty years. It can take one that long to identify with a particular specialty beyond the first degree area of interest. Likewise, the elapsed time between a master's degree and a Ph.D. can be ten or twenty years. There is no law that says an engineer has to get a Master's degree and then a Ph.D. one after the other. In fact, most engineers who do go on for advanced degrees stop at the master's level.

The lifelong opportunity for both intensive and extensive studies in engineering and allied fields is present today to a greater extent than ever before. There have never been as many opportunities or crying needs for the knowledge and experience that engineers can bring to bear on an exploding world of major problems.

You can surpass your competition in many ways. You can surpass your peers in terms of educational credentials, professional registration or licensure, specialty certifications, sheer weight of experience, diversity and range of knowledge from many standpoints, or intensity or depth of expertise, again gained through a combination of education, experience, diverse employment, and so on.

As an aside, don't sell short the opportunity to do high-tech temp-type consulting. Remember that you are not merely a worker filling in temporarily for someone else or until a peak load is worked down. You are a professional. Your expertise is valuable on whatever basis you elect to make it available to prospective clients. Naturally, you can also use this activity style or strategy to identify a permanent slot as an employee or manager. Temporary professionals and managers are hired for flexibility. This has always been a relatively easy way to scale up and down for engineering projects. But other opportunities can develop along the way.

If you can accommodate the high-tech temp career style, you will find a growing market. More and more large companies of all types use temps to avoid the trauma of frequent hirings and firings. It has been found to be good for morale because employers of temporary professionals don't usually get rid of their permanent staffs.

As business increases its reliance on temporary professionals, the size of permanent staffs could decrease. However, this will have a generally beneficial effect on the engineering profession. Business will become more and more dependent upon the independent professional whose primary loyalty is to his profession rather than an employer. Such independence will not lessen the ethical relationship between professional and client. In fact, since the independent professional will be more and more accountable to the client company, the quality of such independent professional consultants will rise.

Therefore, with the foregoing and other trends, opportunities, and ramifications of professional practice, begin to plan now to surpass your competition by surpassing yourself.

CHAPTER 11

UNCOMMON OPPORTUNITIES IN FORENSIC CONSULTING

FORENSIC ENGINEERING AND SCIENCE: YOUR PASSPORT TO CONSULTING PRACTICE

This chapter is about the exciting, even exhilarating, but exceptionally demanding world of forensic science and expert testimony. It is a relatively unknown and, to most people, mysterious world. It involves high stakes, high pressures, and high-powered participants. It is a world that exists exclusively at the very stress points and failure interfaces of society itself. And it is here that the engineer is most visible.

But it is not about forensic medicine or science in the criminal investigation sense, à la *Quincy*, although this subject is indeed one important segment of the overall field. Nor must we dwell on law, legal methodology, procedures, or practices.

In a complex society like ours there are countless opportunities for people to interact, oftentimes in ways that are harmful physically and financially. Personal injury, property damage, and financial or economic loss thereby become the basis for lawsuits to recover damages. This represents an attempt to make the party bringing the lawsuit, the plaintiff, "whole" to the extent feasible. Resisting and opposing a suing plaintiff is his adversary, the defendant. In fact, the entire process is adversarial in nature and somewhat alien to the lay person.

On behalf of clients, the legal community has found it very useful to introduce specialists expert in specific areas. Such experts are utilized by both plaintiff and defense attorneys to assist in preparing their respective cases adequately for trial. Or, at least, expert involvement may help effect an acceptable negotiated settlement in a less charged atmosphere than typically prevails in the courtroom.

The bottom line, of course, is that plaintiff or defendant wins (unless there is a hung jury), sometimes only after an actual trial takes place if both sides cannot come to some mutually acceptable understanding. Sometimes the trial is quite matter-of-fact. But sometimes it is brutal, bitter, emotionally draining—even electrifying, with the expert caught in the middle.

Many kinds of specialists typically serve as experts in litigation. The list, a long one, includes accountants, physicians and psychiatrists, anthropologists, engineers, dentists, chemists, philatelists, pomologists, toxicologists, geologists, architects, industrial hygienists, scientists of almost every sort, handwriting experts, pest control specialists, employment counselors, fire investigators, musicians, psychologists, pilots, firearms and ballistics specialists, metallurgists, photographers, appraisers, ex-car thieves, economists, ex-safecrackers and professional gamblers, city planners, ex-drug addicts, actuaries, plumbers and electricians, ski instructors, athletes, athletic directors and coaches, management consultants, collectors and art specialists of all types, numerous other types of specialists, and even lawyers. The field of application is as diverse as the subject matter of cases, with virtually no limit as to who can serve as an expert witness.

Forensic engineering and science are merely the application of engineering and scientific principles and methodologies to litigation. Legal engineering, as it is sometimes called, is really a misnomer. Forensics is not a separate field of engineering for which one goes to school and gets a special degree, or even takes a special course.

Introduction of products liability laws designed to protect the public safety, health, and welfare against defectively designed, manufactured, and marketed products has created a particularly important and explosive demand for expert analysis and opinion, notably in the fields of science and engineering. Continued development of such laws, even with some inevitable retrenchment and swinging of the pendulum, has provided the public with an increasing number of, and increasingly effective, legal remedies—both civil and criminal—where such unsafe products have caused injury, damage, or other loss. Thus, the American public in particular has acquired, as never before, important opportunities and resources to protect itself and fight back in order to try and obtain equitable jury damage awards and redress in meritorious cases. Although there have been some abuses in the system, by and large products liability litigation has served the public interest well. However, negligence-type lawsuits also require expert testimony.

These next two chapters are specifically designed to reveal just what expert testimony is all about, who needs it and who provides it, how to start and conduct a forensic consulting practice and work with clients, what really goes on in the courtroom, whether there really is such a thing as a "battle of the experts," how to assess and reconcile apparently conflicting expert testimony, and so on. You will gain an appreciation for and working knowledge of the place of expert opinion in the increasingly fragile, sensitive, and complex interrelationships among consumers, corporations, insurance companies, and the federal government.

CREDENTIALS REQUIRED

As noted before, there is currently no school or formal course in forensic practice. There is no degree or certificate qualifying one as a competent practitioner in this area. There is no code of ethics that requires adherence to ethical principles or canons. Nor is there any organization that effectively monitors and enforces ethical, responsible conduct by persons providing expert testimony.

It is true that some professions have established administrative mechanisms, such as peer review boards and panels, intended to deal with complaints and charges of unethical behavior. Attorneys and the courts are also presumably in a position to penalize expert witnesses for illegal or unethical behavior through impeachment, and so on. But the fact remains that, as a practical matter, expert testimony is an essentially and effectively uncontrolled activity in those aspects more likely to affect the outcomes of many types of litigation.

In addition, from time to time executives and businessmen,

professional people, and consumers may themselves become parties to lawsuits, as either plaintiffs, or defendants, or else be called upon by their employers to testify as expert witnesses. Thus, becoming educated in what expert testimony and forensic consulting are all about is a potentially very practical objective for a growing number of people who may find themselves in the role of plaintiff or defendant.

The engineer who possesses a Professional Engineering license is in a superior position, in terms of credibility, to provide expert testimony. However, experience is also important in qualifying one as an expert. Therefore, engineers who do not have their P.E. can still find many opportunities in forensic practice by virtue of in-depth, specialized experience that is relevant to case requirements.

In the engineering and scientific fields, in addition, there are other types of credentials, including certifications by peer groups such as technical associations. Sometimes the Ph.D. is an important credential. In the future, it can be anticipated that higher degree levels, professional licensure, and various peer certifications will become more and more important in terms of credibility in the forensic field.

IDENTIFY YOUR EXPERTISE

Virtually all engineering and scientific disciplines and subspecialties are potentially useful in litigation. Safety, mechanical, civil, and electrical engineering problems probably account for the largest proportion of forensic work. But chemical, metallurgical, and electronic engineering are also very important, with chemistry and industrial hygiene close behind.

Within each major field specific product or process experience will permit you to solicit specialized work. However, you will soon find that your basic training in your chosen major field, such as mechanical engineering, will be transferable from one problem to another. When you are being qualified by a court in which you are proposed as an expert witness, you will find that you are not necessarily expected to have had twenty-nine years of experience designing, manufacturing, testing, or repairing bicycles in order to understand that a quick-release hub is potentially dangerous if it does not incorporate a retaining mechanism that prevents the wheel from popping off if the hub has not been tightened properly or if it loosens up. Nor do you have to be an automobile mechanic or designer of braking systems to understand that a poorly maintained braking system will be potentially dangerous, why this is true, and under what conditions of service.

But also bear in mind that the name of the game in forensic practice is that a cross-examiner will attempt to discredit your testimony. Consequently, you must be very careful to practice within your field of expertise, even though the "technology transfer" aspects of engineering practice broaden tremendously your expert reach.

For example, if you are a mechanical engineer your training can be applied competently to many problems relating to automobiles, power presses, rolling mills, bicycles, table saws, commercial laundry equipment, lawn mowers and snow blowers, amusement rides, toys, swimming pools, and so on. If you are an electrical engineer, you can anticipate engagements relating to fires, explosions, and electrocutions, as well as electrical failures, across a wide diversity of product line, process, and workplace situations. Likewise, with respect to almost every other engineering field and specialty you can imagine.

Identify your strengths and specialties, as already described in previous chapters. Only this time you are focusing upon the litigation environment. Reading the daily newspapers and your technical journals will provide you with a wealth of additional insights into forensic-type engineering problems that are covered in detail or implied. There are more and more reports in the daily press relating to negligence and products liability-type lawsuits and much excellent reporting on general safety, health, and environmental issues. Clip and file those items that interest you, items that suggest consulting "market development" opportunities, or that expose you to important or unusual aspects of the litigation process.

MARKETING YOUR SPECIALTY TO SPECIALIZED CLIENTS

Forensic clients include attorneys, insurance companies, and municipal and other governmental bodies. You will usually not obtain a forensic engagement directly from a plaintiff or a defendant, but rather through their representatives. Your first step should be to contact lawyers whom you know well or even casually. Make it known that you would like to do forensic consulting. Your expertise is valuable. Indicate an intent to diversify your professional interests based upon your specialties.

Not every lawyer does trial work; even fewer specialize in products liability litigation. Most are general practitioners. Thus, you should indicate to lawyers you approach that you would appreciate their referring your name and brochure or C.V. to attorney colleagues or insurance-industry acquaintances who may be interested in expert engineering counsel. C.V. stands for "Curriculum Vitae."

As before, you should make up a nice public relations package, including your C.V. or resumé, and so on. It should be somewhat more general in nature so that you are not viewed as some "super specialist." Lawyers don't always understand that engineering training has broad applications to a wide variety of problems. They don't always comprehend the meaning of "technology transfer." Naturally, of course, if you do have some particular specialties, include writeups on them in a special section of your C.V., but don't overemphasize them.

Doing forensic work is one form of "consulting for consultants." After all, lawyers are consultants, even though they are not thought of in this light. In fact, forensic engineering can be one of the fastest and surest ways for you to obtain consulting assignments. It may also be a relatively less expensive way for the part-timer to enter consulting practice. But be alert to potential conflicts of interest that may involve products designed, produced, or marketed by your employer and its competitors.

Consider advertising in publications serving the legal community in your state or metropolitan area. There is usually at least one of these. For information on what legal newspaper, newsletter, or magazine is read in your area all you have to do is ask a lawyer friend, or else visit a large law firm in town. They will typically be pleased to help you. After all, you are a potential "expert witness." In fact, such publications may be left out in the reception room for visitors to read.

Since forensic engineering practice, like most professional practices, is a relatively local business, limit your initial territory and contacts to a twenty-five- or thirty-mile radius from your office or home. But include at least the nearest major metropolitan area where you can expect to find the greatest density of lawyers, insurance companies, and casework.

THE FORENSIC ENGAGEMENT: NATURE, STRUCTURE, MANAGEMENT, AND LOGISTICS

A comprehensive forensic engineering engagement consists of a number of identifiable stages. Typically, these are as follows:

1. Initial consultation with the client
2. Documentation review (if you are engaged by defense counsel, then you will also have an opportunity to review a plaintiff expert report)
3. Field or product inspection
4. Laboratory testing or analysis
5. Interviews with plaintiff or defendant (as the case may be, depending upon which party engaged you)
6. Case analysis
7. Case research
8. Report preparation and submission with engineering findings and opinions
9. Postreport consultation with the client
10. Depositions
11. Defense report review (if you have been engaged by plaintiff counsel, the defendant expert report will typically not be prepared or disclosed prior to availability and issuing of the plaintiff report during the so-called "discovery" process, depending upon rules of practice within the particular legal jurisdiction)
12. Rebuttal reports
13. Trial appearance and expert testimony (if the case does not settle out of court)
14. Rebuttal testimony

As a practical matter, as already noted, 97 or 98 out of every 100 cases settle out of court. A relatively small proportion of cases require laboratory testing. Taking deposition testimony of engineering experts is common. Rebuttal reports may be required in a small percentage of cases. Rebuttal testimony is rare.

Depositions consist of taking pretrial testimony of both "expert" and "fact" witnesses out of the presence of the judge and jury, although under oath. Deposition testimony can be taken in your office, your house, an attorney's office, or elsewhere, and it will be recorded by a court reporter or court stenographer.

Let us now look at each of the engagement phases in more detail.

1. Initial consultation with the client. Your first client contact will be by telephone or letter. If the client telephones, you should be prepared to get his name, the name of his firm (unless he is a solo or individual practitioner), address, phone number, and so on. You should then send him or her a copy of your C.V. and any other literature you have prepared. Chances are your prospective client will inquire as to your background, experience, and fees (about which we'll talk later on). Try to do a little bit of screening. Get factual data on the case so you can start thinking about it. Or you may want to discuss various aspects of the case with the client even during this first phone contact.

However, if you are just starting out in the forensic field, try not to become too involved with the client over the telephone, since you may convey an impression of lack of knowledge about the type of problem involved. On the other hand, do try to convey at least a sense of mastery of the technical problem or issues that appear to be involved. Then, since you may reasonably indicate that you have not done that much forensic work, you will be in a position to open the door in relation to asking the client to discuss his specific legal requirements to help you grasp the legal ramifications of your engineering problem-solving effort.

If you receive a letter, call the client. Technically, such an initial letter contact will outline the nature of the case and ask you to call to discuss it further, to discuss your fees, and to present your qualifications and availability. Are you available? Of course you are.

Sometimes a meeting may be desirable or necessary. This is fine, but make it understood that you are entitled to be paid for any conferencing. If the case can be discussed and screened in a preliminary way by telephone within, say, a time frame of ten to fifteen minutes, then don't charge. Again, this is part of your client development program.

If an initial meeting is to be held away from your office such as, for example, at the client's office, then you are also entitled to be paid for your travel time necessitated to suit the client's convenience.

Don't hesitate to decline a case or offer a "limited" or "qualified" opinion if you cannot obtain necessary and sufficient data, or if you cannot see the object involved if critical to your analysis, findings, and opinions. But leave doors open for your client.

Besides a full complement of photographic equipment, your bulging field case will probably include binoculars, a Swiss Army knife, magnifiers, various kinds of tape measures and other measuring devices, range finders, safety goggles, hearing protectors, hard hat, safety shoes, safety gloves, plenty of pens, lined- and graph-paper pads, 50- and 100-pound scales, plenty of batteries, a tape recorder with plenty of tapes, portable calculator or computer, a pry bar, plenty of film, and so on. Both Polaroid and 35mm cameras, with a selection of lenses, are indispensable in forensic work.

Depending upon your specialty, you might need various items of electrical testing equipment, fluid pressure-measuring instruments, velocity-measuring devices, inclinometers, surface and remote temperature-measuring instruments, stopwatch, micrometer, awl, some other hand tools, chemical testing and/or air sampling instrumentation and materials, and so on.

Since you usually get only one chance to inspect a site, workplace, machine, or process, make it good. Record complete data from the equipment, make appropriate measurements, take a comprehensive series of photographs from a photojournalistic standpoint, rather than merely close-ups of the offending product or accident site feature per se. You want "overview"-type photographs as well as details of offending accident injury agencies.

Also be certain that you have authorization to make an inspection and enter upon the accident premise. You may have to do this through the party or company that owns or controls the accident site, its insurance carrier, or an attorney. Be certain everybody up and down the line has given their blessings to your presence at a mutually agreeable time. On the plaintiff side, you may find it helpful, if not crucial, for the injured party to be present.

2. Documentation review. Once you have the engagement (and retainer, about which we will speak a little later on), the client will forward various items of documentation.

The burden relating to documentation review can be considerable, depending upon the case. Numerous hours can be consumed on an ongoing basis over a period of months or even years, depending upon the logistics of the engagement. All of this time, of course, should be paid for by the client. The types of materials that will have to be studied and/or reported upon will include a combination of the following:

1. Client correspondence
2. Factual background about the accident and the plaintiff
3. Accident reports by various authorities and agencies (police department, fire department, first aid squad, etc.)
4. OSHA reports and citations for violations
5. Insurance investigation reports
6. Private investigator reports
7. Witness statements
8. Interrogatories and Answers to Interrogatories by both sides
9. Transcripts of Oral Deposition Testimony
10. Selected legal cases and opinions

11. Medical, hospital, and psychological reports
12. Copies of warning and caution labels, and so on
13. Correspondence, memoranda, reports, and so on concerning other accidents
14. Expert reports by engineers engaged by other parties to the litigation
15. Photographic exhibits and folios
16. Abstracts from the technical historical literature
17. Competitive, comparative, interfirm, interindustry, and/or technology-transfer-type analyses
18. Product literature; instruction manuals; safety manuals; maintenance, repair, and operations (MRO) manuals; specifications; and so on
19. Engineering drawings and engineering change documentation
20. Price-cost documentation including invoices, quotations, cost engineering and estimating data, price lists, and so on
21. Sketches, diagrams, technical illustrations, exploded views, and so on
22. Applications engineering documentation, troubleshooting guides, and so on
23. Graphs, charts, tabulations, experimental data, and so on
24. Laboratory tests and analytical or experimental reports
25. Patent documentary or folio collections to establish retrospective state of the art
26. OEM and/or other internal technical or executive correspondence of litigants
27. U.S. Government Agency recall notices
28. Product service histories
29. Published codes, standards, rules, regulations, standard operating procedures, and so on

The list could go on and on. The file materials you receive may consist of a single sheet, be one-quarter inch or fourteen inches thick, or they may take up the better part of four thick and heavy file pockets. Every case is different. You will have to get some idea of how much and what kind of background materials are to be sent to you for your review before you quote a fee for undertaking the case. Ask the client how much documentation there is that you will have to review, and be guided accordingly. Documentation review could take ten minutes, ten hours, or ten days. Know before hand what is going to be involved. The same holds true for any follow-up casework. Always attempt to get a good idea of how much material will need review and analysis. As we will see later on, the name of the game is to obtain an advance retainer for the estimated time to be consumed by the engagement at least through preparation of your expert report, or through some initial milestone. Thereafter, progress retainer payments should be required and received.

3. Field or product inspection. During this stage of the engagement you will have an opportunity to personally inspect the accident site, workplace, product, system, process, or whatever. You will charge for travel time as well as pure examination time, waiting time, and so on. Sometimes a product or part will be delivered to your office and no field trip will be necessary. Your fee structure should reflect this.

In some instances there will be nothing to inspect. The building or workplace may have been changed significantly. Or perhaps it is no longer available at all. Maybe it was totally destroyed, product lost, or transported to some location to which it would be impractical or too costly to the client for you to travel. Or maybe the client only has photographs. In other instances the product or vehicle involved in the accident might be somewhere across the country or the world, constantly on the move, or otherwise not readily susceptible to inspection. In still other instances a replica product may be available although the accident item is not.

Since all your activities are on a "best efforts" basis, you must be certain that you have adequate, necessary, and sufficient data or physicals to be able to proceed with your analysis. If you get to the point where you have done many cases of the same type, you may not actually have to inspect the product or workplace to be able to offer analysis and opinions with "reasonable engineering or scientific probability" that are viable for client purposes. But it is usually better to be able to inspect the physical object or, at least, to obtain good photographs of it. There are many strategic and tactical problems on the legal side if you cannot or do not inspect the physicals involved in the case if they are available on some reasonable basis. Usually, the client will opt to have you inspect. Leave the decision to him unless you feel it is absolutely critical, from an engineering standpoint, to make an inspection. But have him put it in writing if he does not want one. Tactically, this is to the benefit of the client. Otherwise, later on, your client's adversary will make it look like you should have made an inspection but didn't.

4. Laboratory testing. Chances are you will not have the necessary equipment to do anything but "bench inspection" back in your office, even if relatively sophisticated by ordinary standards. If high-powered laboratory testing routines are necessary, you are best off working with an independent laboratory on whatever basis would be most fruitful. The final arrangements should be between your client and the laboratory rather than between you and the laboratory. This must be understood by your client from the start. It should also be understood that you are not a laboratory, but rather a competent engineer who is accustomed to working with laboratory equipment and/or laboratories if and when the need arises.

5. *Interviews.* You may have to personally interview plaintiff or defendant, depending upon who retains you, or other parties, depending upon the nature of the problem. Keep detailed, dated notes and diagrams, and so on. Your case file folder or jacket can serve nicely as your notebook or drawing board in the field. However, oil, grease, water, and dirt can play havoc with note-taking and sketching efforts. Not all interviews will take place in a nice clean, air-conditioned conference room. More often than not field inspections will take place in an oily, grimy atmosphere or outdoors, typically on a frigid or rainy day. Plan accordingly.

6. *Case research.* Typically, this phase of an engagement involves primarily library-type search for "body of knowledge" information or data pertaining to the problem, aspects of your analysis, the search for historical support for tentative findings and opinions, and so on.

7. *Case analysis.* When you have made your field inspection and have reviewed background documentation or materials, you are ready to analyze the case from the standpoint of your specialty. Here again, keep all notes, computations, references, sketches, narratives, rough draft report sections, and so on. This phase may include scientific accident reconstruction and selection of the most probable reconstruction scenario from among the array of those merely possible. Avoid the use of the term "possible" in relation to your opinions. Strike it from your vocabulary. Your opinions must deal in reasonable engineering or scientific probabilities. In forensic work this means that the probability you are correct is over 50 percent, surprising as it may seem. Statistically speaking, this gives you a lot of leeway.

8. *Report preparation.* Whether you draft it by hand, dictate it, type into a word processor, or speak directly into a recorder, prepare your rough draft preliminary report. Several revisions and much editing are typically necessary before you are ready to finalize it. Try to standardize your report format and divide it up into compact, bite-size sections, particularly if you are not a seasoned writer. This way you can prepare different sections as soon as you have adequate information for each one. Due to the short length of each section, you can better control the limited flow of verbiage than if you attempt to prepare a long narrative which may not quite hang together the way you would like. Some sections, naturally, will build upon others. The bottom line will be your set of findings and opinions, plus your valuable signature and seal. Although your P.E. seal is not a legal requirement for forensic work, it reflects your willingness to stake your professional reputation on your work product.

9. Postreport consultation. After you transmit your report to the client, be prepared for a telephone call or a letter asking for clarification or additional information. Typically, you should call your client first to be certain that you both remain on the same wavelength throughout the engagement. If problems crop up, do not hesitate to contact the client and attempt to work things out. Postreport contacts can be extremely valuable to both you and your client in order to keep abreast of new developments, the need for follow-up casework, prospective scheduling of depositions, trial testimony, modifications, and so on.

10. Depositions. The rules of court entitle you to be paid a "reasonable fee" for your oral testimony at depositions. You should also charge for any and all travel time necessitated to suit the convenience of deposing counsel. This is typically payable by deposing counsel at your usual rates. Where a dispute arises over fees, you may find yourself the recipient of a hostile subpoena by the opposing counsel. But as long as your fees are not excessive, there is no reason to panic. As you will see a little later, you should always obtain an agreement on fees before you agree to be deposed, even if your client has to make a motion in court. Sometimes a court order will limit your fees, but that's part of the way the game is played and you have to accept it. In some jurisdictions, any fee in excess of a court-set amount becomes the obligation of your client to pay you.

If deposing counsel desires that you travel to his or her office but declines to pay you for travel time, tell your client that you will be pleased to be deposed in your own office or office-residence, thereby avoiding entirely the need for deposing counsel to pay for any travel time. But do not be intimidated into traveling to some remote deposition location convenient for the deposing counsel, especially where he openly refuses to pay you for your travel time at your usual hourly rates. Throw the problem squarely back into the lap of your client and do not be intimidated. Do not let your own client intimidate you into swallowing travel time.

At the deposition the lawyer(s) requesting your testimony will go into your background and qualifications, and also the details of the case. Remember that a deposition is often a "hunting expedition." Deposing counsel wants to see how you present and handle yourself, how you field questions, and how much of a threat you and your opinions actually are to his or her case. Anything and everything you say can and will be used against you and your client at the time of trial. Count on it. But a good deposition can also help to settle a case reasonably promptly although, on the defense side, the name of the game is deferral, so any damages that will be awarded can go on earning interest for as long as is feasible.

Watch out for gratuitous statements by deposing counsel to the effect that he or she is not trying to trick you. Baloney. He or she is trying to get you to lower your guard and think in a sloppy or overly relaxed manner so that you don't provide crisp, precise, unequivocal answers. The whole purpose of a deposition is to determine your personal weaknesses and limitations, vulnerable parts of your analysis, and weak points in your client's case overall. Keep up your guard at all times, but don't be argumentative; don't be jovial; and don't volunteer information.

11. Defense report review. If your client is the plaintiff lawyer you may have an opportunity in some jurisdictions to see the defense report. Wherever this is feasible and permissible, insist on getting a copy from your client. In some jurisdictions the exchange of reports is not common practice or required, and the case proceeds to trial more like in a Perry Mason story, with surprises all over the place. The defense report will permit you to understand the strategy and tactics of your client's adversary. It will make your continuing involvement in the case "surprise-free." It will also afford you the chance to determine for yourself whether your counterpart on the defense side has been objective, competent, ethical, and professional. Remember that you must take a very statesmanlike attitude. Your client's adversary is not your enemy. You are nothing more than a catalyst who is attempting to shed light where before there was darkness. The fact that your client's adversary is going to try to trap you at every turn simply means that you must be on your guard at all times, rather than hostile. Kill 'em with kindness.

12. Rebuttal reports. Occasionally you will be required to analyze adversary reports, tactics, and strategies and to submit rebuttal reports. Just be sure you have enough data to do so in a sound and professional way. Understand the types of reports that you may run across. We will talk more about them later.

13. Trial appearance. If the case does not settle out of court you may have to testify in front of a jury. In civil and criminal cases, juries are the rule, but not so in municipal or chancery proceedings, arbitrations, and certain other types of specialized proceedings and hearings.

Typically you will not sit through the entire trial. Your client couldn't possibly afford this if you are on the plaintiff side. If you are on the defense side, it could be another matter. Sometimes a large corporate client, who can afford it, will feel it is worth the cost to have you sit through significant portions of the trial. More often than not you will be present only for your own testimony, although on occasion you may have the opportunity to hear other expert witnesses while you are waiting to be called to the witness stand or box yourself.

You will be subjected to both direct examination by your own client and cross-examination by his adversary. But before you are questioned about details of the case and your own inspection, analysis, findings, and opinions, you will be qualified by your client. His adversary will have an opportunity to cross-examine you on your qualifications. This, again, occurs before the main body of your testimony about the specifics of the case. The cross-examiner may elect to "voir dire" you. This is a cross-examination technique that takes place out of the presence of the jury. You will be questioned about the details of your prospective testimony and different issues of importance. Various rulings may be made by the judge on the basis of voir dire as to what you may and may not say or refer to during your examination in open court, when the jury is called back into the courtroom.

After the qualification process and direct and cross-examination, your client may attempt to conduct a redirect examination. His opponent, in turn, may then elect to conduct a recross-examination. Then a re-redirect and re-recross, and so on.

Keep your cool; stay awake. Retain your professional presence and composure. And don't argue with the cross-examiner or the judge. It is the most fascinating, but also most grueling, part of the entire forensic process. You will earn your fee—every penny of it. Luckily, you will not go to trial very much since most cases settle out of court.

14. Rebuttal testimony. Occasionally, although rarely, after you have provided your testimony and have left the courtroom, your client may call you back a day or two later to return to court and provide rebuttal testimony. This will typically occur because your client feels that there is a critical need for his adversary's expert testimony to be rebutted. Naturally, such a court appearance is billable, just like the first. And again, as we will discuss later on, you should obtain payment in advance.

Management and logistics of the forensic engagement are often horrendous, from setting up field inspection schedules, which could involve five or six parties to the lawsuit as well as court-ordered inspection dates, to being on call for a trial that may never take place at all or a deposition that tests your patience if fees become an issue.

Subpoenas and court orders can start floating all over the place. The forensic engagement is a unique experience. From the time you receive the initial assignment to completion of the legal discovery process, the date of a deposition, and the date of an out-of-court settlement or trial, anywhere from two months to five or six years can elapse!

Remember, too, that there is no businessperson whose personal schedules are so subject to disruption, inefficiency, and pressure as your ordinary, everyday, run-of-the-mill trial lawyer. As a consequence, as an engineer practicing in the forensic area your own schedule will be even worse. This is because it is ultimately a function of those of your clients, and theirs are awful.

The average deposition does not take more than two or three hours for a relatively simple case such as a slip and fall. As a practical matter, you will rarely be deposed on such cases.

But the average products liability case deposition could run from three to five hours or more. In some instances it might even consume a full day or even two or three multihour sessions on separately scheduled days. But these are rare for the run-of-the-mill case.

As for court appearances, a half man-day is typical for slip and fall cases. About 60 percent of the products cases will also take no more than one-half day. Another 37 percent will take a full day (i.e., morning and afternoon sessions, typically from 9:00 or 9:30 A.M. to 4:00 or 4:30 P.M.). The 3 percent balance could keep you in court for two or even three consecutive days. Of course, depending upon where the appearance takes place relative to its proximity to your office, you may charge on a per diem basis even if your actual time testifying is only three hours. Travel time could be four or five hours.

Once you get involved in forensic practice you will thrive on it. It will keep you involved in one of the most stimulating and exciting applications of engineering science available to the skilled practitioner. And, of course, due to the unusual pressures that demand and command your attention and divert you from personal pleasures and pursuits, the financial rewards are commensurately great.

THE FORENSIC ENGAGEMENT: LEAD TIMES AND FEES

Whether you value your professional time at $75 or $175 an hour, there are certain rules for survival in the jungle of forensic engineering practice. On the defense side, if your fee is payable either by an insurance company or a manufacturer being sued in a products liability action, as a general rule billing for services rendered is an acceptable and reliable procedure. Few disputes will arise over fees.

Insurance companies and large corporations do go broke on occasion, of course. But the likelihood of your going unpaid is negligible if an insurance company is footing the bill.

But if your defendant client is a self-insured individual or some entity other than a medium-sized to large industrial corporation, retailer, and so on with a national reputation, you are better off requiring an advance retainer to cover all prospective casework, or at least successive engagement milestones. This goes for original investigative casework leading up to your expert report as well as fees for prospective trial appearances at which you are expected to provide testimony if the case does not settle out of court. And beware the "I'm an honest man" syndrome. The more times a prospective client says it, the more careful you should be.

If your client is a plaintiff attorney, make it an invariable, non-negotiable rule to always require advance retainers for pretrial casework, and deposits in anticipation of trial. In the case of fees for trial testimony, any unexpended retainers will then be refunded by you, in whole or in part, depending upon the extent of your actual appearance, if any.

The bottom line is to avoid the necessity to bill plaintiff attorneys for professional services rendered. If a prospective plaintiff attorney insists that you bill him you may want to take a flyer now and then. But the best practice is to be paid in advance. Actually, you are probably better off declining the case if you can't get paid up front and receive a check significantly in advance of conducting casework. You will get to know your customers and you'll develop a "sixth sense" for trouble.

Deposing counsel typically foots the bill for deposition testimony. It is rare that a deposing attorney will consent to pay for any deposition preparation time, however. This is more legitimately billable to your own client. But, typically, it will be impractical to bill and collect for such preparation time unless it involves extensive conferencing with your client. Therefore, make it a general rule to write and format your report in such a manner that you have an absolute minimum of D/P (i.e., deposition prep time).

Over the course of many years of forensic practice I have developed a set of standard letters that have been effective in dealing with clients and their adversaries, and in smoking out prospective problems such as any relating to possible deposition fee disputes. You may find these forms, presented as Exhibits 11-1 through 11-3, instructive and useful.

A fourth form is illustrated in Exhibit 11-4. This is a very simple type of evidence log sheet. Since much forensic casework involves evidence movement, and since the chain of possession is

EXHIBIT 11-1

RESPONSE TO INQUIRY RELATING TO FOLLOW-UP CASEWORK

TO:
DATE:
REF:
SUBJECT: FOLLOW-UP CASEWORK

Thank you for your recent inquiry, which we are acknowledging in this way. In many instances casework of a supplementary or follow-up nature is required by the client. This may take the form of evaluation of new or modified facts and/or data, critical review of laboratory test results, expert reports, transcripts of oral depositions, engineering drawings, product literature and manuals, additional field and/or product inspections, client conferences, preparation of supplementary TSGI expert reports, code and case research, etc.

Fees for such follow-up professional services are at our usual rates on our customary retainer basis. Actual charges will be billed against retainers received, which are *not* minimums. Unexpended retainer balances will be promptly returned to the client or, where follow-up fees are anticipated to exceed retainers, additional follow-up retainers will be required. Retainers may also serve as stop orders. FOR DEPOSITION TRANSCRIPT REVIEW, BASIC RETAINER COVERS STUDYING AND PAGE MARKING FOR CRITICAL DATA ONLY (billed @ 75 pps./hr.). IF YOU REQUIRE THAT A TYPEWRITTEN ABSTRACTING SUMMARY BE PREPARED IN ADDITION, PLEASE DOUBLE THE FEE DUE AMOUNT SHOWN BELOW.

In the present case, above captioned, at our current fee rate of _____ per hour, please forward an initial follow-up retainer of:

$ _____

prior to our performing requested follow-up casework on the basis of approximately:

_____ hours covering _____ transcript pages (where applicable) for professional time preliminarily estimated per your _____ of _____ , 198 .

If there is a compressed time frame where time is of the essence, please indicate the date when follow-up analysis, reports, etc. are desired. If we cannot respond within this time frame, or if there are problems inherent in the request which only come to light as we proceed and which require additional clarification, discussion, time, personnel, etc. we will let you know immediately or as soon as practicable. Some initial expenditure of time may be necessary, however, and will be billed against the initial follow-up retainer accordingly.

Thank you for your courtesy and cooperation. We will be pleased to perform the required supplementary or follow-up casework as soon as possible, consistent with other commitments. Thank you also for the opportunity to be of continuing service to counsels.

R. Matthiew Seiden

EXHIBIT 11-2

DEPOSITION TESTIMONY FEE LETTER

DATE _____

TSGI Report Addressee

REF: _____

Product/Process/Facility

Dear

It is our understanding that deposition testimony is desired from Associates of THE SEIDEN GROUP, INC. in above-referenced matter. Prior to scheduling the deposition, we will greatly appreciate your returning a signed copy of this letter to us constituting an agreement and acceptance with respect to our standard fees.

Our present testimony fee is ____ per man-hour. Total fee due is calculated on an elapsed time, portal-to-portal (including travel time) basis for appearance and testimony. An invoice for professional services rendered at the deposition will be promptly submitted by us. For out-of-state counsels a deposit of ____ is required, to be billed against. Invoice amounts are due within thirty (30) days of invoicing. Counsel requesting the

(continued)

EXHIBIT 11-2 (Continued)

deposition is responsible for forwarding payment in full within the thirty (30) day period trade term aforementioned. A service charge of 1-1/2% per month will be billed on outstanding balances over 30 days old.

Please note that any scheduled deposition is subject to conflicting court appearances, which will ordinarily be given higher priority. We will do our utmost, however, to advise counsels of developing conflicts as soon as possible. As soon as we have received the signed copy of this agreement, we will be pleased to calendar the deposition. Please advise as to desired location. The signed agreement must be received in our office at least ten (10) days prior to the scheduled date. Otherwise, there is a good chance other commitments will have been made for that time. Please also note that your account with TSGI must be current before we can schedule the deposition.

Thank you for your inquiry. We look forward to the opportunity to be of further service to counsels in this matter.

Respectfully yours,

RMS:amm
cc: R. Matthiew Seiden, President

Agreed to &
Accepted by _____ (Signature Above Line)
                                 (Type Name Below Line)
        For _____ (Firm Name)
      Dated _____ 19___

---

sometimes critical and can become a legal issue, it is best to keep track of evidence movements in and out of your office or through your hands. This exhibit may prove helpful to you.

A fifth letter is presented as Exhibit 11-5. It is a request for data and/or documents that may be helpful or critical to the conduct of your investigation. As you can see, it is specifically designed for use in products liability-type litigation.

When you first enter forensic practice you will not be able to estimate engagement time frames too well. But there are certain

EXHIBIT 11-3

TRIAL FEE DEPOSIT LETTER

DATE _____

TSGI Report Addressee

Product/Process/Facility

REF: _____

Courthouse Location

Dear

Thank you for your trial notice re above-captioned matter. All appearances are contingent upon receipt of a deposit in our office, prior to trial, of a full day appearance fee. IN THIS CASE, A TRIAL DEPOSIT OF $ IS REQUESTED. Notice of at least TWENTY-FOUR (24) hours is required for all appearances.

At present our local appearance fee schedule is   per half man-day or portion or   in total for a full day including Bergen, Essex, Hudson, Morris, Passaic, and Union. Other NJ, NYC, and nearby Pennsylvania appearances are billed on a per diem basis of   and up. More remote out-of-state appearances are billed from   per diem plus expenses. If the case settles out of court you will be immediately refunded the full deposit amount as long as we have been able to abort our trip to the courthouse, but less preparation time. Unapplied trial fee deposits will be promptly refunded to you. We do not charge for "stand-by" because we cannot stand by. Appearance should commence in the morning or TWO (2) half-day or full day fees will be applied if there is a holdover. Appearances are on a first-come, first-served basis. But we will accord highest priority to federal, peremptory, and criminal matters. Summertime and foul weather logistics can make appearances infeasible or impossible.

Any required pretrial nontestimony casework is separately billable at   per man-hour. This fee and outstanding casework balances are payable in advance of any prospective appearance. No fees are acceptable at the time of trial. Outstanding balances and the above trial deposit MUST BE RECEIVED BY TSGI AT LEAST FOURTEEN (14) DAYS PRIOR TO TRIAL DATE. Please note that we CANNOT accept your client's personal check. Payment must be by your own attorney check,

(continued)

EXHIBIT 11-3 (*Continued*)

or else a bank or certified check or money order from your client. If trial deposit cannot be received at least two weeks prior to trial date, PAYMENT MUST BE BY BANK/CERTIFIED CHECK OR MONEY ORDER, but the prospective appearance can be reserved and calendared only AFTER receipt of the requested trial testimony fee deposit in either case. Nominal pretrial file review is included in trial fee if appearance is required.

Please accept our sincere best wishes for a favorable out-of-court settlement in this matter. Thanking you for your continued courtesy and cooperation, and for the further opportunity to be of service, we are

Respectfully yours,

R. Matthiew Seiden, President

RMS/amm

---

guidelines you can follow that will be helpful in estimating dimensions of the forensic engagement for different classes of cases:

1. Slip, trip, and fall (ST&F)-type accidents resulting in negligence lawsuits. In this category, generally allow up to five hours through preparation of your expert report, if your travel time is relatively local (i.e., up to one to two hours of round-trip travel) and there is minimal documentation review. Allow up to three hours if there is no travel time but you are working from photographs or other background documentation only. Naturally, these figures also assume that the case does not become hopelessly open-ended. If it does, advise the client and request follow-up fees. Always advise the client in advance that fees will depend on whether the case is routine or standard (three to five hours) or special, and that more intricate problems will most likely require more extended casework. Also note that sidewalk and building stairway fall-down-type cases, for example, will be of shorter duration than cases involving falls from machinery, trucks, and so on that may result in products liability-type litigation. Time frames and fees for such cases are

EXHIBIT 11-4

EVIDENCE LOG

Evidence Description	Evidence Movement		Date	Reason For Evidence Movement	R/F D/T P/U	Name*		
	In	Out				Received From (R/F); Delivered To (D/T); Picked Up By (P/U)	Init-ial	
1								1
2								2
3								3
4								4
5								5
6								6
7								7

(*) Type or print name & sign name or initial.

TSGI Associates are required to return evidence in their possession to *THE SEIDEN GROUP, INC.* office after final report is completed and transmitted to client.

Client Name _____ Case I.D. _____

EXHIBIT 11-5

REQUEST FOR DATA AND/OR DOCUMENTS

DATE: _____

REF: _____

Dear Atty.

As authorized by your office in connection with above-captioned matter, TSGI is proceeding with casework on a "best efforts" basis.

It is recommended that the following information be obtained, where feasible and as relevant, and transmitted to TSGI for evaluation.* Not all such data necessarily will be critical to formulation of TSGI findings and opinions, of course. However, for purposes of completeness they may be of some interest. In addition, some of these data may become useful as the litigation proceeds.

1. Accident reports. ☐ 1
2. Selected medical reports. ☐ 2
3. Photographs of accident product and accident site. ☐ 3
4. Product catalogues, specification sheets, and price lists. ☐ 4
5. Instruction, safety, maintenance, installation, parts, and other product-specific manuals prepared by the manufacturer (OEM). ☐ 5
6. Sales invoices and proposals. ☐ 6
7. Maintenance and repair records. ☐ 7
8. Engineering drawings, experimental data, sketches, and diagrams. ☐ 8
9. Product safety audit data including hazard foreseeability analyses and actions taken, and failure mode and effect analyses (FMEA), etc. ☐ 9
10. Patent history. ☐ 10
11. Field experience in relation to accidents and injuries. ☐ 11
12. Records of customer complaints and dealer-distributor reports of accidents. ☐ 12
13. History of product recalls and factory modifications to eliminate/mitigate safety defects/hazards. ☐ 13

14. Corporate organization chart and organization chart for accident product-producing division or other organizational entity.   ☐ 14

    * Items with boxes containing an "X" would be of particular interest at this time. However, other items also may become significant at a later time, as noted.

15. Copy of corporate/divisional product safety manual with standard operating procedures, corporate and divisional product safety policies, product safety department/activity organization charts, job descriptions/position profiles, biographical profiles/ resumés of product safety department personnel, history of product safety programming at the company, and product safety engineering techniques utilized.   ☐ 15
16. Description of applications engineering forms and procedures utilized, generally, as well as specifically in connection with the accident product and/or actual order for the accident product/equipment/process/system, plus order engineering-focused safety documentation.   ☐ 16
17. Abstracts and/or summaries of oral depositions, interrogatories, complaint, and other legal documentation.   ☐ 17
18. Records of user/customer/employer/plaintiff modifications.   ☐ 18
19. Records of start-up activities in relation to the accident equipment, when OEM personnel were present, and records of subsequent OEM customer contacts, for whatever reason, including servicing and modifications.   ☐ 19
20. Comparative competitive product analyses, market research studies performed, and product line RD&E to determine necessary and/or desirable product safety attributes.   ☐ 20
21. Records relating to human factors/human engineering studies of the accident product/process/system and safety-in-use.   ☐ 21

(continued)

EXHIBIT 11-5 (*Continued*)

22. List of field installations in NY-NJ-PA area, with dates of original sale, genealogy data (if any), OEM servicing, modifications, etc. ☐ 22
23. Names, qualifications profiles, professional licenses, and/or certifications of engineers, supervisors, managers responsible for accident design. ☐ 23
24. Address of product-designing division or state of design origin, if different from division address, and dates of design efforts. ☐ 24
25. Names, addresses, and dates of engagements of independent outside product safety management consultants responsible for setting up product safety management and/or auditing programs and/or procedures. ☐ 25
26. Copies of product safety management reports and recommendations by independent outside product safety management consultants, if any, and internal corporate documentation, memoranda, etc. relating to reliance on same with respect to corporate product safety assurance. ☐ 26
27. Accident product/system genealogy; date of manufacture, date sold new (and to whom, if used), date shipped new from factory, date resold (by whom), date refurbished/rebuilt/modified (and by whom), original and resale price history, product movements/relocations with dates and locations, etc. Detailed chronology from point of original product inception/design/development could be critical. Provide comprehensive details on source/acquisition through disposition/sale/resale/brokering, etc. for every single entity over the entire life cycle of the accident product/system/subsystem/component/OEM part/etc. ☐ 27
28. Particularly if a subsystem or component OEM part is involved, be certain to provide detailed data as to whether it came direct from the OEM, from a captive or independent distributor/dealer, from the stock or warehouse of the final product producer, etc. ☐ 28
29. Particularly if a subsystem or component OEM part is involved, be certain to provide detailed data as to what role the OEM played in design, ☐ 29

development, manufacturing, marketing, testing, etc. with respect to knowledge of the ultimate application and customer type.
30. What specific knowledge did a subsystem or component part OEM have with respect to the particular order involving original or spare parts/subsystems involved in the accident, and to what extent did such OEM participate in order/application engineering. ☐ 30
31. Copies of case-specific litigation policies, memoranda, letters, operating procedures, instructions, etc. communicated to defense counsel by the defendant manufacturer for purposes of handling the case, interfacing with the public, the press, the government, etc. ☐ 31
32. Is the accident product, product type, model, series, variation, or similar product type exported abroad from the USA? Manufactured abroad by subsidiaries, associated or affiliated companies, subcontractors, licensees, etc. for sale abroad? For export to the USA? ☐ 32
33. What added safety features, warnings, instructions, etc. relevant to the present case are incorporated into units exported from the USA? Into units made abroad for sale abroad? Into units made abroad for export to the USA? ☐ 33
34. If additional safety features are provided on exports, imports, or non-USA-made units for sales abroad, why? What are the specific codes, standards, and/or regulations controlling the requirement for such added safety features compared with those supplied with (or absent from) the accident product? Obtain copies. ☐ 34
35. Where a dealer, distributor, marketing franchisee, etc. is involved, was the original selling entity the same as the servicing or reselling dealer, etc.? If not, determine identity of all marketing entities in the commercial chain subsequent to manufacture. ☐ 35
36. Were any relevant OEM safeguards removed or bypassed or otherwise modified? By whom? Obtain engineering documentation of all such OEM ☐ 36

(continued)

EXHIBIT 11-5 (*Continued*)

safeties. Among other things, it must be determined therefrom whether, in fact, such safeties would have been capable of preventing, mitigating or otherwise accommodating conditions leading to subject accident and injury.

37. ☐ 37

38. ☐ 38

39. ☐ 39

40. ☐ 40

41. ☐ 41

42. ☐ 42

43. ☐ 43

44. ☐ 44

45. ☐ 45

Thank you very much for your cooperation and courtesy in this matter. We appreciate this opportunity to be of service to counsels.

Respectfully yours,

R. Matthiew Seiden, PE, CSP, CPSM

RMS/amm

in a different category altogether. But straight ST&F-type accident investigations can be an excellent entry point for mechanical and civil engineers desiring to do forensic work.

2. Machine guarding cases resulting in products liability lawsuits. In this category, mechanical and electrical engineers will typically find a ready market for their skills, with some opportunities for chemical engineers as well. Safety engineers, human factors specialists, and related professionals are also much in demand in this category. A typical machine guarding case with relatively local travel time should not take more than eight to twelve hours through the preparation of your report, if the case is not open-ended. Add unusual travel time requirements and corresponding fees to suit. When the problem is not merely absence of a suitable guard but malfunction, time frames could escalate to fifteen to twenty-five hours or more. Be careful that you know what you are getting into. Try to size up the case before you quote an initial retainer, and work on a milestone basis thereafter if you are not able to grasp the problem right off the bat. Don't let the prospective client intimidate you into quoting a flat fee. You'll be a big loser most of the time if you do, and he will merely go on to another expert for his next case, with the same intimidating bad "deal" in his hip pocket. As you gain more experience with particular classes of cases, of course, you will be better able to quote a realistic fee right at the start.

For instruction and warning-type cases, figure three to five hours through report preparation without a field inspection. Note that all of the foregoing time estimates assume that you have ready access to a good technical library including safety references, codes, standards and regulations, and so on.

3. Consumer products cases (except foods, drugs, and cosmetics). Time frames can vary widely, depending upon the product, where the inspection is to take place, and what the problem is. But a time frame of about five to ten hours is probably typical where there is a guarding problem, or where a part breaks or malfunctions and contributes to injury, but where no lab work is necessary.

4. Flammable substances, fires, explosions, and electrocutions (FE&E). These can be extended assignments. You have to figure a going-in retainer at ten to twelve hours minimum with local travel time, unless it involves a simple kitchen range or oven flashback-type fire from accumulated gas, or a flashback fire

from aerosols, open solvent containers, gasoline, simple electrocution cases, and so on. For cases involving industrial fires and explosions, metallurgical fires, building fires, process fires, marine fires, utility power line electrocutions, or the like you could wind up spending twenty-five to thirty hours or more on a case, and you might have to bring in consulting associates, which will escalate the time still further. Be very cautious in screening FE&E-type cases with clients. Scientific accident reconstruction can become very extended and problemsome. Arson work, in particular, is very tricky. As a practical matter you may be better off keeping away from arson-type investigations until you have had more experience. There are plenty of opportunities to do forensic consulting relating to other types of FE&E categories.

5. Toxic chemicals, health and environmental hazards. Chemists and chemical engineers, industrial hygienists, toxicologists, and other occupational health and health science specialists will typically be involved here. Casework through the expert report could run from three to thirty hours or more and become multidisciplinary. Screening the case on the telephone with the client is an absolute must. You can even invite him to send you a one- or two-page letter for your review without charge, but do not do any extensive casework in order to determine whether or not you can do anything for the client. The other alternative is to charge a flat rate or three hours for initial documentation review and casework screening, or up to three hours for conferencing as a self-standing abbreviated engagement. In some instances you may have to conference with the client and bring along one or more of your consulting associates. All such time must be paid for on an advance retainer basis, including travel time that is necessary to suit client convenience.

6. Automotive accident reconstruction. Mechanical, civil (including traffic), safety, and other related engineering specialties can be useful for purposes of scientific accident reconstruction. Your minimum time will probably be ten to fifteen hours, unless casework involves merely computing minimum speeds from skid marks using any one of a number of specialized slide rules or nomographs that are available for this specific purpose. Automobile work can become products liability-focused, so screen cases very carefully. A serious problem that sometimes occurs is determining whether vehicle damage observed *after* the accident is indicative of a defect that *caused* the event, or

merely breakage and so on that was caused by the incident. Also remember that although events and/or damage patterns may be consistent or not inconsistent does not establish a cause-and-effect sequence. Just because two opinions are not inconsistent with each other does not necessarily establish that one event followed or derived from the other. Watch your logic and the pitfalls of determining causal sequences.

7. Construction defects. Architects and engineers can both be helpful in building failure cases. Time frames can be anywhere from five to twenty-five hours, depending upon the extent of field work, plan review, reconstruction, and so on. If computer time is necessary, casework costs can escalate tremendously; there is no particular upper limit to the time that might have to be expended in such cases. But for run-of-the-mill-type casework, this range is probably realistic in 85 to 95 percent of the cases you will see, and five to eight hours for perhaps 75 percent.

8. Construction accidents. Mechanical, civil, safety, and electrical engineers can be very helpful here. Engagements typically consume five to eight hours unless they involve products liability-type design or manufacturing defects. If so, then figure eight to twelve hours through report preparation, including field inspection and local travel time. On rare occasions, accident reconstruction may increase casework time requirements. Here, again, if metallurgical failure is involved then laboratory analysis may also be required. Most construction-type accidents that are litigated, however, involve improper and unsafe management or supervisory practices and procedures, lack of equipment guarding, or lack of construction-in-progress safeguarding means.

9. Safety management cases. Here the industrial engineer and safety specialist will be effective. Typical time frames can vary from five to eight hours.

10. Combination cases. Some cases will involve one or more of the foregoing types as well as others. A complete list would be impossible to prepare. However, an index to selected product, subject, and professional specialty casework categories handled by the author's firm, The Seiden Group Inc., follows (Exhibit 11-6). It is not complete by any means, nor does it represent the full breadth or depth of work that is actually out there but has not been encountered by TSGI. However, the list is extensive enough to give you a good idea of the wide diversity

EXHIBIT 11-6

TSGI: The First Decade . . . an Index to Selected Products and Other Subjects Investigated From 1973–1983

Acids & alkalis
Active versus passive safeties
Active versus passive auto seat restraints
Adhesives
Adventurous, nonmalicious behavior
Advertising misrepresentation
Aerial baskets & ladders
Aerosol containers
Aesthetics & safety (form versus function)
Agricultural machinery
Air compressors
Air-conditioning & heating systems
Air pollution studies
Air sampling studies
Alarm & intrusion systems
Aluminum wiring-caused fires
Amusement & carnival rides
Ant killers/traps
Appliances
Applications engineering
Armored cable
Asphyxiation & choking
Athletic & recreational equipment & supervision
Attractive nuisances
Audible alarms
Automatic controls, valves, & instrumentation
Automatic filing systems
Automobiles, buses, trucks, & motorcycles
Automobile accelerator-brake foot pedal system design
Automobile accident reconstruction
Automobile fuel tank explosions
Automobile parking bumpers/wheel stops
Automobile seat covers
Automotive service lifts
Baby carriages & strollers
Backup/reverse motion alarms
Bag-making machinery
Bakery machinery
Ball-bearing-making machinery
Band saws
Barricades
Barrier guards
Baseball pitching machines
Basement flooding
Basic performance standards & rules of practice for safe design
Bathtubs & showers
Bending & forming rolls & mills
Bicycles & tricycles
Battery explosions
Beach chairs
Beer keg explosions
Blenders & mixers
Blood alcohol content
Blow dryers/hair dryers
Boat trailers
Boiler explosions
Bottle cap flyoffs
Bottle explosions/bursts
Bottles, jars, cans, & other containers, closures, & packaging
Bottling machinery

Uncommon Opportunities in Forensic Consulting

Box making, forming, slotting, gluing, & folding machinery
Brakes & clutches
Brake monitors
Breakaway & nonbreakaway signposts
Brittle fractures and ductile failures
Building codes, standards, & regulations analysis, compliance, & violations
Built-in versus built-on safeguarding means
Bulk handling & loading machinery
Bulldozers
Bullworkers

Cables
Cameras
Camping equipment
Can-making machinery
Canes, crutches, wheelchairs
Carbon monoxide poisoning
Carbonated beverage bottle explosions & cap liftoffs
Cargo containers & "igloos"
Carpeting
Cardboard-tube-making machinery
Catching devices
Ceiling plaster falldowns
Cement mixers
Centrifugal clutches
Centrifugal governors
Chain saws
Chain & wire rope
Chairs, tables, benches, drawers
Chemical hazards
Chemical reactors & process equipment

Chemical spills
Chemical storage drums
Christmas tree & ornamental lights
Circuit breakers
Circular saws
Classroom & school shop supervision
Clothing, fabrics, seams, & stitching
Clutches & brakes
Codes & standards analysis, compliance, & violations
Coffee carafes & decanters
Collapsing walls & other building structures
Collision attenuation systems
Color coding
Combustible gas detectors and indicators
Combustible liquids & solids
Commercial freezer cabinets
Commercial garbage compactors, stationary
Compressed gases, compressors, & compressed gas cylinders
Compression, shear, tensile, & torsional failure
Concrete burns
Concrete mixers
Concrete-pipe-making equipment
Construction defects
Construction machinery, all types
Contractor failure
Control handles & knobs
Controlled failure-type safeties
Controls & instrumentation
Converting machinery
Conveyors, all types

(continued)

EXHIBIT 11-6 (*Continued*)

Corks
Corrosion & corrosion-inhibiting systems
Corrosive chemicals
Corrugated paper mills
Cost-benefit analysis
Cranes, hoists, & winches
Cribbing & blocking
Crown versus screw cap safety problems
Crushers, grinders, & mixers
Crutches, canes, & wheelchairs
Curling irons
Cutoff saws

Dance floors, built-in & portable
Dart games
Deadman controls
Defeatability proneness of product functional & safety features
Demolition operations
Department store & supermarket design, construction, & housekeeping hazards
Design defects
Detergents
Diaper-making machinery
Die-cutting machinery
Dispensing equipment
Diving accidents
Do-it-yourself products (assembly required)
Dockplates/dockboards
Doors & door openers/operators
Drain cleaners
Drainage & drainage systems
Drill presses
Drilling, milling, & boring machines
Drop lights/utility lights
Drumsticks
Dry-cleaning chemicals
Dumbwaiters
Dump trucks/dumpsters
Dust explosions
Dusts, fumes, gases, mists, & vapors

Economics of science & technology
Elderly, handicapped, & infirm, safety hazards confronting
Electric extension cords
Electric motors & motor-generator sets
Electric heating blankets and pads
Electric power lines
Electric space heaters
Electric switch spark-initiated explosions
Electrical appliances
Electrical power failures
Electrical circuit breakers
Electrical switch gear
Electrocutions
Electronics industry production equipment
Elevated workplaces
Elevators, lifts, & escalators
Embossing presses & mills
Engineering psychology
Envelope-making machinery
Environmental stress corrosion
Error-provocative products & workplaces & error-forgiving versus unforgiving product designs & constructions
Escalators
Eyeglasses

Uncommon Opportunities in Forensic Consulting   239

Eyeletting machines
Excavating operations & machinery
Exercise bicycles & apparatus
Exhaust & fume hoods
Exploding bottles & bottle cap flyoffs
Explosion-proof motors
Explosions

Fabrics
Face masks & shields
Failure analysis
Falling objects
Falling object protective structures (FOPS)
Falling, slipping, & tripping hazards
Fans
Fasteners, connectors, & joints
Feasibility: ethical, technical, economic, managerial, & political
Fences
Ferris wheels
Film splicers
Filter presses
Fire engines
Fire escapes
Fire extinguishers
Fire hydrants
Fire- & smoke-damaged machinery & merchandise
Fire & smoke detectors & alarms
Firefighter turnout gear
Fireplaces & chimneys
Fires & explosions
Flammability of materials including flammable fabrics & ignition points

Flammable vapor detectors/ "sniffers"
Flat-type amusement rides
Flexible couplings & shafts
Floor mats & runners
Floors, slippery & otherwise hazardous
Fluid power system failures
Flume rides
Food, drug, & beverage machinery
Food choppers, grinders, & slicers
Foot pedal & treadle controls
FOPS (Falling object protective structures)
Forklift trucks
Forming rolls & calenders
Foundation defects & failures
Foundry equipment
Fracture analysis
Front-end loader-backhoes
Frozen pipes
Fuel tanks
Fume hoods
Furnaces
Furniture

Garbage-refuse collection trucks
Gates
Gaskets, packing, & seals
Gear-hobbing machines
Genealogy tracing/product life cycle monitoring
Generic protective, safety, & control features
Giant slide-type amusement rides
Glass fractures
Glass laboratoryware
Glazing

(continued)

EXHIBIT 11-6 (*Continued*)

Go-karts
Golf-ball-making machines
Golf carts
Grain bins & silos
Graphic arts & printing machinery & equipment
Grinders, crushers, & mixers
Grinding wheel bursts
Guardrails & guide rails
Gym stands
Gymnasium workout equipment & mats
Gymnasium supervision

Hammers, mallets, & other striking & struck hand tools
Handrails & guardrails
Hazardous substances handling, labeling, storage, transportation, & disposal
Heat-sealing machines
Heating, ventilating, & air-conditioning equipment & systems (HVAC)
Heating pads & blankets, electric
High-pressure system failures
High-temperature system failures
Hoists & winches
Home building products
Hot tar cookers & luggers
Hooks
Hot water heaters & heater controls
Housekeeping, workplace
Human factors engineering & engineering psychology
Human error, failure, & oversight (negligent & nonnegligent)
Hydrostatic pressure failure

Ice & snow control
Icemaking machines & systems
Illuminating levels
Inching & jogging controls
Incinerator flashups
Independent testing laboratory liaison
Industrial espionage & trade secrets
Industrial hygiene studies
Industrial process guns, spraying equipment, & systems
Ink & dye processing equipment
Instructions & instruction manuals
Insurance inspection failure
Integral analytical instrumentation-type safeties
Interlocks, all types
Intersection accident reconstructions
Ironworkers

Jacks, automotive & industrial
Jig borers
Joints: metal, wood, plastic, & multimaterial
Jungle gyms & other playground climbing apparatus
Juvenile furniture

Kerosene heaters
Kettles & vats
Kitchen appliances & equipment

Laboratory apparatus & glassware
Ladders & stepladders
Laminating machinery

Lapping machines
Lathes
Laundry presses, dry-cleaning & other equipment
Lawn mowers, tractors, & brush cutters
Lead poisoning from paint chips & flakes
Lift trucks
Light dimmer switches
Lighting systems
Limit switches
Liquid level controls
Load & moment monitors
Loading docks & terminal yard operations
Local ordinance compliance & violations
Locks & latches
Log splitters
Loose clothing & jewelry around machinery & electrical appliances
Loss of earnings appraisals

Macaroni & noodle machinery
Machine tools, all types
Machinery access features
Machinery brokers, dealers, & rebuilders
Machinery falloffs
Machinery guarding, all types
Machinery turnovers & upsets
Maintenance-related problems
Management & supervisory error, failure, & oversight
Manuals, all types (operator instructions, installation, maintenance, safety, replacement parts)
Manholes & manhole covers
Manufacturing defects

Marine cargo handling operations & equipment
Marine fires & explosions
Masonry nails
Materials handling equipment & operations, all types
Meat choppers, grinders, & slicers
Mechanical linkage failures
Mechanical power transmission equipment & drives (MPT)
Merchandising displays & fixtures
Merry-go-rounds
Metal failure & fatigue
Metallurgical process fires & explosions
Microwave ovens
Mischievous nonmalicious behavior with physical security breaches & hazards
Milling machines
Minibikes/mopeds
Mirrors: automotive, factory, & household
Mixers, grinders, & crushers
Mobile equipment upsets, upset protection, avoidance, & control
Mobile homes
Mobile work platforms & lifts
Motorcycle helmets
Motorcycle exhaust system burns
Multilingual instructions & warnings
Multidisciplinary, multiprofessional investigative teams & engagement management
Multivehicle accident reconstructions

(continued)

EXHIBIT 11-6 (Continued)

Nailing machines/guns
Noise levels

Occupant restraint systems
Occupational health studies
Octopus/monster/spider/tilt-a-whirl-type amusement rides
Office furniture, equipment, & systems
Oil burners
Ordinary carelessness
OSHA violations & citation appeals
Outriggers
Oven tipovers
Overpass bombing
Overspeed safeties

Packaging failures, including merchandise damage & bottle explosions
Packaging/wrapping/palletizing machinery
Paint spraying guns & nozzles
Paints & paint jobs
Papermaking machinery
Parking lots
Passive versus active auto seat restraints
Passive versus active safety devices
Patent infringements
Pedal boats
Pedestrian hazards
Perfume production machinery
Personnel lifts
Petroleum terminal operations & equipment
Pharmaceutical processing & production machinery & equipment

Piggyback cargo containers
Pipe-bending & -forming machinery & rolls
Pits & holding tanks
Plastic grinders
Plastics injection molding machinery
Plastics manufacturing processes
Playground equipment
Plumbing & sanitary fixtures
Pollution control
Pop-out windows
Portable power tools: drills/saws/nailers/other
Potholes
Power lawn mowers
Power presses: punch presses/press brakes/can-making machines/die cutters/plastics injection molders/heat sealers/balers/platen type printing presses/other "power press effects" (pneumatic bladder-type cargo positioners/load dividers, etc.)
Power saws, portable & stationary
Power shears
Power staplers & nailers
Powered lift tables & work platforms
Practical limitations of published safety codes and standards as design guidelines
Pressure cookers
Pressurized container failure
Printed circuit manufacturing equipment
Printing presses
Product modification, retrofit/backfit, & modernization

Uncommon Opportunities in Forensic Consulting 243

Product warnings & instructions, all product types
Property maintenance code analysis, compliance, & violations
Pumps & compressors
Pump jacks
Propane explosions
Prosthetic & orthotic devices
Psychological testing
Pyrex glassware

Quick-opening & tamper-resistant closures
Quick-release & captive fasteners
Quick-release bicycle hubs

Radial saws
Radiator caps
Radio transmission interference
Railroad crossings & rights-of-way
Railroad passenger car doors & stairs
Railroad passenger station platforms & access facilities
Ramps
Rear-end auto collision-generated fires & explosions
Recalls
Recreational vehicles
Refuse containers/bins
Regulators
Restaurant & dinner theater design, construction, & housekeeping hazards
Retaining walls
Rigging operations
Risk-utility analysis
Road plates

Road rollers
Roadway/highway hazards/signalization/signing/pavement markings/guide rails/guardrails/geometry/defects/roadside hazards/"booby traps"/maintenance/winterizing/potholes/construction safety/channelization/etc.
Roller coasters
Roller & ice skating rinks
Roller skates
Roller tanks/roller dollies/caterpillar dollies
Rolling mills, ferrous & nonferrous/slab mills/tube mills/sheet & plate mills
Roll-up doors/reversal devices
Roll-over protective structures (ROPS) & falling object protective structures (FOPS)
Roof failures & leaks
Rope & cordage
Roundup & trabant-type amusement rides
Rubber/linoleum/plastics mills & calenders
Rupture/frangible discs

Safety belts, lanyards, & safety lines
Safety goggles
Safety by location/distance features
Safety & relief valves & vents
Safety manuals
Safety serendipity
Saws, power, portable, & stationary
Scaffolding & ladders
Scales & weighing equipment

(continued)

EXHIBIT 11-6 (Continued)

Scientific accident reconstruction (SCAR)
Scientific laboratory apparatus
Scrambler amusement rides
Scrap- & waste-baling presses
Seating systems, automotive & industrial
Second collisions, automotive & aircraft
Service manuals
Shearing presses
Shock cords
Shoes/shoe heels/stitching
Shopping carts
Sidewalks, stairways, ramps, landings, floors, driveways, yards, construction openings, sidewalk cellar doors, sidewalk openings, utility covers, guardrails, handrails, catwalks, work platforms

Ski bindings
Slack cable safeties
Sliding ponds
Slings & spreader bars
Slip clutches
Slippery floors & other walking/working surfaces/constructions
Slips, trips, & falls
Slitting mills
Small business environment & product/workplace safety
Small elevation differences
Snap hooks
Snow blowers & throwers
Solvents
Spray bottles & atomizers
Spray guns & nozzles
Stairs & handrails/guardrails

Stapling & tacking machines/guns
State of the art versus state of the industry/state of the trade
Stationary power tools
Steel mill machinery
Stoves, domestic & camping
Static electricity explosions
Steering linkages
Storage racks & shelving
Structural defects and failures
Subcontractor failure
Sweepers, power
Swimming pools/holes & lounge chairs
Swing & gang saws
Swings, playground & backyard
Swinging cage-type amusement rides
Switchgear

Table saws
Tabletting presses & machinery for pharmaceuticals
Tamper-resistant controls & interlocks
Tamper-evident/fail-evident features
Tanker trucks
Teapots
Technological forecasting
Technology transfer
Television set fires
Tennis courts, nets & net winches
Tension control systems
Textile calenders & other textile machinery, processes, & products

Thermal failure
Three-wheeled vehicles
Timers
Tire defects, blowouts, other failures, explosions, & safety cages
Toilet bowls (water closets)
Towing equipment
Toxic chemicals
Toy guns
Toys, novelties, & projectiles
Tractors
Tractor-trailer rigs
Trade codes/standards/practices
Trade secrets & industrial espionage-type security breaches
Traffic signalization
Trailer hitches
Trailer homes & travel trailers
Tramways
Transformers
Transportation-damaged machinery & merchandise
Trash compactors
Treadmills
Tree-removal machinery
Tree trimming, pruning, & limbing
Trenching machinery & operations
Truck brakes
Truck cab access systems
Truck cargo lifts
Tube & pipe mills
Torque control
Turret lathes
Two-hand trips

Unfair competition & trade practices

U.S. Government codes, standards, regulations analysis, compliance, & violations
User-hostile/unfriendly/unforgiving designs
Utility lights
Utility vehicles

Valves, controls & instrumentation
Vapor & gas explosions
Vacuum cleaners & wet vacs
Vending machines
Vendor failure
Ventilating equipment & systems
Visibility studies
Vocational/employability/rehabilitation/psychological assessment

Walking hazards
Warnings, signs & labels
Water-damaged machinery & merchandise
Water pollution control
Web tension control systems
Welding, brazing, & flame-cutting processes & equipment
Welding spark ignition-caused fires & explosions
Wheel blocks/chocks
Wheel bumpers/stops
Wheelchairs & wheelchair lifts
Wheels & rims
Winches
Winding machinery
Windows
Window cleaning equipment
Window latches & locks, including vehicle

(continued)

EXHIBIT 11-6 (*Continued*)

Wire extruding machinery	Zero mechanical state (ZMS) safeties & lost motion
Wood chipping equipment	
Woodworking machinery	Zoning studies & zoning ordinance violations
Work feeding & holding tables, manipulators, positioners, lifts & platforms	
Wrongful birth & wrongful death financial & economic appraisals	YOUR INQUIRIES ARE INVITED IN STRICTEST CONFIDENCE

of assignments where engineering expertise has been critical in litigation involving personal injury, property damage, and economic loss.

Engagement time frames can vary all over. However, 97 percent of all the work you will probably ever see in a typical forensic practice falls within a range of three to fifteen hours. Lead times from five to ten hours will probably account for approximately 85 percent of the typical practice. Note that elapsed time from start to finish of an engagement might well be measured in days or months between making the original field inspection and completing your expert report. But the forensic assignment usually has a relatively shorter lead time requirement, making it manageable even for a full-time, predominantly design-oriented engineering practice, as long as forensic work does not become too large a percentage of the total activity. Otherwise, forensic demands can interfere seriously with other types of engineering projects.

The following summary of rules relating to plaintiff cases should be followed faithfully to avoid financial problems:

1. Require advance retainers for the full number of hours estimated for the case through the preparation of your expert report. Your estimates will grow more reliable as you gain more experience in forensic work.
2. Require that all plaintiff attorney checks be received at least two weeks prior to rendering of any and all engineering services. Insurance carriers and defense attorneys can usually be billed, with some exceptions.

3. Advise the client up front that if the case should become a "can of worms" and is open-ended, you will require follow-up retainers for prospective follow-up services to be rendered whenever you reach an appropriate milestone.
4. Do not accept personal checks from plaintiffs themselves. Require that any direct payments from plaintiffs be certified checks or money orders. In fact, as a general practice, do not accept an engagement from a plaintiff directly but only through his legal representative.
5. Require trial fee deposits for all prospective court appearances for plaintiff cases, to be received at least two weeks prior to trial. If a check cannot be forthcoming and delivered by your client within this time period, then require that it be a certified check or money order, as above, unless it will be an attorney trustee or trust account check. Treat an attorney business check (as opposed to a trustee account check) as an ordinary personal check. Payment can be stopped within a time period as long as two weeks after it is deposited by you. Don't get stung by one of those few unethical lawyers who make it tough for all the others.
6. Generally speaking, attempt to establish flat casework fees for work that can be more or less routinized or standardized, and set fees accordingly. Obtain advance retainers in all cases, as noted above.
7. Bill for all travel time, conferencing time, and so on, but attempt to build certain allowances into your basic fee and/or retainer. You will find it impractical and virtually impossible to invoice and collect for isolated bits and pieces of time expended on a case. Therefore, carefully assess casework requirements in advance and require follow-up retainers for prospective follow-up casework.
8. Do not bill for deposition preparation time. You won't collect. Design your reports so that they are easy and quick to review for depositions and trial appearances.
9. Always return unexpended portions of trial fee deposits promptly. You must respect the fiduciary relationship you establish with clients when you require and receive such advance retainers and deposits.
10. Since the first trial date will usually be adjourned one or more times, perhaps over a period of weeks or months, you should make it known to clients that trial fee deposits should be forwarded to you only when they have a trial date that looks

reasonably firm. A peremptory date, for example, is in this category. This is a court-ordered trial date, set by a judge, usually to avoid any further delays in bringing the case to trial. However, peremptory trial dates are not common in civil litigation.

11. Do not bill for standby time when you are placed on call for a prospective trial appearance. Don't stand by—you can't afford it. Just stay in touch with the client office so you can respond reasonably promptly, and go about your business. Normally, your client and the court will try to work around you, as with physicians.

12. Keep a trial fee deposit log as part of your bookkeeping system. Due to the use of the trial deposit method, you must have a perfectly effective and efficient handle on these transactions, otherwise these checks will get lost in the shuffle and you can wind up with a lot of ill will on your hands. Use the type of form shown in Exhibit 11-7. Since trial fee deposits are not yours until earned, you might want to keep them in a special escrow-like non-interest-bearing checking account. The total amount in this account will never be very large, but matters could become hopelessly complex if you don't keep the kind of record shown in the exhibit.

If an attorney's trustee account check bounces, whether it is made out to you or anybody else, only God can help him. But if his commercial account check bounces (and this can happen up to two weeks after issuance), call the client to find out what happened. More often than not he will tell you to redeposit it and that will be the end of the problem. As a matter of fact, if and when you receive a client check back, it will usually be because he has forgotten to sign it.

But if a client's commercial check bounces a second time, call him and warn him you will file an ethics complaint. If you cannot obtain satisfaction from the client, acquire a grievance form from the appropriate agency in your state, fill it out, and mail it in.

As an example, Exhibit 11-8 shows an Attorney Grievance Form available to complainants in the state of New Jersey. Note particularly Section E regarding your own immunity relative to non-disclosure of the ethics grievance to parties outside the attorney disciplinary system, excepting your own attorney and/or witnesses you may call to testify in your behalf.

If you are careful in the conduct of your practice, and do not expose yourself to the potential for financial losses, then your down-

EXHIBIT 11-7

TRIAL DEPOSIT LOG

CLIENT/CASE	DATE			AMOUNT			REMARKS
	Received	Returned	Earned	Received	Returned	Earned	

EXHIBIT 11-8

ATTORNEY GRIEVANCE FORM

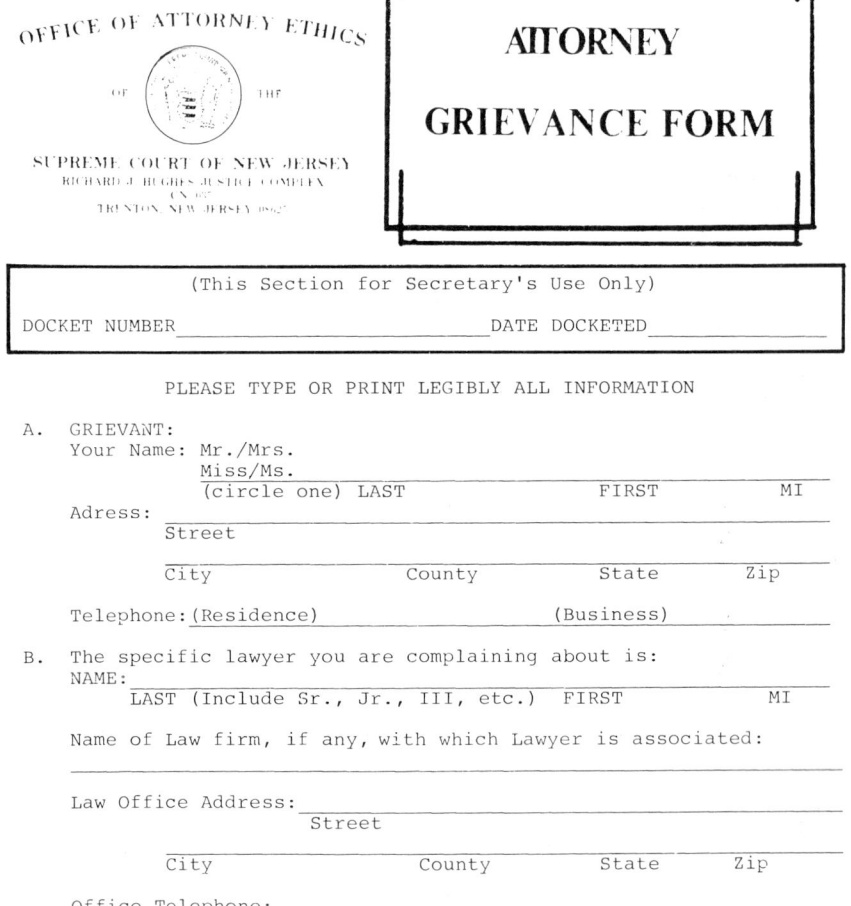

(This Section for Secretary's Use Only)
DOCKET NUMBER_____DATE DOCKETED_____

PLEASE TYPE OR PRINT LEGIBLY ALL INFORMATION

A. GRIEVANT:
    Your Name: Mr./Mrs.
               Miss/Ms.
               (circle one)   LAST   FIRST   MI
    Adress: _____
            Street

            _____
            City   County   State   Zip

    Telephone:(Residence)_____(Business)_____

B. The specific lawyer you are complaining about is:
    NAME: _____
          LAST (Include Sr., Jr., III, etc.)   FIRST   MI

    Name of Law firm, if any, with which Lawyer is associated:
    _____

    Law Office Address: _____
                        Street

                        _____
                        City   County   State   Zip

    Office Telephone: _____

C. NATURE OF GRIEVANCE: State what the lawyer did or failed to do
    which you believe may subject the lawyer to discipline. Include
    all relevant <u>FACTS</u> surrounding your grievance including dates,
    times, places, and names and addresses of important persons
    involved in your case, the amount of fees or retainers paid, and,
    if litigation is involved, give the name of the court and docket
    number. Attach additional pages as necessary.

    _____

    _____

    _____

◄CONTINUED ON NEXT PAGE►

C. (continued)

_____
_____
_____
_____
_____
_____
_____
_____

D. The TYPE OF CASE handled by the lawyer was: (check one)

__(A) Adoption
__(B) Bankruptcy/Insolvency
__(C) Criminal and quasi-criminal (Municipal Court)
__(D) Domestic Relations (Matrimonial, Support, Custody)
__(E) Estate/Probate
__(F) Federal Remedies/Civil Rights
__(G) Government Agency Problems (Local, County, State, Federal)
__(H) Collection Matter
__(I) International Law
__(J) Juvenile Delinquency Matters
__(K) Contract
__(L) Labor
__(M) Immigration/Naturalization
__(N) Negligence (Personal Injury/Property Damage)
__(P) Patent/Trademark/Copyright
__(Q) Landlord/Tenant
__(R) Real Estate
__(S) Small Claims Court
__(T) Tax
__(V) Admiralty/Maritime
__(W) Workers Compensation
__(X) Corporation Law
__(Y) Other Litigation
__(Z) Other Non Litigation

E. CONFIDENTIALITY AND IMMUNITY:

You are advised that, under Supreme Court Rule 1:20-10, once you file this Attorney Grievance Form you are REQUIRED thereafter to keep all communications regarding this ethics matter CONFIDENTIAL. You may not breach this confidentiality by disclosing your ethics grievance to persons other than members of the attorney disciplinary system, except to discuss the case with other witnesses or to consult an attorney. So long as you maintain the confidentiality of these proceedings, Supreme Court Rule 1:20-11(b) grants you immunity from law suits with this state as a result of filing your grievance. <u>If you breach this rule of confidentiality</u> YOU WILL LOSE THIS IMMUNITY.

_____   _____
         Date   Signature

OAE-G3
8/1/84

side risk should be minimal. Over a ten-year period, for example, your losses should be less than 1 percent, if that. The trick is to control the quality of your clients and the quality of the engagements. Don't be afraid to decline engagements that appear potentially problematic, or to turn off prospective clients who approach you in a less than totally open and professional manner. A willingness to meet your reasonable fee requirements and arrangements can be one important factor in assessing whether the prospective client is approaching you in good faith.

As a general rule, charge by the half day for nearby court appearances, and on a per diem basis for more remote appearances. The morning session should be valued more than the afternoon due to preparation time, if nothing else. For example, if you charge $1,000 for a full-day trial appearance, then charge $750 for a half day, whether morning or afternoon session. If the trial spills over from the prior day's half-day appearance to the next day, but only for a half day, then the second half day is worth $750 also. Therefore, encourage your client to start you in the morning. A single full-day appearance, consisting of morning and afternoon sessions, will cost him less than, for example, an afternoon (at $750) and the following morning (also at $750). Overnight, incidentally, you will be boning up some more and trying to review the prior day's testimony, so you will have additional prep time. It also gives your client's adversary an excellent opportunity to review your testimony.

This method of charging tends to load round-trip travel time into the first half day of testimony, whether it is the morning or afternoon session, which is entirely realistic for the more local appearances that lend themselves to half-day billing.

Incidentally, one of the important reasons for requiring retainers up front, trial fee deposits, and other advance payments is that as you become busier, it will be a good way to establish priorities as to who gets serviced. You will have many clients with current needs. Your priorities will be easier to establish if you simply provide services first to those clients who have reserved your time on a serious basis by forwarding retainers, trial deposits, and so on.

Another reason for requiring trial fee deposits in advance of any and all court appearances (and in sufficient time for the checks to clear), is as follows. As mentioned previously, probably only 3 to 5 percent of all cases wind up going to trial; the rest settle out of court. But of those that are tried, the chances are only fifty-fifty that either side will "win." Therefore, the percentages are not overwhelming that your own client will be on the winning side, no matter how good your testimony may have been.

If you do not have your trial fee in hand by the time of your appearance, and if your client should not prevail in court despite your professional, competent, ethical expert testimony, you will be fair game for negotiation. That is, your client may be one of those attorneys who will attempt to beat you over the head in an effort to have you either reduce your fee or even "forgive" your fee because he lost and his client cannot really afford to pay your trial fee, or because he would have to foot the bill himself and he would like to feel you will "work with him" so he can come to you on future cases. Stand firm. You earned your fee. And when you have a retainer or trial deposit in hand, this game is precluded.

Lawyers often are, by training and temper, inveterate and incorrigible negotiators. They don't always recognize this and you don't want to hurt their feelings by bringing their attention to it. Therefore, be gentle but firm when it comes to your fees; simply refuse to negotiate. They may wonder why all their negotiating expertise isn't working, but they will come around to your way of doing business if you are insistent and persevere through their expert efforts to negotiate your fees downward. As usual, you must use judgment; but if you submit, once in a while you may be stung when you least expect it, and when the stakes are high. Negotiation may also take the form of arrangements as to *when* you get paid. Here, too, keep your guard up.

TWILIGHT ZONE

There are some less well known aspects to the high-powered world of litigation and forensic practice. For any one of a number of reasons, plaintiff attorneys may decline to pursue cases after spending a significant amount of time screening them. In some instances an engineer is consulted to determine whether or not the client attorney should proceed at all. My associates and I personally decline or discourage an average of maybe 5 to 10 percent of all the cases presented to us every year due to such a screening process. In most instances a full-fee basis is entirely in order.

There are other little-known classes of "engagements." In one type of situation the defendant will be an attorney or law firm. Perhaps the individual or firm did not file a case properly within the time period allowed under the statute of limitations, or perhaps the case just got lost between the cracks. Or maybe it is believed by the plaintiff that the lawyer or firm did not handle the matter competently or prepare the case diligently, or didn't name an important defendant.

In these instances an engineer will be consulted by the plaintiff attorney to determine whether, in fact, there had been a bona fide cause of action from an engineering standpoint if there was personal injury, property damage, or other economic loss. If so, then the plaintiff will attempt to show, through his present attorney, that there originally had been a viable case but that the opportunity was lost or forgone to sue the original defendant responsible for the injury or other loss. The first attorney or law firm then becomes vulnerable to a malpractice action, although obviously there is no culpability for having caused the injury or economic loss per se in the first place.

In another type of situation an attorney may want to reinforce his decision to decline a case due to some inherent engineering-related weakness. In such a circumstance he may not really need an engineering expert, but he does not want to take the risk of unilaterally telling a prospective client, who may even be a friend or member of the family, that he recommends against proceeding. So he gets the blessings of his engineer to buttress his own opinion and decision. In effect, he "blames" his engineer for the negative assessment of the case and gets the client out of his hair and himself off the hook. This is fine. In such instances he will probably tell the prospective client that he would not be hurt in the least if the client wished to go to another lawyer for a second opinion, but his engineer said there was no case.

Finally, there may be what can be termed "fortuitous convergence." Your client forgot to file within the statute of limitations; however, it turns out that it really wasn't a good case. To fend off his client's threats with respect to a possible malpractice action, the attorney engages you to evaluate the case from an engineering standpoint. In some instances you may come back with findings and opinions supporting his suspicion that there was no case to start with, thus, there was no damage done even though the attorney erred. Here again, he will probably advise his disgruntled client to get another attorney and another expert, if only so the client can be convinced of the nonmeritorious potential of the original case.

At the limit, the client may sue his attorney. Then, with you as his own expert, the attorney will be able to show (it is hoped) that there really was no case to start with despite attorney error in not having timely filed, and so on.

Some of the aforementioned problems can be handled by a simple letter that doesn't involve any significant time on your part. If so, don't charge. It's just a little more client development, though unique in nature. Naturally, you will keep a file on the case. I have personally handled many cases of this type over the years. A simple letter from an engineer can sometimes help a lawyer get out of a tight spot quite gracefully, yet competently, ethically, and professionally.

In all of the above instances the engineer is being called upon to provide expert input. As long as you never compromise your ethics or professionalism, there is absolutely nothing wrong with accepting "mini-engagements" of the types described. Again, in some instances it could be a full-fledged engagement with respect to structure, inspections, reports, testimony, fees, and so on. In other instances you probably should not charge but merely consider a short letter or memorandum public relations. An appreciative client might well be quite willing to pay you a fee of a few hundred dollars, on the other hand, for your trouble. Just be sure you keep the game honest, as always, if you accept a fee. At either extreme, your expertise can be invaluable and can help resolve a dispute or even preclude a lawsuit.

CHAPTER 12

THE FORENSIC CONSULTING BATTLEGROUND

THE BATTLE OF THE EXPERTS: FACT OR FICTION?

Expert testimony plays an increasingly pivotal role in both negligence and products liability litigation for both plaintiff and defendant. But is there really such a thing as a battle of the experts?

Application of engineering science expertise in products liability litigation is an important and growing field of activity, and one in which significant contributions to human well-being and quality of life may be made by the engineering community. It is also one of the more visible activities of engineers.

However, for the typical professional engineering practitioner—and for most people—the courtroom environment is alien and intimidating because it is fundamentally adversarial in nature. The so-called "battle of the experts" remains a favorite phrase erroneously used by the uninformed. It is also tactically applied by laypersons and even by many lawyers to describe apparent differences in technical testimony and to convey the impression that one side is going to prevail and, therefore, is "right" while the other side is "wrong."

Differences of opinion among experts are, in reality, merely the results of one or more of the following factors, where the Greek letter delta (Δ) is used as usual to denote a change, difference, increment (increase), or decrement (decrease):

$\Delta 1$ Differences between ethical practice and unethical or other unprofessional motivations of some people undeserving of the term "expert," "professional" practitioner, or engineer.

$\Delta 2$ Differences in actual, assumed, or given facts, including limitations on background information available to each expert.

$\Delta 3$ Differences between competent and incompetent analysis of identical facts.

$\Delta 4$ Differences between competent and incompetent testing and experimental procedures dealing with "known art."

$\Delta 5$ Differences between competent and incompetent analysis of test and experimental data dealing with "known art."

$\Delta 6$ Differences between analytical procedures, taking into account both facts and theoretical considerations versus those based solely upon theory or mathematical analysis, and so on, to the essential exclusion or omission of factual or "real world" bases for the analysis.

$\Delta 7$ Differences between state of the art (intended to refer to the applied or known art as opposed to experimental art at the "cutting edge" of science and technology) and state of the industry or state of the trade.

$\Delta 8$ Differences between technically and economically feasible product designs based upon body of knowledge and performance standards, guidelines, philosophies, practices, and procedures, and so on, versus designs complying solely with published codes and industry or trade specification-type standards, and so on.

$\Delta 9$ Differences between correct and incorrect assumptions relating to retrospective contemporary ethical mandates controlling underlying, conflicting design philosophies at issue.

$\Delta 10$ Differences in amount of preparation by the various experts. This may be due to engagement or casework constraints,

The Forensic Consulting Battleground 259

or limitations on analysis or investigation scope imposed by client financial or other limitations or directives, resulting in necessarily limited or qualified opinions.

Δ11 Differences in admissibility of expert opinions as to limitations on scope or content imposed by the court, thereby artificially creating apparent net opinion differences or even ruling out some opinions altogether.

Δ12 Differences among interpretations and applications of R&D (Research & Development) data or new technology and experimentation at the "cutting edge" of science with minimal "known art" content.

Δ13 Differences in scientific accident reconstruction scenarios in cases involving product "failure," component breakage, malfunction, "mysterious" actuation of control circuits, and so on, where competent scientific, engineering, or mathematical analysis, laboratory testing, and the like lend themselves to alternative interpretations, or where insufficient or fragmentary background data and witness statements do not permit conclusive opinions to be offered by any expert involved. In such instances there may well be alternative reconstruction hypotheses that are consistent with available data and/or factual information which lead, in turn, to possible findings and opinions at variance with one another as to either or both causality determination and responsible entity identification. In such cases, also, product or system "guarding" per se may not be an issue, so that product failure is, by itself, the critical issue.

On balance, with the exception of expert opinion variances arising from Factors 12 and 13, and with all other things being equal (which they almost never are), expert opinions should be, of necessity and rigorously speaking, identical in any particular case. The fact that opinions may be made to appear different in a specific instance for plaintiff versus defendant is an illusion having more to do with legalistic legerdemain and virtuosity than inherent engineering or technological substance.

"Battle of the experts" is thus an insidious turn of phrase. If not altogether designed to confuse and confound laypeople and juries, and to undermine subtly the inherently superior position in order to gain an equal hand for the inferior, it is fully capable of doing so. However, if sufficient information is available so that seemingly divergent expert opinions can be analyzed using the foregoing thirteen factors as criteria for evaluation, more intelligent and balanced assessment of frequently illusory disagreements can be made. Recognition of the "quality" of variances can then be utilized to formulate more rational and considered viewpoints, facilitating selection or rejection of one expert opinion as opposed to another.

If we were to create a mathematical expression or formula to express the apparent difference between two opposing sets of testimony, it would read something like this:

$$\text{Total apparent testimony difference/conflict} = \text{The sum of factors } \Delta 1 \text{ through } \Delta 13 = 100\%$$

where the value of some factors would be zero, while the values for others would be given other weights.

The phrase "battle of the experts" is a misnomer that does grave injustice to the cause of science. Science is not biased or prejudiced. It has no adversary, no politics, no morality. It is merely the cumulative, organized natural knowledge of the ages. It illuminates where before there may have been little light.

However, engineering science can be purposefully misused or incorrectly and ineptly applied in practice. The lay observer cannot always recogize the difference between competent and incompetent analysis, ethical or unethical practice, and real or illusory expert opinions. Unfortunately, this provides opportunities for abuse and unprofessional behavior.

A juror, upon hearing two apparently different sets of testimony from two evidently well-qualified "experts," may well conclude that the real difference is merely the result of two opposing philosophies of problem solving. The juror is simply not in a position to realize that one expert is upholding the highest ideals of his or her profession and ethically applying the single body of knowledge that exists, while the other may actually be providing unethical, nonobjective, and/or incompetent testimony.

There really is no such thing as a battle of the experts. There is only a private war in a public arena, fiercely waged between plaintiff and defendant under the rules of court and legal practice. Ideally, this contest objectively utilizes ethically presented engineering science to shed light on issues otherwise too intricate or specialized for lay jury evaluation.

In this sense, the expert engineer should be no more than a catalyst whose obligation and mission is to educate and dispel darkness in the search for truth. The tactic of pretending that there is some battle of experts is merely a transparent ploy designed to benefit a questionable, unmeritorious, or even sophistical argument. Ideally, the observer or analyst should not be swayed by mere force of dialectics or dramatics. He or she must be alert and on guard at all times to assure that the objectively most meritorious position pre-

The Forensic Consulting Battleground

EXHIBIT 12-1

EXPERT OPINION RATING AND RECONCILIATION WORK CHART

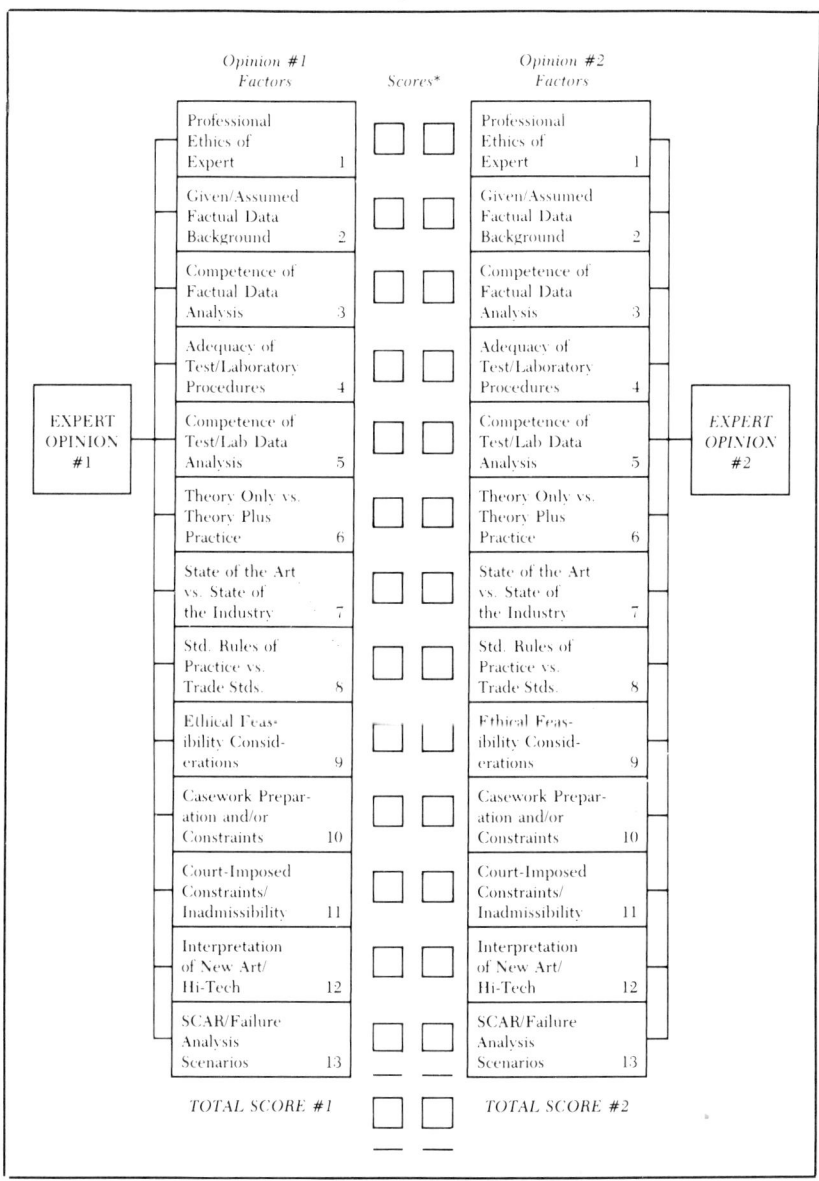

(*) Score 0–10 or N.A. (Not Applicable) for each factor. Only score opinion elements actually/potentially critical to court/jury deliberations. This chart is *not* intended for non-verbal communication forms such as body language.

vails. Unfortunately, the rules of the game are defective, and this does not occur often enough.

Exhibit 12-1 is a work chart you can use to rate and reconcile testimony of plaintiff and defense experts. It follows the formula factors presented earlier in this section. You can score each factor for each expert in any way that is meaningful for you, such as from 1 to 10 or "Not Applicable" (NA).

THE "COURT-WISE" EXPERT: IMAGE VERSUS SUBSTANCE

The engineer who engages in forensic practice to any significant extent will sooner or later be required to provide testimony at oral depositions and testify in court. As a practical matter, probably 97 or 98 out of every 100 cases settle out of court. Therefore, on average, actual court appearances will be required for only about 2 or 3 percent of total forensic casework load. Of the cases that go to trial, probably about 50 percent win and 50 percent lose.

A large individual forensic practice may well result in an average of perhaps one or more court appearances a month. Over a period of years, an active forensic practitioner can acquire a great deal of useful experience and knowledge as to how to conduct and present himself acceptably, credibly, and professionally in the courtroom.

Unfortunately, such experience is sometimes derided by the press, the lay public, and even lawyers as being "court-wise." There is the intimation that the engineer is automatically unethical; that he or she can be "bought"; that he or she will offer opinions on anything, even if it is out of his or her field of specialization; that the engineer will say whatever he or she thinks the client wants to hear; that he or she will try to "snow" the court and the jury; that he or she is unprofessional and dishonorable.

In over a decade of forensic practice, I have seen many unethical engineering reports. On excellent authority, I have heard of unethical testimony and been confronted with engineering arguments that are so incompetent as to verge on unethical practice by reason of totally irresponsible or unprofessional work product alone. I have also personally witnessed engineers practicing well out of their fields.

In the forensic engineering field, in my opinion, a significant part of the criticism sometimes leveled at the legal community with respect to encouraging nonmeritorious products liability lawsuits and products logjams in the courts is due to engineering reports and opinions that never should have been written or offered in good conscience. Such reports only prolong and increase the costs of litigation.

Thus, the image of the engineering community has become tarnished. The expert engineer is sometimes viewed automatically as being corrupt, spoiled, or unscrupulous.

What worsens matters, as the experienced trial attorney will quickly tell you, is that a "court-wise" expert without the highest credentials can be vastly superior and preferable to the impeccably credentialed witness without courtroom charisma.

Meanwhile, how has the engineering profession responded to all this? Although forensic practice requires an extraordinarily high degree of professional expertise when ethically and competently practiced, the forensic engineer is not well regulated. His or her nonforensic consulting colleagues are better controlled. Even lawyers cannot always recognize unethical practice, or may prefer not to! They believe that a true "battle of the experts" is in progress.

But the ethical forensic engineering practitioner takes a pragmatic and unruffled viewpoint. He knows the ground rules and applies the accepted body of knowledge in his particular specialties to the full extent ethically feasible. His court-wise demeanor permits him to retain his professional equilibrium and integrity even when forcefully and rudely attacked by cross-examining counsel. His court-wise manner permits him to cut through and neutralize the irrelevancies, smokescreens, sophistical arguments, insults, condescensions, acrimony, attempts at entrapment, dramatics, bluffing, bullying, impertinence, and other unpleasantries sometimes leveled at him by a cross-examiner who, by careful design, may be merely exhibiting his killer instinct in an effort to destroy the expert's credibility under questioning. He wants to win for his own client and, at the least, to unnerve the expert sufficiently to prevent cool-headed, collected, professional responses. The lawyer, of course, is merely practicing his art, and he is accorded wide latitude in doing just this.

The true substance of ethically applied court-wise demeanor, when combined with competent testimony, permits professional engineering services to be provided on the highest level. Since the reputation of the entire engineering profession is, in a real sense, on trial at every court appearance, the ethical and competent engineer in forensic practice who is also court-wise is actually in the best position to abide by his professional code of ethics and present his arguments fairly and squarely with dignity, confidence, and pride of accomplishment. He will not permit himself to fall prey to the sometimes ruthless tactics artfully utilized during cross-examination. He will always strive to be in control of himself and his testimony, despite the fact that he is at all times actually playing in someone else's ball game. Cross-examining counsel will be unable to demean him or, through him, his profession, or impugn the integrity of being a true professional in his field and the technical proficiency he has in his subject matter.

Depending upon the individual, of course, the character of the court-wise engineering expert or witness can be one of two kinds:

1. A true professional upholding the highest ideals and standards of his chosen field, or
2. An unscrupulous rogue who is no more professional than a mercenary, despite technical competence that could be used ethically, but is not.

On balance, achieving court-wise demeanor is in reality an enviable accomplishment that facilitates and ennobles, rather than impedes and discredits, both the judicial process and forensic engineering practice. It is an achievement that permits the experienced and ethical forensic engineer to serve his own profession honorably, competently, and with distinction, thereby serving the best interests of public safety, health, and welfare as specified and mandated by the Professional Code of Ethics for Engineers.

In addition, there remains a significant need for the engineering profession to practice more effective self-regulation with respect to forensic engineering practice. The legal community simply has not achieved a sufficiently effective measure of control over engineering involvement in the litigation process to assure that the highest levels of professionalism in forensic engineering practice are brought to bear at all times by all parties to a controversy.

Therefore, the expert engineer must be ever alert for lawyers' sometimes repugnant cross-examination tactics, which can seriously undermine and detract from an engineer's sound arguments and testimony if not correctly and promptly perceived and brought to the attention of the court and the jury.

On the other hand, professional persons who serve as expert witnesses must not permit the prospect of occasionally repulsive tactics and public behavior of cross-examining counsels to deter them from providing the benefit of their training and experience to the court and jury in the pursuit and administration of justice. The legal process cannot be successful without the presence and counsel of the expert witness. Expert testimony is a social and legal process. The merits of many cases can rise or fall due to engineering arguments.

Therefore, when the occasion arises, professional men and

women have a solemn obligation to participate in the legal process to the best of their abilities, notwithstanding the sometimes distasteful but artful conduct of counsels on cross-examination. The public must not be deprived of the engineer's indispensable services.

An important principle of forensic engineering practice is that the only way to protect expert opinions so that they survive the courtroom jungle is for the expert to protect himself or herself. Thus, he or she must learn to be court-wise. Remember, however, that the expert witness is not a lawyer. Even the engineer experienced in forensic practice may be vulnerable to making a slip now and then if he or she is assumed to possess a higher degree of legal knowledge than is really the case, even though a certain amount of legal "information" is acquired by osmosis.

The expert engineer has a profound obligation to help keep the litigation process honest. By understanding the difficult environment in which testimony is proffered, the prospective expert can better protect the integrity of his or her actual testimony. To do this, he or she must understand how cross-examination is specifically and expertly calculated to destroy expert testimony—sometimes by attempting to destroy the expert in the bargain.

Thus, it is not enough for the forensic consultant to be competent, responsible, ethical, and forthright. He or she must also be court-wise if expert testimony is to be meaningful and effective. The expert witness must constantly and diligently strive to overcome a wide variety of artful cross-examination tactics and strategies. Expert testimony must come through loud and clear, liberated and extricated from often brutal attempts to impugn the credibility of both testimony and testifier alike. These attempts are intended to mask the constructive essence of expert testimony, thus confounding, diverting, and, it is hoped, swaying the jury.

A court-wise forensic engineer will be better able to afford the court and jury an equitable opportunity to hear and understand the testimony as intended. He or she will no longer be at the mercy of the cross-examiner. Even the client may not always be in total and effective control of a cross-examination situation. A cross-examining attorney has many ways to take advantage of an unsuspecting witness. The forensic consultant must learn how to deal with such cross-examination tactics while remaining totally honest, impartial, professional, courteous, and unperturbed during his or her presentation.

Becoming court-wise is a necessary skill acquired primarily through experience, and includes understanding the forensic environment. What has been said so far, and what follows in the rest of this chapter, will give an excellent head start and foundation upon which to build an ethical, successful, effective, and exciting forensic practice.

WHAT YOU ALWAYS WANTED TO KNOW ABOUT CROSS-EXAMINATION BUT WERE AFRAID TO ASK (AND NOBODY WOULD TELL YOU)

For the engineer engaged in forensic practice, in a very real sense the bottom line is cross-examination. It is here that all aspects of your casework and professional efforts converge, are laid bare, and are subjected to the most critical scrutiny. Consequently, it is absolutely essential to understand and anticipate the types of questions that will be leveled at you.

Your expert report must be viewed as a trial report; it should really be written for cross-examination. If you have succeeded in preparing a report whose contents can withstand the rigors of cross-examination, it will possess utility in facilitating out-of-court settlement. But if your report raises more questions than it answers, or reflects inadequate analysis, preparation, understanding of the problem, inadequate background, and so on, you will make yourself vulnerable to sharp and effective attack at either or both deposition and trial. Naturally, of course, if you are vulnerable you put your client in an equally weak position.

An effective and efficient report that has high settlement value is only one important step toward resolution of the case. You may still have to defend your position at time of trial. Without an adequate trial report designed to withstand cross-examination, you and your client might just as well call it quits at the very beginning. But the game is not over just because you have produced a viable report. This is only your ticket to the next lap of your journey through the forensic thicket.

There are a number of general categories of cross-examination questions; you must be prepared for all of them, no matter how distasteful and impertinent they may be. They are as follows:

1. Questions that attempt character assassination—of *your* character.
2. Questions that attack your professional credentials, educational background, and the extent of your participation in professional and trade associations.
3. Questions attacking quantity, quality, or relevance of your experience. You will be maligned, if you are a university profes-

sor, for having no practical experience. But you will also be attacked if you have spent your whole career in industry without teaching. You can't win.
4. Questions imputing bias or prejudice to the expert (you) and attacking your ethics.
5. Questions attacking you as a "professional" witness, "mercenary," "whore," "gunslinger," or "hired gun."
6. Questions attacking or testing your competence, expertise, opinions, computations, analysis, and even report format.
7. Questions attempting to demean your professionalism, style, and dress.
8. Questions attacking your casework methods, procedures, amount of preparation, adequacy of supporting materials, documentation, and sources.
9. Questions attacking expert opinions or positions as avant-garde, outdated, superficial, irrelevant, illogical, overly intricate, and so on.
10. Questions attacking a plaintiff expert opinion or position where a product complies with all published codes, standards, and laws but the expert maintains that the product is unsafe anyway. Conversely, questions attacking defense of product hazards based solely on code compliance, if the defense expert relies on code compliance as the ultimate test of product safety.
11. Questions subjecting you to general ridicule, disparagement, intimidation, rudeness, and impertinence in order to unnerve and provoke you into an emotional response or outburst. Typically, such an uncontrolled behavior pattern permits the cross-examiner to pounce upon and attempt to discredit the testimony.
12. Questions that attempt to set you up, keep you off balance, catch you off guard, wear you down, or put words in your mouth.
13. Questions attacking the factual background upon which you depended, and forcing you to analyze alternative accident reconstruction scenarios, even if not in evidence or of very low probability.
14. Questions that nibble away at your expert opinion by moving testimony in the direction of the cross-examiner position. This strategy can be characterized as "incremental opinion reconciliation."
15. Questions of a "fishing expedition" nature, although these are usually found only at deposition proceedings.

16. Questions attacking you on the basis of nonsubstantive legal technicalities and multiple meanings of words and their rule-making power. You are accused of usurping legal prerogatives of the court. For example, you should never use the term "negligence," which is reserved for lawyers, but rather, "dereliction." Semantic subtleties can be a formidable ally when the cross-examiner is out to get you.
17. Questions built upon an array of fallacious arguments, twisting the meanings of your words or thoughts so they can be criticized by the cross-examiner to suit his or her purpose.
18. Questions attacking the contents of materials contained in your file. Always look through your file prior to deposition or trial to be certain that you have only critical materials. Prior to any proceeding at which you are going to testify, ask your client to check your file for materials that he or she might not wish to have readily accessible to adversaries. In some instances your client may have sent you background information that represents his or her legally privileged, private, nondiscoverable "work product," or other types of materials that he or she can legally withhold from adversaries.
19. Questions designed to retain or regain control of the cross-examination. The last thing an attorney can afford is to have control of the examination wrested from him or her, for whatever reason, by the expert witness.
20. Questions designed to grasp an opportunity, or even momentarily forgo an advantage, until a more opportune or vulnerable point is reached during cross-examination of the expert.
21. Questions designed to attempt to impeach expert testimony due to alleged contradictions between testimony obtained at a prior deposition (as documented in the deposition transcript) and what has been said during either direct or cross-examination at trial, or even what you have testified to in prior depositions in totally unrelated cases.
22. Compound questions that really consist of two or more separate questions rolled into one, and to which a simple "yes" or "no" is not feasible, even though the cross-examiner attempts to put you under great pressure to answer with a simple "yes" or "no."
23. Questions that are based upon facts or assumptions not in evidence, or that are incorrect or irrelevant, or represent special cases and which, therefore, are really a special kind of compound question with multiple parts that should be addressed one by one. Such questions cannot be answered with a simple

"yes" or "no," but may require considerable development before you can offer an answer.
24. Questions that should elicit an "I don't know" answer from the expert may be so skillfully designed that they lead you astray and into areas outside your specialty. Naturally, the cross-examiner will summarily swoop down on you and tear you to shreds by trying to show that you habitually practice outside your specialty and, therefore, are incompetent and not credible.
25. Questions otherwise skillfully designed to trick you. And so on.

It is very important that you steel yourself to the fact that nothing personal is intended in any of the attacks made upon you by cross-examining counsel. He or she is merely attempting to protect the best interests of his or her own client, and is accorded great latitude by the court in doing precisely this, as already noted. You just happen to be in the way. You will become accustomed to getting caught in a cross-fire. If you can develop a very thick skin and hard stomach you will be better able to cope with the abrasive tactics of cross-examination. Do not take things personally. As observed above, one cross-examination tactic will be to attempt to unnerve you into becoming defensive and hostile so you will become emotional and try to lash back. *Don't.* Stay cool.

Compile a catalog of questions from your personal experience or that of your colleagues and any other sources you may come upon over the years. If you have any friends or acquaintances who are trial attorneys, ask them point-blank what are their favorite questions that are designed to unnerve or discredit experts on cross-examination. You will probably be shocked at the immediacy and calculating nature of their reply, if they will tell you the truth. This indicates that the subject is common knowledge among attorneys and very important to the trial process.

The only way you will be able to present your opinions fairly is to first understand the rules of the game. You must learn how to protect and defend yourself because, on cross-examination, you are susceptible to being attacked relentlessly, ruthlessly, and repeatedly.

This is the way the game is played. Repugnant as it may be, it merely reflects the ultimate law of the courtroom jungle. Unfortunately, right does not always make might, even in the courtroom, especially if you don't know how to defend yourself and your opinions. Cross-examination is, by its very nature, a quintessential example of predatory behavior since it is part of a "war game" where the judge is arbiter of the law, and the jury is arbiter of the facts.

Remember that only part of your effectiveness depends on your qualifications, competence, and experience. An equally great asset in forensic engineering lies in being court-wise, as we have discussed. This permits your opinions to get through to the jury the way you intended. One of the distinctions and satisfactions of forensic practice is that you are afforded a continuing and highly visible opportunity to defend findings and opinions that you believe in enough to protect in "open combat."

Questionable ethics and competence will ultimately be exposed for all the world to see. The forensic practice of engineering occurs in the real world, in a predatory arena where survival of the fittest is the first principle. Your opinions, first and foremost, must survive. But, if they are to survive, you must survive. You must become court-wise to protect your opinions, as well as yourself.

There are thirteen (and more) deadly sins of giving expert testimony. Make a copy of this list, paste it on a card, and keep it in your wallet whenever you go to court.

The First Thirteen Deadly Sins of Giving Expert Testimony

1. Don't be argumentative.
2. Don't interrupt.
3. Don't smile, laugh, or joke.
4. Don't be verbose or preachy.
5. Don't volunteer information.
6. Don't exaggerate.
7. Don't use jargon.
8. Don't lose your cool.
9. Don't be unresponsive.
10. Don't open mouth before engaging mind.
11. Don't second-guess your cross-examiner.
12. Don't underestimate your cross-examiner.
13. Don't be an advocate.

When in court you should dress in a neat and businesslike manner. Be well-groomed. It is best to be unobtrusive and restrained, and to avoid eccentricities and extremes in attire and behavior. Naturally, you should be prompt, courteous, modest, and maintain a professional demeanor. You should speak clearly and forcefully, be respectful and dispassionate. You can be cordial without being flippant. Pay attention to the proceedings. Don't engage in conversations or whispering with persons around you. Be aware of your "body language." Get a good night's sleep before the day of your testimony. Relax. Do not talk to jurors, although you can be courteous to them in chance encounters, such as in the elevator, lunchroom, restaurant, or washroom. Try not to enter or leave the courtroom too many times. Speak *to* jurors rather than *at* them during your testimony. Try to maintain reasonable eye contact with both your client and the cross-examiner. If you see a juror sleeping, maybe it's your fault—but let the judge wake him or her up and do the scolding. And don't be afraid to say "I don't know," if you don't know the answer to a question.

Above all, you must learn to be court-wise and have an acceptable courtroom presence. Otherwise you will be "eaten alive" by cross-examining counsel. Your client can sometimes resurrect or rehabilitate problem situations and testimony later, such as during a redirect examination. But there will have been a significant lapse of time with the potential for loss of continuity, focus, emphasis, relevancy, and concentration by counsels, jury, and judge alike.

Another way you can train yourself in courtroom practices and procedures is by sitting in on trials now and then. If you know of a colleague who does forensic work, ask to be informed when he or she has to testify at a trial. If you can, take a day off to attend. You'll learn a great deal from such an experience. You will observe what the courtroom and courthouse atmospheres are like, and more important, you will have the opportunity to hear a live proceeding and the question-and-answer process leading up to the jury decision.

As a practical matter, the trial question-and-answer process is nothing more than a specialized, qualitative form of problem solving for purposes of legal decision making. It has analogs in engineering methodology. The trial process is simply less rigorous than its engineering cousins, even though it is controlled and circumscribed by many rules. And it is less dependable, effective, and efficient in terms of exposing all necessary and sufficient issues and arguments. Also, of course, the adversarial atmosphere precludes realization of the benefits of collaboration among experts. Instead, controversy is promoted, which is not necessarily the most fruitful methodology where engineering problem solving is involved.

After three trial appearances you will have acquired roughly 20 to 40 percent of what you need to know about cross-examination. After five to eight trials you will have roughly a 60 to 85 percent working knowledge, and after ten or more trials you will be qualified to protect the integrity of your testimony and yourself in the future, for all practical purposes. This doesn't mean that you won't make an occasional blunder; everybody does.

The same thing can be said of depositions. After three to five depositions you will be able to handle yourself so that you won't readily be taken advantage of by deposing counsel. Remember, again, a deposition is largely a "fishing expedition."

When you first start out in forensic-type consulting, have your client give you some pointers. This will be invaluable. And after each appearance, ask your client to critique you. These exercises will fill in the gaps and alert you to problems you might otherwise have never even known existed.

There are some additional pitfalls that you should learn to avoid. Learn to differentiate between the "information only" content of every cross-examination question and its "noise level" or noninformation content. As alluded to before, that twelve-part question just asked by your cross-examiner is an example of a potentially meaningless "nonquestion," especially when a "yes" or "no" answer is demanded in three seconds or less and the question is ingeniously framed in less than ten words. Chances are that when you bring this to the attention of the cross-examiner, he will simply agree and withdraw it. Interestingly, even during the "heat of battle" there is a certain amount of collective team effort that takes place. Although the process is far from perfect, everybody is trying to make it work, even while attempting to bend it a little in favor of the client's best interests, protecting the integrity of professional opinions, and generally attempting to develop and hold onto the fragile threads of logical, technological, qualitative, and quantitative consistency and of truth.

Remember that your cross-examiner will attempt to ask you as many leading questions as possible. On the other hand, your own client cannot ask you leading questions during his direct examination. Keep in mind constantly that your cross-examiner will attempt to ask you only questions to which he knows, or believes he knows, the answers. He is very skillful at doing this.

Fortunately for you, and unfortunately for him, however, unless he is also an engineer and has some familiarity with your specialty, you have an automatic advantage. Of course, in a pinch the cross-examiner can have his expert by his side to prompt him. But don't let this worry you; chances are this won't happen. And if it does, you know as much as the next engineer in your specialty.

Remember that your cross-examiner doesn't want to give you an opportunity to explain anything. It is not to his advantage to permit you to sound authoritative or gain the dependency of the jury, and certainly not its respect if he can help it.

As you gain more firsthand experience in forensic work, particularly through cross-examination episodes, you will discover a rather disquieting fact. What makes for good law can engender really terrible engineering. Conversely, what you may consider excellent engineering can create horrendous, unworkable law. This is one problem that is guaranteed to drive you and your client up a wall every time.

However, since there is a good chance that you will wind up learning more about law and the legal process than your client will ever know about engineering, this problem will probably bother you a lot more than it does the client. For example, making a jury understand what human factors engineering is all about in ten minutes or less is a virtually impossible task. The judge is breathing down everybody's neck to move the case. Each attorney is sweating out every passing minute that diverts a jury from issues he or she feels are more germane to the case. Thus, the jury really never learns, for example, that reasonably foreseeable human error, failure, and oversight should be design conditions during the product development process. What appears to you to be nothing more or less than good engineering, and a "body of knowledge" background that the jury must be taught to comprehend and apply, becomes an impossible legal presentation problem that few attorneys are willing to address. Good engineering can thus be bad law (or, more correctly, untenable legal process).

You don't want to display an erratic or eccentric personality on the witness stand. Your client must recognize, of course, that you do have a personality, and that it may be different from somebody else's. There is nothing wrong with this. You don't necessarily have to come across as being insipid and nondifferentiable from the rest of the world. However, whatever your personality type and however it comes across (except that it should be natural), you must be predictable and predictably professional.

In a certain percentage of cases your client, and therefore you, will have been dealt what looks like a losing hand. But there may be something salvageable. Consequently, the only thing left for your client to do is to attempt to play that losing hand brilliantly. There may be, in fact, something meritorious about your client's case. He must bring out its strengths. And it is up to you to address them within the constraints of objective, competent, ethical engineering practice. There is nothing wrong with being vulnerable; just be certain that your casework is as surprise-free as feasible, within the constraints of casework funding available to you and your client. Then, if the case loses, at least you know that you and your client have done your level best to bring out its strengths. It is up to the jury to make the final decision. Remember that it is not you who has lost; you have done your best.

Of the cases that go to trial—which, again, are roughly three out of one hundred (the rest settle out of court)—approximately 50 percent are decided in favor of the plaintiff and 50 percent in favor of the defense. Keep these figures in mind. In fact, extract all of the percentages given in these pages and use them as standards against which to measure the characteristics of your own forensic activity. It is like flipping a coin. The more cases you do, the more closely the percentages in your own practice will approach those that are generally encountered.

Only take to court materials you won't mind being taken from you, marked into evidence, and thereafter lost forever. A client may promise you faithfully that you will get back that $100 handbook or that irreplaceable, out-of-print, 1905 treatise. If you ever do get either back they may be damaged beyond repair. Since you have absolutely no control over anything you take to court, the best practice is to leave everything valuable back in your office. Make copies of anything you need including title page, copyright page, table of contents, and relevant text. Don't let your client intimidate you into taking valuable books to court. Stand absolutely firm. Make extra copies of your report, file folder, and notes. Better to be safe than sorry.

As your travels through the marvelous, murky world of forensic practice expand, write your own book. Treat your experiences as a "laboratory course." Keep notes. Keep a diary. Analyze each experience. Again, make a list of cross-examination questions so you can prepare your own catalog or inventory of them. This is the way to fight back against the otherwise potentially devastating and demoralizing onslaught of your client's adversary. But you must not be an antagonist or adversary—your sole objective is to protect the integrity of your opinions at all times.

THE EXPERT REPORT: "CHAMBER OF HORRORS" FOR THE UNINITIATED

In this section you'll gain an important understanding of the various kinds of expert reports that are generated in forensic engineering, including those written by the attacking plaintiff's engineer and by defense experts.

The first rule is to account for differences in expert opinions, as discussed in the previous section. Remember that there is only one body of knowledge in any field. There is not one body of knowledge for the plaintiff and another for the defendant; the principles and rules of practice are the same for both.

Therefore, when you encounter two (or more) opposing sets of findings and opinions, try to account for the differences. No matter what the legal tactics or strategy may be, you will be better fortified to understand the outcome of the case. You will also be able to understand why some party won who should have lost, and vice versa. Expert reports and analyses can be categorized as follows:

Type 1—Sound and acceptable analysis. Reports that are ethical and competent in their attack upon bona fide defects relevant to accident and causal sequences, on the one hand, or with respect to their defense of designs, constructions, practices, and so on, on the other.

Type 2—The irrelevant defect. Reports that attack bona fide defects and/or violations of codes, standards, regulations, practices, and so on relating to design, construction, conditions, and management which, however, have absolutely nothing at all to do with actual accident causation or else are inconsistent with specific background causal data. Alternatively, identified defects are not demonstrably contributory because they are neutralized or offset by other features of design or construction, and so on.

Type 3—The "Superman" requirement. Reports that state that specific safety features were present which should have prevented the accident had they only been used when, in fact, such features may represent safeguarding means which assume that every victim will have the presence of mind to actuate them while in the very process of having the accident. Or, alternatively, such safeguarding means require that the victim be a veritable machine or "Superman," and not be susceptible to human error, failure, or oversight. In the latter case, of course, there is an entire subsidiary body of knowledge in the field of human factors engineering or man-machine systems design which treats this specific issue. Human factors analysis is conveniently ignored by this report type. Such reports do not deal with "real world" human behavior. Active versus passive safety devices are frequently a focal point here. Furthermore, in this report type the presence of warning labels and instructions are frequently erroneously accepted as adequate when, in fact, higher-order safeguarding means were technically and economically feasible.

Type 4—The code compliance fallacy. Reports that state that the product, process, structure, facility, workplace, and so on, has met all published safety codes, standards, regulations, and so on, and, therefore, is ipso facto "safe." Typically, this type of report places the total burden for personal injury or property damage on the shoulders of the plaintiff and, as with the Type 3 report, does not acknowledge the existence or applications relevance of human engineering analysis. Such reports totally ignore consideration of any relevant product features or defects which go beyond the scope of published safety codes and standards, and so on. They also neglect the basic philosophical and doctrinal foundation of the safety engineering field, aside from practical and relevant safeguarding modes which were ethically, technically, and economically feasible. Also, they may confuse state of the art with state of the industry or state of the trade.

Type 5—The "merry measurement" syndrome. Reports that describe in meticulous detail and highly technical and competent terms how a product, process, structure, or the like, functions. The report type may be replete with highly intricate engineering computations, formulas, diagrams, charts, and tables. But it then stops dead. It does not address safety at all. It does not address the alleged defect and does not place responsibility. It merely concedes that the accident occurred.

Type 6—The "bumblebee can't fly" analysis. The "bumblebee" report bases analysis solely upon theoretical scientific-mathematical models, textbook data, engineering calculations, laboratory testing and experiments, or field observations and measurements, and so on, that do not reasonably take into account—in fact, even ignore—the stated facts of the accident as reported by the victim or eyewitness. Under such conditions the results of this investigative procedure insidiously deny the very existence or occurrence of the accident event itself. In a sense it conclusively and scientifically "proves" that the accident never could have happened! Indeed, on the basis of such "scientific" manipulation and calculation, truly the bumblebee cannot fly! However, this type of report admits into its analytical process possibilities or hypotheses that, on the basis of factual data, should have been discarded and eliminated as improbable or impossible. In some instances, the absurdity of "scientific" findings and opinions offered by logic and analysis not constrained and tempered by "real world" facts or events is intuitively obvious even to the casual observer. This type of report is frequently rich in the presentation of either or both laboratory and field measurement data. Such data are sometimes elaborately tabulated and statistically analyzed. Unfortunately, this profusion of numerical data and computations is usually irrelevant. However, the "bumblebee can't fly" effect is almost always quite transparent when subjected to scrutiny by the demands of competent technical analysis. This report type is frequently combined with Type 5.

Type 7—Unsound analysis. Reports incorporating incompetent and invalid analysis, pure and simple, including improperly conclusive opinions from insufficient facts. Sometimes the "logic trap" or deficiency is as simple as erroneously concluding that, merely because two events or arguments are not inconsistent, one necessarily follows from the other in a rigorous cause-and-effect sequence.

Type 8—Unethical practice. Reports of an unethical character which, however, may only be identifiable by another qualified engineer involved in the case—you!

Type 9—Design for confusion. Reports that are calculated and intended solely to confuse, confound, and complicate issues without necessarily being critical and without offering new or useful insights. This report type is usually competent as far as it goes but invariably begs the critical issues. This report type will typically also include a good deal of extraneous or irrelevant material and discussion. Analysis may appear impressive to the uninitiated.

Type 10—Critiquing the twisted meaning. A report from one expert purposely or inadvertently misinterpreting the report of another expert. It attempts to put words into the expert's mouth. Then it proceeds to attack the incorrect and improper misinterpretation. But, while the misinterpretation is fundamentally incorrect or untenable, the critical arguments sound plausible and superior. This is one form of sophistry. This report may even erroneously try to show that the second expert opinion really agrees with the first, after all.

Type 11—Inadequate foundations and net opinions. Reports that do not present reasonable or adequate support for findings and opinions offered and/or are not self-teaching. Such reports are of the "net opinion" type. Net opinions are usually not admissible. Thus, such reports do clients no good at all when the chips are down. They sometimes are purposely prepared this way to throw up a smoke screen.

Type 12—Noncomplying formats and/or content. Reports that do not otherwise comply with ANSI/ASTM Standards E620-77 and E678-80 regarding standard practices for the reporting of opinions and evaluation of technical data.

Type 13—The limited scope analysis. Reports that are limited in scope, depth of analysis, or otherwise constrained by client, financial, tactical, time, or strategic factors. Therefore, findings and opinions will be more limited or qualified than more comprehensive investigations. These reports may well be totally competent and objective as far as they go. But, unless their limitations are spelled out, such as through clear statement of the problem addressed, specific hypothetical questions answered, or otherwise limited scope, conclusions may well be misleading and bordering on the unethical. These reports, of course, lend themselves to sharp cross-examination and deposition questioning with respect to alternative fact sets and

SCAR (Scientific Accident Reconstruction) scenarios. Reports may also be purposely limited in scope to avoid altogether discussion of what are perceived to be complicating and/or compelling issues that would be injurious to the findings and opinions presented and expressed.

Type 14—The ICBM syndrome. The ICBM syndrome will be more familiar to all of us in its popular form, the "It Can't Be Made," "It Can't Be Done," or "Not Invented Here" argument. The latter is customarily abbreviated to "NIH." Reports of this type are, at root, antiscience or antitechnological in their character. This is due mostly to the defensive, if not unethical, imperative to pooh-pooh all ideas, positions, and arguments that are contrary to their own or that are not helpful to their cause. As a practical matter, of course, if you locked up all the experts in a room for a few hours or a few days, depending upon the intricacy of the task set down before them, no doubt they would come up with a multitude of fresh and ingenious new ideas for ethically, technically, and economically feasible safety devices that would fill the bill admirably. That is, of course, presuming all experts are mandated to put client pressures aside and simply solve the problem, pure and simple, instead of just arguing over whether or not a safety device is possible or feasible without really working at a solution. Negative thinking is the worst killer of unborn ideas in the world. This is why the brainstorming approach is so useful and constructive as an idea generator in so many applications. The ICBM report type, incidentally, is frequently combined with others listed above, such as the "bumblebee can't fly" variety. However, where the ICBM syndrome reflects a way of corporate life and general management thinking, it can be just as devastating. For such a management its "LRP" program, usually an abbreviation for "Long-Range Planning," probably better also include a lot of "Long-Range Praying"!

Type 15—The avant-garde assault. The incremental safeguarding means proposed in the plaintiff engineering report may have been technically and economically feasible but more in the nature of an invention rather than merely an innovation or imitation at the time you designed, produced, and sold the accident product. Or it may not have been in use in your industry, general industry, or special services, although progressive engineering application of technology transfer principles and safety engineering rules of practice would have identified its potential utility in conjunction with your product. Perhaps it represented a concept or physical configuration at the very cutting edge of then-contemporary technology that had not yet really graduated from the laboratory. But be careful in your thinking here. Just because something did not represent the state of the industry and was not generally used or was not codified does not mean it did not have practical and relevant merit with respect to accident prevention. There is a big difference between state of the industry and state of the art. By the way, we are specifically excluding the realm of experimental laboratory developments when we talk about state of the art in this context.

Type 16—The "MLC" report. This type report is a bona fide response of a defendant to a legitimate plaintiff attack on a defective and unreasonably dangerous product that is the subject of litigation. In this case, however, the defendant has lots of company since there are other derelict parties to the action and, therefore, multiple defendants. And since "misery loves company" (MLC), a defendant is properly defending himself by laying a portion of the blame on somebody else's doorstep as well as resigning himself to an adverse judgment. He knows he can't wiggle away as the manufacturer of the accident product, but he also knows he has a reasonable defense in this case because he was not the only one to blame and, therefore, shouldn't be made to shoulder the whole burden. Maybe a vendor, distributor, outside design consultant, subcontractor, prior owner of the product or machine, or someone else also screwed up badly and in a manner that identifiably and significantly contributed to the accident incident or severity. The MLC-type report thus gives a defendant an opportunity to "spread the risk" and minimize awarded damages to the extent feasible. Many perfectly good defense-type reports prepared for carriers are of this type. They play an important role in products liability-oriented accident investigations.

Type 17—The "UFO" report. This report type expertly develops the "Unessential Failure Origin" for the accidental occurrence. While scientific accident reconstruction is always of interest, it may not be critical. In this report, however, the UFO becomes the focal point. Analysis never proceeds beyond the highly technical, competent reconstruction. Its approach is reminiscent of the "merry measurement syndrome" of report Type 5. The writer presents the technical discussion and then stops cold. There is an implication that the product did not fail due to any design or structural defect in most cases of this class where, in fact, no breakage or the like occurred. In most instances there was human failure. The UFO-type report totally ignores human factors ramifications. Here, again, the report also ties in somewhat with the Type 3 "Superman" report. This report type may also present the failure as though it were an inevitable, unavoidable event, given the "reasonably safe" product design. For example, a glass coffee decanter that breaks due to thermal shock or physical impact is presented as a "reasonably safe" design and construction. The "failure origin" is thus presented as an inevitable and unavoidable consequence of an accepted material of construction being used in traditional and accepted applications. Therefore, it is argued or inferred, the product was not defective and the scientific accident reconstruction "proves it." However, the report totally ignores the technically and economically feasible alternative use of stainless steel, which would have made totally irrelevant SCAR-based opinions relating to acceptability of glass. Thus, the failure origin based upon competent and meticulously documented SCAR analysis is totally unessential. Had stainless steel been used, neither thermal shock nor physical impact would have resulted in the failure or the accident, and human error, failure, or oversight would have been neutralized or adequately accommodated by the safer design and construction of stainless steel. The glass decanter, on this basis, is seen to be unreasonably dangerous notwithstanding the beautifully presented scientific accident reconstruction that "proves" the glass decanter could not possibly have behaved any differently under the circumstances and that the glass was not itself improperly produced or imprudently applied.

Type 18—The old switched defect trick. This defense-oriented report type is particularly insidious. It attacks a modification to prevent occurrences or recurrences of the accident type, made by a product owner either before or after an accident, as being impractical, too costly, or even unsafe in other ways. It makes it appear that the retrofitted item is of the very design and construction that the plaintiff says was lacking, and which lack the plaintiff expert claims constituted a design defect that made the product unreasonably dangerous and unfit for the service intended. In fact, this product modification, intended in good faith by the owner-user to preclude accident occurrences or recurrences, may really be impractical, too costly, or hazardous itself. However, these very problems with many owner-user modifications are widely known and reasonably foreseeable by manufacturers. They comprise one very important reason why manufacturers should incorporate adequate safety features into product designs and constructions in the first place, instead of delegating product safety to owners or users. The plaintiff expert might even agree that the actual modification is too expensive, impractical, or dangerous. But the focus of the plaintiff expert is most probably on a practical, economical, and acceptably safe design that is different from the one built onto the product as a modification by the owner-user, which is the one being attacked quite justifiably by the defense report. However, it is the intent of the defense expert that is repulsive. This type of defense analysis merely tends to mislead, confound, and confuse the jury and the court. It is a thinly veiled attempt to obscure the competent, ethical, and objective thrust of the plaintiff expert position. No doubt the plaintiff expert will have ample opportunity to rebut the erroneous assumptions and incompetent, if not unethical, smokescreen being thrown up by the defense expert. But this defense report type complicates the true issues. It adds absolutely nothing to a critical understanding of the case. It merely tries to undermine and sabotage professionally supportable plaintiff expert analysis and opinion. It is an unethical variant on report Type 9, a "design for confusion", as well as Type 10, which twists the meaning of the plaintiff expert position. In effect, the defense expert has managed to switch defects on the plaintiff before our very eyes. It is, of course, but an illusion. The defense expert has analyzed the *wrong* design alternative, *by design*. But if the plaintiff expert is not careful, this report type can be troublesome. It is best countered by observing, as above, that an unsafe product modification is oftentimes the result of a sincere effort by an owner-user, unfortunately nonexpert in product safety, to compensate for the lack of adequately safe original design or construction features that should have been built in by the manufacturer.

Type 19—The "GIGO effect." A variant on the technical cliché, this report type documents inspections, tests, analyses, and so on that prove, beyond the shadow of a doubt, that the product, process, and so on met all manufacturer's engineering and production specifications; that there was no malfunction, breakage, or other failure that caused the accident; and that in fact the product did exactly what it was intended to do under the design conditions. Unfortunately, however, this is precisely the problem. The design and/or construction was not capable of performing safely under reasonably foreseeable conditions of service and commerce. It did exactly what it was designed to do; nothing more, nothing less. Thus . . . GIGO.

Type 20—The "open, obvious, and warned-against hazard" cop-out. This is a variant on the "Superman" type report. It assumes that human beings will invariably behave in a risk-aversive manner and never make mistakes. It ignores human error, failure, and oversight as a bona fide source of accidents and injuries. It fails to acknowledge the existence and meaningfulness of human factors methodology and the experience of the ages. It fails to recognize that even the "open, obvious, and warned-against hazard" must be properly guarded with highest order safeguarding means feasible if accidents are to be avoided. It fails to recognize the limitations of short-term memory, the habituation phenomenon, production pressures, human forgetfulness, zeal, distraction, fatigue, and so on.

Type 21—The strange case of the unforeseeable hazard. Pure and simple, this report type simply denies that the hazard was or should have been foreseeable by the OEM. Therefore, since the hazard was not reasonably foreseeable, the designer-manufacturer was under no obligation to guard against it in any way, shape, or form. In this case we have to apply the tests as to whether the hazard was, indeed, reasonably foreseeable. These tests are incorporated into the checklist presented as Exhibit 12-2, titled "Practical Techniques and Sources for Discovery and Determination of Reasonably Foreseeable Product Safety Hazards." The tests must be applied on a retrospective basis, of course, to ensure state-of-the-art relevance.

Type 22—The adverse risk-utility-feasibility gambit. Somewhat like the "unforeseeable hazard"-based report, this type denies that the proposed incremental safeguarding means (PRISM) is utilitarian or that it will eliminate or mitigate the hazard. It denies that the proposed safety device would be economically or technically feasible. It may even conclude, in some cases, that addition of the proposed safety feature would render the product less useful or even useless, or that elimination of the hazard would do the same. Or it might state that the proposed safety feature would create still a new hazard. This report type is usually contrived and nothing more than a smokescreen.

Type 23—The old "passing the buck" trick. This is a close relative of the "MLC" and "Superman" report types. It blames other parties and may emphasize mistakes of everyone except, naturally, the manufacturer. For example, it may blame parts or materials suppliers or subcontractors, the employer, co-workers, bystanders, and so on. Of course, blaming the plaintiff stands by itself and is thoroughly covered. But the main thrust of this report type is to show that the employer or product owner, generally, for example, was in error and improperly operated, applied, maintained, transported, stored, modified, or set up the product, process, system, and so on.

Type 24—The unfit and unsafe modification. This is a special case of the old "pass the buck" trick. Here the emphasis is upon the employer or owner or even plaintiff product modification that is, indeed, unsafe. The problem is that the particular modification attacked by the report has absolutely nothing at all to do with how or why an accident occurred. There is absolutely no causal connection with the modification and the accident on the basis of background factual data or competent, responsible, objective, ethical engineering analysis. The attack on the modification and, therefore, on the plaintiff position is really groundless. The report is, again, merely a smokescreen. But sometimes, depending upon the particular modification selected, safe or unsafe, interesting arguments can be advanced that may well confuse the poor, innocent jury. This report type is also a special variation on the "irrelevant defect" and "old switched defect trick" formats.

EXHIBIT 12-2

PRACTICAL TECHNIQUES AND SOURCES FOR DISCOVERY AND DETERMINATION OF REASONABLY FORESEEABLE PRODUCT SAFETY HAZARDS*

1. Existing published safety codes, standards, and regulations.
2. Company, competitor, and general industry field experience including misuse, abuse, and error patterns; product failure and malfunction patterns; aftermarket and downgraded service patterns; and so on.
3. Products liability case law and legal publications (periodicals and legal association special services for trial attorneys, etc.)
4. Technical reference manuals, textbooks, data sheets, manufacturer trade literature, professional and trade association proceedings and periodicals, and so on.
5. Product safety checklists, cookbooks, scientific accident reconstructions and malfunction and failure analyses of reported incidents.
6. Consumer usage testing under simulated, reasonably foreseeable conditions of service and commerce by independent testing/polling organizations, industry trade associations, manufacturers, and so on.
7. Human factors analysis and application of the rules of practice for safe design (performance standards).
8. Hazard foreseeability, technological forecasting, and technology transfer techniques and procedures (including the Delphi method, brainstorming, comparative analysis [interfirm/interindustry/interproduct], etc.).
9. Other market research and commercial intelligence, including overt sources as the press, dealers and distributors, customer complaints, U.S. government and industry trade association accident statistics, commercial clipping services, other media coverage.
10. Applications engineering, defeatability proneness, safety serendipity, and form versus function analysis.
11. Laboratory experimentation, analysis, testing, mockup and general modeling (including dimensional analysis and simulations), quality assurance and reliability analysis.
12. Covert intelligence (industrial espionage methodology).

* Also facilitates discovery and determination of reasonably foreseeable conditions of service and commerce.

Type 25—The defeatable, bypassable safety device. In this report the proposed safety feature is attacked because it is capable of being bypassed, defeated, disabled, altered, or broken. One problem is that it takes the viewpoint that all safety devices can be defeated and, therefore, it just doesn't pay to try and do anything about the hazard due to the perverseness of human nature. Another problem is that it assumes that an operator, employer, or product owner will attempt to defeat the safety device—any safety device—as an immutable law of human nature. A third problem is that it assumes that almost anyone who gets near the product will take it apart and rewire it, rebuild it, tear it to pieces, and otherwise take whatever extreme steps are necessary to defeat the poor safety device that is proposed by the plaintiff. It takes the "man against himself" standpoint that is, of course, totally absurd in the ordinary course of human conduct. This report type may also make the point that the proposed safety device may wear out, malfunction, break, and so on, and that, additionally, to build in redundant safeties would be uneconomical or ridiculous.

Type 26—The "big lie." This report acknowledges that the accident happened but denies that it happened the way the plaintiff said or testified it did. It is a variation on the Type 6 "bumblebee" report, but in reverse. It basically calls the plaintiff a liar, although not in so many words. This report typically advances a new accident reconstruction scenario or even more than one. In its most insidious form, it presents rank speculation without any foundation or connecting logic. The reader is essentially asked to take the alternative scenario(s) on faith. Sometimes it may sound plausible or even technically elegant to a layman. But it falls short when subjected to engineering scrutiny. It is ultimately found to be groundless, without engineering merit, and without an acceptable causal sequence as its basis. As with certain other report types this format is usually an unethical or incompetent, irresponsible smokescreen designed to confound and confuse rather than enlighten.

A sort of "composite indefensible defense report" can be constructed as a distillation of the substance of the foregoing types of incompetent, irresponsible, unprofessional, and/or unethical report types. It takes the form of a checklist presented as Exhibit 12-3. This checklist does not apply to those sound and ethical defense reports which may very well make some of the same points. For example, in some cases hazards really are not reasonably foreseeable. In others it really might not have been technically or economically feasible to eliminate or mitigate the hazard, and so on. But when an expert report proves to be indefensible on the defense side, its thrust will

EXHIBIT 12-3

THE COMPOSITE INDEFENSIBLE DEFENSE REPORT

1. The product complies(ied) with industry and/or government codes and standards (i.e., ANSI, ASME, ASTM, SAE, OSHA, DOT) (4).
2. The product meets(met) all OEM engineering and manufacturing specifications, does not malfunction, and, in fact, works just fine and as intended (19).
3. The plaintiff, employer, co-worker, and so on erred and improperly operated, applied, or maintained the product, and so on (3, 16, 20).
4. The owner, employer, or user-plaintiff improperly modified the product (16, 18, 24).
5. The defective part or hazard was supplied or designed by an outside vendor or subcontractor, and the manufacturer thus had no control over hazard elimination (16, 23).
6. The hazard was not reasonably foreseeable (21).
7. The hazard was open, obvious, and/or warned against (20).
8. The hazard cannot be eliminated or mitigated without reducing or destroying product utility (22).
9. It is not economically feasible to eliminate or mitigate the hazard (22).
10. It is not technically feasible to eliminate or mitigate the hazard, and the proposed safety device is not practical, can't be made, or won't work (22).
11. A safety feature would only be defeated or bypassed anyway (25).
12. The safety feature proposed would be unsafe itself (18, 24).
13. The safety feature added by the employer, owner, plaintiff, and so on is unsafe but typical of any safety feature that would be added or proposed (18, 24).
14. The accident didn't happen the way the plaintiff said (16, 26).
15. The proposed safety device had extremely limited use at the time of design and manufacture of the product or, in fact, was not even on the market so that it could not be purchased, adapted, or otherwise incorporated into the product (22).
16. The plaintiff was trying to commit suicide (26).

(continued)

EXHIBIT 12-3 (Continued)

17. The product had plenty of safety devices built into it. All the plaintiff had to do was use them (3, 16, 20, 23).
18. The whole industry has been designing the product the same way for the last fifty years. How can the plaintiff come along and reasonably say the whole industry and its cadres of degreed engineers and research scientists have been wrong, incompetent, or uncaring (19)?
19. We designed the product the best way we could. Honest. We tested it, continually modified it, and did our best to eliminate or mitigate the hazard. We just couldn't find a good way to make the product safer. Honest (22).
20. There's no law that says we have to put that safety device on our product. There certainly wasn't anything that said we had to put it on when it was designed, manufactured, and marketed (4).
21. Our product doesn't become a "living, breathing machine" until somebody sticks it into a system or applies it within his overall product design. Why should we have to worry about the applications safety of our product? It's not a self-standing or useful commercial item until somebody else does something with it or to it. Why blame us? Our product is just a "component" and can't be used alone. How should we know how it's going to be used (23, 24)?

usually be in one or more of these directions. Numerals in parentheses key in the item to the foregoing report types.

RETROSPECTUS

A major difference between forensic and conventional design applications of engineering science is one of emphasis. Traditional design engineering should take past field experience into account during the design and development process. This is important in attempting to satisfy present and prospective applications criteria with respect to reasonably foreseeable conditions of service, including reasonably foreseeable patterns of both use and misuse or abuse.

But forensic engineering is predominantly retrospective in nature; it attempts to reconstruct the retrospective state of the art and state of the industry. This is necessary to determine to what degree a product or other engineering work involved in litigation may have deviated from either or both when it was designed, manufactured, and/or marketed. It is the "then-contemporary" state of things that is of current interest in the forensic environment.

Critics of plaintiff expert attacks on alleged defective or dangerous products frequently decry findings of variances between accident-causing designs and either or both state of art and state of industry as merely "twenty-twenty hindsight." However, this is an unwarranted oversimplification. In fact, frequently it is nothing more than a thinly veiled smokescreen.

Sound forensic analyses of products involved in accidents will attempt to carefully reconstruct and document the then-contemporary, retrospective state of art and/or industry. It is only through careful, competent retrospective, reconstructive engineering analysis that a comparative study can be made to support findings of variances, noncomplying, or poor, unsound, and unsafe designs or practices.

In fact, the approach of the engineer in forensic practice is akin to that of the accident reconstructionist in the nonforensic safety and health environment. Careful retrospective analysis is critical to future accident prevention, redesign, or other remedial efforts, as well as for purposes of placing responsibility for failures or defects in connection with litigation.

Scientific accident reconstruction is a very useful tool for progressive, responsible manufacturers as part of their ongoing efforts to eliminate or reduce future accident potential, thereby, also minimizing the probability of future products liability lawsuits.

Therefore, in your own consulting travels, whether forensic or nonforensic, be alert to the historical background of your assignment from the safety engineering standpoint. And recognize the need, in forensic engagements, for developing the retrospective state of art and state of industry at the time the accident product or other design was first introduced.

Search out older safety books, codes, and standards in your specialty. You will need them to develop and support your findings and opinions, whether your client is a plaintiff or defense attorney. Frequently the client has his or her own historical library resources and sources. They can be crucial to the case.

A WORD OF CAUTION

In the last section a distinction was made between "state of the art" and "state of the industry." The term "state of the art" has two meanings. On the one hand, it may mean an experimental gadget or concept that has not yet graduated from the scientific laboratory. It is more a research or laboratory "curiosity" or ivory tower concoction than it is a proven and accepted, commercially available, or feasible device or product.

On the other, the term is also used to describe a concept or product that was technically and economically feasible at the time the particular item involved in litigation was designed, produced, and/or marketed.

In other words, we are talking about the then-contemporary or retrospective state of the art or available technology. This latter common usage is very pragmatic. It is an important definition and distinction. An entire industry may be doing the same thing but doing it wrong. We have all heard the expression "Forty million Frenchmen can't be wrong"—but they built the Maginot line and the Germans attacked them from behind.

On the other hand, "state of the industry" or "state of the trade" reflects that segment of the state of the art that manufacturers elect to apply for whatever reasons. Technology transfer considerations are significant in defining state of the art or available technology in relation to any particular problem.

As has been correctly recognized by the courts, state of the art at any given time may well be partly a function of the diligence with which industry has invested in safety research. The ultimate view is then simplified of what should have been knowable and done in time to have permitted elimination or mitigation of the product hazard involved in a particular accident. That is, a manufacturer should not be excused from liability if, in fact, managerial error, failure, or oversight, or inadequate investment in safety virtually assured that hazards in a product would remain unknowable or undiscoverable right up to the time the accident product was shipped. The manufacturer should not be exempted from blame if inadequate investment in product safety research and development effectively precluded identification, evaluation, and control of the accident hazard so that the problem remained unaddressed and unsolved.

This argument is totally consistent with rules of practice for safe design. It is also consistent with the safety engineering mandate to apply hazard foreseeability procedures.

Companies or whole industries may effectively disregard progressive product safety thinking. That is, on the basis of industry-wide economic characteristics over which, in actuality, they collectively are capable of exercising control, they in fact collectively do nothing to promote progress.

A plaintiff report attacking unsafe design may do so on the basis that it was merely a "state-of-industry" design, although a "state-of-art" alternative was feasible. This is a bona fide attack on what is alleged to be an unreasonably dangerous product.

But there is a problem. Juries are composed of sincere, dedicated, and thoughtful men and women who are doing their share in serving the public interest as jurors. But you can't make engineers or safety professionals out of them in one hour or even eight hours of even the best expert testimony.

If you find yourself in a situation where the product involved in an accident was designed in compliance with all state-of-trade standards, even though they might be inadequate and even though it would have been feasible to design the product more safely, you and your client could be in for big trouble. Or if your case research shows that the state of the art was in advance of state of the trade, but no manufacturer in the world has ever incorporated the safety device you know would have been feasible and would have precluded the accident, you are also potentially in deep waters. Similarly, if there is no code or standard that addresses an alleged design defect, but the product is like everybody else's in the industry, you could have a problem. Finally, if you know that the technology transfer technique would have been feasible and that, if used, the designer-manufacturer of the accident product would have discovered the defect and/or found a solution, but he didn't use it and didn't find it, you might again be in trouble.

Juries actually tend to be relatively conservative. They like to see a piece of hardware, something concrete—even a product catalog to show them that somebody someplace made it better and safer at the very time the accident product was being sold. Juries tend to believe that if a book or other published work represented as being authoritative says something, then it is acceptable. Most people's expectations are molded by the written word.

If you cannot show a jury a piece of hardware or product literature, you will lose some credibility no matter how viable your engineering arguments may be from purely technical and professional standpoints. And it is not always feasible to march into court with a working model. Economic and other factors may weigh heavily against this.

So, on a scale of 1 to 10, even if your arguments are excellent and viable from a purely engineering perspective, without hard support your credibility before a jury is probably only a 7 or even a 6 or 5, depending upon the intricacy of the problem and that of your solution. It may be even less if your solution is particularly creative, even though any engineer with ordinary skill in the art would have been fully capable of coming up with it if he was resolved and committed to solving the problem. In fact, you might even know in your heart that the basic problem, from an engineering standpoint, is relatively trivial in terms of the applied technology that the OEM would have had to bring to bear to solve it, if only he had worried over it enough in a responsible, thoughtful, competent, and objective manner.

All of this does not mean that you shouldn't write an engineering report to present your findings and opinions, or that you should fear defending them in court, or that you should be intimidated into advising your client against proceeding with the case. But you should alert the client to the problem. You must both go in with your eyes open. It may be an uphill fight. It may even be one of those potential losing hands that you were both dealt.

In some instances a more intrepid client may want nothing more than for you to get him to a jury. You may be able to do this readily in an ethical, objective, professional, competent, responsible way. But the deck may be stacked against you. Another client may be afraid of the case, especially since plaintiff attorney fees are contingent upon the outcome of the case (unlike yours). He may be afraid to invest the time.

This does not mean that one attorney is right or wrong, that you are right or wrong, or that your client is being unethical if he wants you to get him to a jury despite the obvious problems ahead. It just means that there may be a real struggle down the road.

The plaintiff must also be made aware of the prospects. No matter how good your arguments are, no matter how practical and how credible, you never know what a jury will do even under the best of conditions. A judge's instructions to the jury concerning the applicable law could be faulty or confusing, which can confound an already problematic situation. Many things can get in the way of a proper trial outcome as you see it, based upon your own impartial, professional, competent, and responsible analysis and presentation. We have seen some of the problems in previous sections.

Finally, a word on something you may have wondered about. Why do cases drag on for so long until they come to trial or are settled? Obviously, one of the problems has to do with severe pressures on an already overloaded court system in the United States. But this is not the problem I am referring to.

On the defense side, if the plaintiff has a strong case that is recognized as such by the defense, the name of the game is deferral. If the defense sizes up the case as one where a jury award for damages is likely, the case will be dragged out as long as is possible. Trial dates will be adjourned time after time for any one of a number of both good and bad reasons. After all, why shouldn't a defendant try to earn interest on his money for as long as he possibly can? This is really the ultimate name of the game. So don't be surprised if a trial date is adjourned numerous times. Naturally, the plaintiff wants his case to come up as early as possible.

CROSSFIRES

By definition, forensic practice is embedded in a litigious sector of human affairs. It is adversarial in nature. The principals in litigation—plaintiffs and defendants—are at each other's throats since the civil justice system is the solution of last resort in a civilized society. Otherwise unresolvable conflicts and personal injury, property damage, or other economic loss are the raw materials of the adversarial legal system.

The engineering expert thus functions at the center of conflict. Involvement in forensic casework positions the expert in a veritable crossfire. Each lawyer is doing his job to protect the best interests of his or her client, and there you are—in the middle. The plaintiff wants to be made "whole" to the extent feasible. The defendant disclaims any responsibility or obligation to do so. So they fight . . . in court.

Primary problem potential for the expert revolves around getting paid the professional fees that have been earned for professional services rendered. Typically, such services involve casework leading up to expert reports, or deposition and trial testimony.

The rules presented earlier will eliminate most, if not all, fee payment problems that can arise having to do with report and trial testimony-related casework. In a word, get paid in advance. And get paid sufficiently in advance to avoid having checks stopped or bounced by those few members of the legal profession who make it tough for all the rest. Advance payment will also assure that you do not have to spend six months or more trying to collect fees for which you bill clients, and even then possibly to no avail. And, of course, retainers, deposits, and other advance payments will permit you to establish priorities as to which clients deserve your valuable professional attention first.

Unfortunately, there are lawyers around (as there are debtors in every line of business) who will defer paying you as long as they possibly can. And there are some who will conveniently assume that you will wait until the case is settled or goes to trial for payment—which could be two or three *years* down the pike. Therefore, you have to be specific about financial arrangements. You simply must get retainers or trial testimony deposits paid up front as a matter of good business, if not business survival. As pointed out previously, trying to collect payment on an invoice for trial testimony where the client has lost the case can be quite a feat. Ask around; any engineer who does a significant amount of plaintiff work will tell you the same thing.

Luckily, the author was coached before he went into forensic practice by both experienced forensic engineers and lawyer friends and acquaintances. So I started out doing all of the right things. Consequently, I have been stung very few times, and for very small sums of money. When it happens it is invariably because I have not followed my own rules and have let down my guard when I should not have done so.

Those of us in forensic practice have had clients with marital problems who have become financially strapped. Or they go into personal bankruptcy. Or they have a drinking or drug abuse problem and can't raise the dough. Or they lose one too many cases and blame it on you. Or they "fly the coop" for various and sundry reasons. Or they leave one firm and join another, but nobody knows anything about anything and the buck gets passed back and forth. In a few cases lawyers have been known to be dishonest and unethical: They have ordered casework for which they had no intention of paying, or they come to you for a super-rush emergency field inspection and report before, they claim, they can possibly get a retainer to you. Better to decline such cases unless you really know the client.

The "crossfire clincher" will be the situation, thankfully rare, where you have been summoned to a deposition to provide expert testimony. Unknown to you, the attorney who called the deposition can't make it or doesn't really want it, and doesn't show up. Instructions and/or paperwork manage to get fouled up as to whether the deposition is on or off. But there you are. You, your client, and other lawyers are already present—all of you, that is, except the attorney who called the deposition in the first place.

However, due to an impending trial date, your own client hesitates *not* to have the deposition proceed as originally scheduled, just in case your testimony proves valuable to him at time of trial. Therefore, he permits the deposition to go ahead and you wind up providing testimony. Then you bill for your services.

All the while, you have no idea as to what has been going on behind the scenes, why one of the attorneys isn't present, why there is some legal discussion back and forth about scheduling, and so on. You are present in good faith. Nobody told you anything.

Needless to say, everybody thereafter disclaims responsibility for having authorized the deposition. Your client says it was the original deposing counsel who, in turn, says he had cancelled it. Nobody seems to be interested enough or willing to give you a straight story. The worst part is that your client settles the case shortly after you bill for your fee. But the complexion of his practice may have changed, and he no longer is going to need experts in his newfound legal specialty. He will probably never have to call you again, so he could care less whether you get paid or not. Anyway, the record shows who ordered the deposition, right?

Now, of course, it would be a relatively simple thing to resolve the matter if only your client would make a motion in court to determine who is to pay you, or how payment is to be distributed if it is found that all counsels must share the burden. This is the usual procedure where there is a dispute. And ninety-nine times out of a hundred this is what will be done, to everyone's satisfaction.

But your client wants no part of it. He justs wants to be rid of the case—and you. After all, he settled the case and got his fee out of it. And the amount is too small—a few hundred dollars, at most—for you to sue over, even in Small Claims Court. Motions cost time and money. And your client would need a court order to split the fee with the other firms because he wasn't authorized to make such a disbursement. He would have to go back and make changes in all the settlement papers. And on and on and on. You still don't know or understand all the legal ramifications of passing the buck in this instance, and your client tells you you're just being jerked around by deposing counsel (whoever he may be by this time).

So what do you do? You *don't give up.* You remain persistent. You nag and pester a lot. You may eventually wear them down. After all, they should not be the ones to get angry. You are the one who is owed money. Remind them of this simple fact if anyone gets testy with you. If all else fails, you can always sue them all in Small Claims Court, even just for the experience. Remember that it was not your fee rate that was at issue. Everyone is simply trying to sweep the problem under the rug and not pay you at all. It becomes a matter of principle after a while.

While we're discussing it, you should send away for information about Small Claims Court procedures in your state and county. You never can tell when you might have to avail yourself of this useful resource. In New Jersey, for example, the Small Claims Court is a division of the County District Court in which one can sue another person or a business for smaller sums of money, currently up to $1,000. The Small Claims Court process operates quickly. Most of the time your case will be heard within two weeks after the action is filed. In New Jersey each of the district courts, with some exceptions, has a Division of Small Claims.

The real advantage of the Small Claims Court is that you can handle the whole matter yourself. Divisions of Small Claims are designed to be used without the assistance of a lawyer. Fees are very nominal.

However, remember that when you sue in Small Claims Court for an amount over the legal limit, any amounts due you over such limits are automatically and legally waived or given up by you due to the fact that you have elected to sue in Small Claims Court. In any other court, of course, you will need a lawyer to represent you. But the defendants must show up or lose the case by default. This alone can help you, since such high-priced legal talent should not be wasting its time in Small Claims Court.

One of the ways to prevent getting caught in crossfires is to follow a simple rule: Don't depend upon your client to protect you unless your best interests are identical with those of his client or, naturally, his own. Although it is reasonable for you to expect protection from your client under diverse conditions that can arise during the course of your involvement in a forensic engagement, you should always be ready, willing, and able to protect yourself to the extent the law permits as a last resort.

If all of this sounds scary, don't let it throw you. In fifteen years of forensic practice, after thousands of cases and thousands of clients, I can count the number of times I have had to take someone to court on the fingers of one hand—and I have always won. The losses I have sustained total less than $1,000.

It is not only possible, but probable that if you conduct your practice on a businesslike, professional, ethical, and mannerly plane you will have good reason to be proud not only of your own conduct of the practice, but also of your clients and even their adversaries. With all of the potential problems and harrowing moments, your forensic practice will prove to be worth the time, effort, and dedication that you give it. You will have served society and the public earnestly and sincerely in fulfilling the ethical mandate to the engineering profession of promoting and protecting, first and foremost, the public safety, health, and welfare.

MOVING AHEAD

From the foregoing blow-by-blow description of the forensic engineering environment, you now know what to expect, what to avoid, and how to avoid it. The next step is up to you. Engagements are available almost for the asking, if you do your homework and follow the suggestions offered.

Attorneys are always looking for new faces, backup capabilities, multiple opinions, and access to expanded and diversified resources. There is plenty of work to go around. The field is expanding, and even those engineers who practice essentially full-time cannot possibly hope to service all the clients who depend upon competent engineering counsel and who are theoretically within reach through active client development activities.

Following the guidelines offered in this chapter can pay off both financially and in terms of your professional development. There are nominal risks associated with competent forensic practice, unless you happen to lose or destroy evidence entrusted to you or forget to show up for a trial. Once you learn the ropes, your practice is capable of becoming as large, as fast, and as diverse as you desire. Your three worst enemies will be time—the limited hours available in which to fit all the work that comes your way, "forensic burnout" due to the unique and ongoing rigors of forensic practice, and the temptation to be undisciplined in relation to requiring advance retainers. You must not permit yourself to be intimidated or pressured into providing engineering services without advance retainers from noninsurance-carrier clients.

One of the reasons you must go to some trouble to present yourself in a totally professional light, and must come across as competent, dependable, and ethical, is that some attorneys have been burned in the past by experts who do not deliver. Attorneys and other forensic clients are typically quite willing to pay top dollar for expert engineering services. But word gets around about bad experiences. Nobody wants to get stung. So you may well find that prospective clients are cautious, though also eager to work with you if only you can convince them of your professionalism. Word travels fast about top performance, too.

Of course, it is precisely because of the failure of some "experts" to deliver adequately and professionally forensic engineering services that you have an additional opportunity to capture certain engagements. The prospective client has been looking around for new professional contacts. But you must not let client mistrust and caution, due to some bad experiences, get in the way of his or her relationship with you. You must do your homework from a public relations standpoint. Once you have done this, you have no reason to compromise your professional or commercial posture with respect to fees, for example.

With all of its unique, demanding characteristics, you will find forensic engineering practice one of the most challenging and interesting activities on your professional agenda. You will be dealing with problems at the other end of the scale—at the failure points and interfaces where products, machines, workplaces, and structures are ultimately used by human beings. Its potential for professional development is virtually unlimited.

Naturally, the discussion in this and the preceding section cannot begin to expose you to all of the intricacies of forensic practice. But if you follow the suggestions and rules presented, you will be off to a running start when those first few forensic engagements present themselves. You will not get caught up in the many traps that lie in the path of the uninformed engineer who does not understand the realities of the forensic engineering services marketplace.

There is a somewhat less obvious but extremely beneficial potential result that can come from forensic consulting practice even if you only engage in it part-time while working full-time for an employer. Your forensic experience—obtained, naturally, with your employer's blessings—can be very useful to him. Your courtroom experience and general knowledge of the litigation environment can make you very valuable in the event that your employer is sued. In a products liability action you may well be able to serve as a defense expert. If nothing else, this will set you apart from your colleagues and peers who may be totally ignorant of the forensic side of engineering practice. You will gain instant visibility. Top management will come to depend upon you. If they're really savvy, they may even ask you to set up a forensic engineering department, although not necessarily by that name. In any event, your consulting experience in the forensic field will give you a competitive edge in terms of advancement opportunities within your company. Your experience will be absolutely unique and not readily acquired in the ordinary course of engineering employment.

In fact, since many large companies have internal forensic engineering activities, you may expand your career opportunities the next time you decide to switch jobs.

CHAPTER 13

A CRASH COURSE IN SAFETY ENGINEERING THAT CAN SAVE YOUR NECK

THE BASIC PROBLEM

Toward the end of Chapter 8 we noted that financial and economic analysis relating to safety, health, and environment is a critical ingredient of engineering problem solving. In this chapter and the next we will zero in on the key concepts and methods that every engineer must know to protect his client, himself, and the public.

Whether in product design and development, workplace layout and process analysis, or facility and/or systems design and construction, engineers must be thoroughly familiar with the basic rules of practice for safe design. For the consultant particularly, a design error that creates an unsafe product, workplace, or facility can become costly both in financial terms and in human suffering due to accidents, injuries, and/or fatalities. A negligence or products liability lawsuit may not be far behind which could financially ruin an engineer or businessperson who does not adequately incorporate safety into his or her designs. An unsafely designed, mass-produced product can adversely affect the lives and/or careers of thousands of unsuspecting men, women, or children across the nation. And a poorly thought out process could release toxic substances affecting a few people or an entire community.

The objective of this chapter and the next is to help you keep out of trouble as a consultant. The important principles and practices are not difficult to learn or apply, but they must be learned well and applied faithfully—*every time.*

Before we move ahead, however, let us view safety engineering and product safety from still another viewpoint. The opportunities for the consulting engineer in today's world are virtually boundless. Any attempt to cover every field of specialization, or even a limited few, in detail would be impossible. And to treat them all in a reasonable amount of space would result in broad generalizations that would not do justice to any one of them. Consequently, this book deals with principles and practices of consulting engineering that will be of value to all engineers.

However, it so happens that safety engineering is a specialty that can also be effectively applied by all engineers. Therefore, these next two chapters may be used as a guide not only to principles and practices that all engineers need to know, but also as a basis upon which to practice specifically in the field of safety engineering. Using the material in these chapters in conjunction with your own specialty—be it civil, chemical, electrical, or mechanical engineering—you will be equipped to address safety engineering problems from the design, as well as defensive, standpoint. And remember that, to a significant degree, defensive engineering is really no more than good engineering. The specific safety engineering approaches presented, with appropriate variations for application to different engineering specialties, can be used across the entire spectrum of consulting opportunities that are within your reach.

SAFETY ENGINEERING

Safety engineering is a specialty within the overall engineering field, with its own subsidiary body of knowledge going back almost to the turn of this century, that deals with the identification, evaluation, and control of hazards and risks. Through engineering, education, and enforcement, hazards and risks are eliminated, mitigated, or otherwise accommodated.

For example, companies with the most successful safety records typically articulate a corporate policy, and embrace and implement the philosophy, that:

1. *All* injuries can be prevented.
2. Management must accept its responsibility in preventing personal injury.
3. It is possible to guard against *all* exposures.
4. Safety is good business, both efficient and economical.
5. Both superior engineering and training are crucial to safety achievement.

PRACTICAL LIMITATIONS OF SAFETY STANDARDS AS DESIGN GUIDELINES

Our survival course in safety engineering begins with a look at safety standards. A bewildering array and almost staggering number of standards typically confront the design engineer and business executive, who may well rely upon them for guidance in relation to safe design. But there are important shortcomings in their content, particularly in the area of safety. Limitations must be recognized if codes and standards are to be applied meaningfully and effectively in the public interest, and in a way that permits the engineering profession to overcome shortcomings and defects.

For example, almost all published safety codes and standards consist of compromise, consensus, maximum acceptance, minimum level content, voluntary documents. They do not necessarily reflect the technically and economically feasible state of the art or best practices, or even a state of the industry more current than at some point in time when they were last updated and revised.

Thus, mere compliance with published safety codes and standards does not automatically guarantee that a complying product, workplace, or facility is reasonably safe. It may only reflect state of the trade or state of the industry, rather than some more fundamental or more encompassing set of standards, rules, guidelines, and so on.

State of the industry may be vastly different from technically and economically feasible state of the art. Published codes and standards comprise but one portion of the body of knowledge in the safety engineering field. And the record will show that state of the industry is characteristically well behind state of the art.

At least twenty-six defects and limitations in published safety codes and standards can be identified readily. They are listed following, and discussed in greater depth afterward.

Defect 1. Incomplete coverage
Defect 2. Some major hazards not addressed
Defect 3. Minimal safeguards acceptable
Defect 4. Hazard reduction rather than elimination
Defect 5. Near misses not considered
Defect 6. State of the art and economic feasibility not built-in
Defect 7. Genealogy tracing not recommended
Defect 8. Generic hazards and safeguards, including the rules of practice for product safety, not built-in
Defect 9. Some classes of product users not covered
Defect 10. Some classes of hazards excluded
Defect 11. Updating and supplementing the standard not built-in
Defect 12. Cross-referencing to sister standards not built-in
Defect 13. Technology transfer not recommended as necessary
Defect 14. Standards for preparing standards not offered
Defect 15. Two types of conflicting standards may be present
Defect 16. The body of knowledge mandate not built-in
Defect 17. Nonrepresentative content due to dominating standards-setting entity
Defect 18. Standards compliance labeling and certification problems
Defect 19. Apportionment of safety responsibilities and responsible entities may be unrealistic or self-serving
Defect 20. Maintainability safety not guidelined
Defect 21. Special-purpose products not addressed
Defect 22. Human failure and the error-provocative, accident-conducive design not addressed
Defect 23. Managerial trade-offs not disclosed
Defect 24. Dissenting opinions not disclosed
Defect 25. Retarding progress and judgmental requirements problems
Defect 26. Nonexistence of a safety standard

DESCRIPTIVE CATALOG OF CODE AND STANDARD DEFECTS

Let us now describe each of the limitations and/or defects listed above so we know what to look for when we use codes and standards. That will be half the battle in effectively applying them. You will have a better shot at staying out of trouble both with respect to protecting a client from the potential for customer accidents, as well as yourself against an injured plaintiff suing you personally due to a design defect that harms.

Defect 1—Incomplete coverage. Safety standards seldom treat the comprehensive range of reasonably foreseeable hazards. They are generally incomplete and fragmentary, although generally helpful as far as they go. They are guides reflecting desirable and acceptable practices and incorporate specific aspects thereof.

Defect 2—Some major hazards not addressed. Certain reasonably foreseeable hazards under reasonably foreseeable conditions of service, not considered by the standard as written, may be as or more severe than any hazard specifically incorporated. For example, accident frequency with respect to a particular hazard may be foreseeably low but the associated accident severity may be foreseeably high. Yet the standard may impose safeguards only against the more frequently recurring incidents in some cases or, by whatever criterion or measure, the "major" hazards.

Defect 3—Minimal safeguards acceptable. The typical safety standard is the product of a compromise or consensus procedure. That is, segments of the affected industry represented in the particular standards-setting body reach some agreement through negotiation. Thus, such a standard is also referred to as a "negotiated" or "maximum acceptance" standard. It is also a "voluntary" standard and minimum safeguard. It may even be subliminal, so that responsible design and safety engineers should not rely upon it exclusively in assuring user safety. However, most standards do not contain a statement to the effect that provisions represent minimum acceptable safety attribute levels.

Defect 4—Hazard reduction rather than elimination. The conscientious application of many standards will result in risk or hazard reduction rather than elimination. Hence, injuries may be mitigated rather than precluded. This is a desirable but unacceptable result if hazard elimination would have been technically and economically feasible within the existing state of the art at the time the standard was drafted.

Defect 5—Near misses not considered. Actual accident frequency, prompting inclusion of a particular standards provision safeguarding against the associated hazard, will not usually reflect the far greater number of near misses or close calls wide experience shows will have occurred, but which went unreported. Therefore, a low actual accident frequency count may be a distorted and inaccurate indication of true product hazardousness content, and but an apparent or illusory measure compared with the real exposure.

Defect 6—State of the art and economic feasibility not built-in. Technically and economically feasible state-of-the-art safeguards are not always incorporated into standards.

Defect 7—Genealogy tracing not recommended. Standards are seldom designed or intended to apply retrospectively to product units already in use in field applications although genealogy tracing, life-cycle monitoring, product traceability, or recall procedures and the like are feasible for purposes of retrofitting. In fact, such procedures may actually be quite essential and in general use as a managerial control tool within a particular industry for nonsafety-related reasons. These include, for example, ensuring operating performance improvement upgrading through retrofitting, effective field servicing, and implementation of marketing strategy. Standards should facilitate application of cumulative technological know-how.

Defect 8—Generic hazards and safeguards, including the rules of practice for product safety, not built-in. Standards are generally very specific as to design and construction features. But they seldom incorporate sections on generic hazards and safeguards, the general design philosophy to be followed with respect to product design safety, the authoritative treatises to be respected and applied, or the methods to be employed in identifying and evaluating hazards that facilitate their control—that is, hazard foreseeability techniques such as fault-tree analysis. Also, most standards do not approach the problem from the epidemiological viewpoint (i.e., through application of Haddon's theory of abnormal energy transfers and the like), although this technique is very powerful and universally applicable.

Defect 9—Some classes of product users not covered. The standard may not state that not all classes of product users are intended to be protected when, in fact, the standard as written has this net effect.

Defect 10—Some classes of hazards excluded. The standard, by design, may not cover certain classes of hazards, but this intended limitation is not communicated to the user.

Defect 11—Updating and supplementing the standard not built-in. No provision or attempt may be made to update or supplement the standard when warranted due to financial or other administrative constraints upon the standard-producing entity. Therefore, the standard obsolesces as time passes.

Defect 12—Cross-referencing to sister standards not built-in. Standards seldom are cross-referenced to sister documents within the national or international engineering communities at large even though, taken together, the coverage would be so much more complete as to hazards treated and safeguards recommended.

Defect 13—Technology transfer not recommended as necessary. Standards seldom charge the user-designer with the obligation to apply the technology transfer concept on a disciplined and rigorous basis to avoid overlooking important classes of hazards and safeguarding means generally available to the engineering community at large. Likewise, the "commercial intelligence" concept, which has been effectively utilized by top managements in other connections, and which has both overt and covert or clandestine ramifications, is seldom a factor in standards setting and writing.

Defect 14—Standards for preparing standards not offered. There is, at present, no generally recognized standard for the preparation of standards. However, ANSI published its "Guide for Consumer Product Standards—ANSI Consumer Council Publication No. 1." This is stated to be ". . . a list of considerations for use in preparing consumer product standards." The Standards Association of Australia has also published "SAA Guide for Consumer Standards—SAA Misc. Publication MP 29-1975," also based on the ANSI document.

Defect 15—Two types of conflicting standards may be present. There are two kinds of potential conflicts. For example, different standards-setting bodies and groups of experts may incorporate different requirements into their standards. Such differences may be only in degree but may also be in quality, so that the alternative standards are actually inconsistent with one another. Without a comprehensive understanding of the circumstances, test methods and conditions, rationale, trade-offs, personal and professional backgrounds of the drafters of the standards, whom they officially represent, it may be impossible to reconcile the differences.

In addition, be on the lookout for code provisions that may actually conflict with the rules of practice for safe design to be covered in the next section. This might not be easy to accomplish, but it is absolutely critical that you identify and evaluate such conflicts if you can. If you don't, your code-complying product may be violating basic rules of practice and be unreasonably dangerous. Don't necessarily depend upon the standards writers themselves for this kind of information—they may not be able to help you. Talk with some people on the original standards-writing committee in order to determine if any conflicts do exist and were taken into account, even though the code does not call out provisions relating to them. Also determine compromises and trade-offs. But, again, don't expect too much in this direction. You'll probably find out that most of the people you speak with have never even heard of the rules for safe design, let alone be able to recite them. It is much more likely that your inquiry will get the reaction that the code can't cover everything and that the standards consumer must use his or her own judgment to a significant degree. This is true. So why didn't the code say so? In any case, it is better that you should know about the problem in advance, before your "complying" product attacks the marketplace and everyone in sight.

Defect 16—*The body of knowledge mandate not built-in.* Safety engineering logic as a critical ingredient of design philosophy is unassailable. The highest feasible level of product safety content should prevail, lower-order safety provisions of published standards notwithstanding.

Defect 17—*Nonrepresentative content due to dominating standards-setting entity.* Occasionally, a standards-setting body or committee may not be representative of the total affected producer-marketer-consumer or buyer-seller universe. Representatives of a segment of that universe may dominate. Thereby, the effectiveness of the standards-setting process is impaired or destroyed, as is the utility of that standard as a truly independent, objective, professional, meaningful document. Such conflicts of interest may, as well, undermine the credibility of the standards-setting entity to the point where the entire standardization process comes under criticism as being self-serving. There has been, for example, a great controversy as to effectiveness of the voluntary or consensus standards system in the United States.

Defect 18—*Standards compliance labeling and certification problems.* Labeling methods typically employed by certifying bodies for communicating compliance with standards (for example, through "standard" test procedures) may be misleading to product users. Yet standards seldom specify acceptable or unacceptable labeling techniques for certification purposes.

Defect 19—*Apportionment of safety responsibilities and responsible entities may be unrealistic or self-serving.* Some safety standards distinguish among safety-related responsibilities of manufacturer, rebuilder, modifier, owner, installer, employer, or employee. In some cases this distinction is arbitrary and clearly contrary to good engineering practice. Standards involved could legitimately be considered self-serving with respect to those who drafted them and misleading to the user.

Defect 20—*Maintainability safety not guidelined.* Most standards do not separately treat maintenance or maintainability hazards but focus upon operational or so-called "point of operation" hazards. Maintainability engineering is as important as other aspects of design from a safety viewpoint due to specialized problems and exposures typically associated with maintenance, setup, in-process cleanup, and other operations support activities. A major body of knowledge in this field has existed for a number of decades in both the private and military sectors.

Defect 21—Special-purpose products not addressed. Where special-purpose machinery or complex equipment systems are being designed, there may not be any specific safety codification. Consequently, the designer is particularly vulnerable if he or she does not conscientiously seek out related codification and other bases for assuring product safety.

Defect 22—Human failure and the error-provocative design not addressed. Safety standards seldom admonish the designer to apply human factors engineering principles and practices to the extent feasible. Certain product safety standards do incorporate the results of this approach as specific design and/or performance features. The human factors approach is invaluable in accommodating reasonably foreseeable human failure, for example, thereby facilitating elimination of error-provocative or error-inducing design which seduces an operator into committing an error. That is, the unsafe condition contributes to, if not precipitates, the unsafe act. Interestingly, however, "ANSI Consumer Council Publication No. 1" includes an appendix that specifies guidelines for use in preparing safety performance provisions in safety standards. In a paragraph of the appendix relating to human effects analysis, it is recommended that safety codes for consumer products take into account the ways in which human beings will probably interact with the product that is the subject of the code. It also recommends that safety codes consider product applications that are reasonably foreseeable, control configurations, instructions, and so on. This ANSI guide further recommends that safety standards address both the promotion of "beneficial interactions" and the suppression of "detrimental ones."

Defect 23—Managerial trade-offs not disclosed. Standards seldom, if ever, spell out built-in trade-offs between operating and maintenance safety and performance efficiency, range of suitable ambient conditions, portability, economy, manufacturability imperatives, merchandising effectiveness, durability, operational ease, maintainability ease, and other factors considered in drafting the standard, or even present them in checklist form. Thus, the anatomy of decision-making and risk-taking processes inherent in the standards-setting process is not always evident. The designer is precluded from assessing the quality of the standard in respect to specific provisions or product features in relation to such trade-offs.

Defect 24—Dissenting opinions not disclosed. Consensus standards do not present information on dissenting opinions, rationales, or alternative proposals advanced during the standards development process. Thus, the design engineer cannot assess the quality of the final product with respect to such dissent.

Defect 25—*Retarding progress and judgmental requirements problems.* Established standards may actually present a deterrent to change. The designer must not be lulled into a state of complacency by the mere existence of a standard. An existing standard is no substitute for good professional judgment. Improvement must be made if progress is to continue. Standards should not be maintained or complied with for their own sake. They are of value only as long as they are consistent with performance, safety progress, or continued operating economy and lowering of production or purchasing costs. The designer must not permit an established safety standard to stifle ingenuity or sense of responsibility.

Defect 26—*Nonexistence of a safety standard.* To the competent, diligent, and responsible designer mere absence of a published code or standard poses no problem whatever. It has been said that all over the world there are extraordinary opportunities brilliantly disguised as insurmountable obstacles. In fact, the absence of a specific safety standard in connection with a particular design project represents such an opportunity. The engineer is intensively and extensively trained in specialized, highly disciplined techniques of problem solving. Trained to be an innovator, he has the cumulative, distilled wisdom of the worldwide engineering community at large at his complete disposal.

Where there is no specific standard as a guide, the engineer well knows he is professionally bound to make his own literature search to find applicable concepts, mechanisms, practices, codes, standards, and so on. He thereby draws upon the experience and creativity of other engineers in diverse fields. In the vast majority of cases, the engineer is not called upon to be inventive but to be innovative or imitative with respect to safeguarding. He need not "reinvent the wheel." It is seldom necessary to resort to avant-garde, unproven designs to solve most product safety-related problems. It is generally not even necessary to push the state of the art to its limits. In short, the absence of a safety code or standard relative to a specific design problem most certainly does *not* mean there are no guidelines or relevant resources available to the design engineer. It just means he must function on the highest professional level.

Even if a product or other engineered item has been traditionally designed and manufactured in a certain manner, from a strictly engineering viewpoint historical precedent is inadequate justification for perpetuation of any particular design feature. Merely to have survived is no index of excellence. Such a proposition applies to safety no less than to other aspects of machinery of other product performance.

From the above viewpoint, we might conclude the foregoing discussion of standards limitations and defects with the following revealing, if somewhat caustic, commentary in the form of . . .

An Epitaph

Here lies a man who had an accident. He was maimed, cruelly disfigured for life, and later died of his injuries. He made a mistake. He was unable to maintain a constantly high level of awareness of, and attention to, potentially hazardous situations over a prolonged period of time. The machine did not have an adequate guard. The accident would not have occurred if the machine had been properly guarded. Adequate safeguarding means were technically and economically feasible when the machine was designed, built, and sold. Contemporary ethics of the day provided an enabling philosophy which facilitated, if not mandated, application of available and/or appropriate technology and its translation into engineering design conditions and/or criteria relating to safety in use. It would have been ethically, technically, and economically feasible to accommodate, neutralize, counteract, and otherwise eliminate and/or mitigate reasonably foreseeable human error, failure, or oversight through more attentive and responsive original equipment design. It would have been ethically, technically, and economically feasible to design around human frailties under reasonable foreseeable conditions of service and provide superior design features. But the safety code didn't require them and the manufacturer did not incorporate them. Tough luck, man.

THE UNIVERSAL PERFORMANCE STANDARDS FOR SAFE DESIGN

The following performance standard, titled "Rules of Practice for Safe Design" or, alternatively, "Performance Guidelines for Product Safety," is not new. Its roots and content, in different forms, can be readily traced back to the early 1900s and even earlier in the evolution of engineering science and technology.

Since at least World War II, safety engineering methodology has also encompassed the contributions of human factors engineering and engineering psychology. Major statements and recommendations relating to a significant portion of this body of knowledge or generic standard have appeared in National Safety Council and mechanical engineering handbooks for almost half a century, as well as in diverse texts and references on accident prevention. It has also found wide application in forensic engineering applications, notably in products liability litigation.

These Standard Rules of Practice (SRP) should be considered "controlling" in all product design and development efforts. Where product-specific standards do not exist, or existing standards provisions conflict, the SRP are superseding or controlling criteria. However, as a general rule, the SRP are intended to supplement, not supplant, existing codes and standards. Also, all "specification"-type standards and standards provisions should be interpreted and applied using the "performance"-type SRP as their foundation. Thereby, priorities, shortcomings, conflicts, and needs will be more efficiently and competently identified, evaluated, and accommodated.

The SRP will permit you to "read between the lines" of codes and standards that you use daily, whether as an engineering employee or consultant. They will also permit you to look behind the scenes and evaluate the intent, competence, and methods of code and standard preparers.

For example, complying with a standard provision that is self-serving for the promulgating and/or publishing organization can get you into a lot of trouble if a product user gets amputated, killed, or incinerated. Your dependence upon the code or standard may have resulted in minimal product safety content well below the level that was technically and economically feasible even though, technically speaking, you did indeed comply with the code.

PERFORMANCE GUIDELINES FOR PRODUCT SAFETY
(The Twelve Rules of Practice for Safe Design)

1.0 All product types covered within the scope of these guidelines shall be designed and constructed in accordance with generic or standard principles and practices of safety engineering which are, in order of highest priority and greatest effectiveness, as follows:

　1.1 To the maximum extent ethically, technically, and economically feasible, hazards existing under reasonably foreseeable conditions of service and commerce, including intended use and reasonably foreseeable use and misuse, shall be eliminated or designed out of the product through engineering means at the earliest feasible stage in the product life cycle.

1.2 To the maximum extent ethically, technically, and economically feasible, hazards existing under reasonably foreseeable conditions of service and commerce, including intended use and reasonably foreseeable use and misuse, shall be enclosed or otherwise physically guarded through physical design features of safety "hardware" at the earliest feasible stage.

1.3 To the maximum extent ethically, technically, and economically feasible, hazards existing under reasonably foreseeable conditions of service and commerce, including intended use and reasonably foreseeable use and misuse, shall be eliminated or mitigated through warnings, instructions, training, administrative routines, procedures, packaging, and/or other safety "software."

1.4 To the maximum extent ethically, technically, and economically feasible, hazards existing under reasonably foreseeable conditions of service and commerce, including intended use and reasonably foreseeable use and misuse, shall be eliminated or mitigated through application of personal protective gear, appliances, or accessories which, however, are ordinarily to be considered temporary or interim safeguarding means (if higher-order safeguards are capable of being installed and until installation of such higher-order safeguards is effected), with the qualification that if such personal protective devices are the only safeguarding means technically and economically feasible in addition to item 1.3 above, such personal protective devices shall be specified in addition to utilization of level 1.3 safeguards; likewise, if they are applicable in any event.

2.0 No safeguarding means or feature should, itself, constitute an accident hazard by creating a new hazard or defeating an existing safeguard, and so on under reasonably foreseeable conditions of service and commerce, including intended use and reasonably foreseeable use and misuse. This follows from 1.0.

3.0 Utilizing accepted principles, methods, and procedures of human factors analysis, all product design, construction, instruction manuals, and standard operating procedures should take into account, to the maximum extent ethically, technically, and economically feasible: (a) reasonably foreseeable unsafe acts due to human error, failure, and oversight, and (b) reasonably foreseeable human physical, physiological, and psychological limitations. These design conditions should be applied to setup, operating, service, and other classes of reasonably foreseeable user personnel. They should include both safety hardware and safety software.

4.0 Any product hazard capable of being eliminated, guarded, warned against, and so on through ethically, technically, and economically feasible means, according to the foregoing guideline provisions and priorities, will be considered to be unreasonably dangerous.

5.0 Although the present performance guidelines are intended to apply to all types of products and product systems, the following portions of "ANSI Consumer Council Publication No. 1: Guide for Consumer Product Standards" (American National Standards Institute, N.Y., 1972) are hereby adopted with respect to safety performance and analysis:

5.1 Section 5: Safety Performance (p.6).
This section of the guide relates to the need to perform a safety analysis (covered by Appendix A of the guide) and also spells out some broad product safety requirements. These include safety aspects of design and construction, materials, noise and vibration, reliability and durability, applications limitations, sanitation and health, physical contact with the user-operator, energy supply and control, stability, fire prevention, end-of-life product discard and disposal, and so on.

5.2 Appendix A: Safety Analysis (p.9).
This appendix lists a number of analytical approaches to determine and assure minimum product safety performance levels. They include compliance with existing codes and standards, epidemiological analysis relating to injury frequency and severity from like hazards in similar or other products, hazard foreseeability and identification techniques, hazard and risk assessment procedures, human effects or human factors analysis and hazardous interaction prevention ways and means, failure analysis, fault-tree analysis, and other hazard prevention and avoidance procedures, etc.

A Crash Course in Safety Engineering That Can Save Your Neck 311

6.0 In addition to the foregoing, product safety analysis/audit and safeguarding means effectiveness and application should take into account "defeatability proneness" or "defeatability resistance" of safety features or safeguarding means. As a general rule, the most "defeatability-resistant" or least "defeatability-prone" safety feature should be adopted, consistent with ethical, technical, and economic feasibility constraints. This follows from 1.0.

7.0 Risk-utility, cost-benefit, or other comparable engineering economy techniques shall be expertly applied during the product safety assessment process, product safety audits, design review, and so on.

8.0 Where warnings are utilized, "hazard intensity level" shall determine signal, key, or impact word utilized, as follows:

    8.1 Danger - Immediate hazards which *will* result in severe personal injury or death.

    8.2 Warning - Hazards or unsafe practices capable of resulting in severe personal injury or death.

    8.3 Caution - Hazards or unsafe practices capable of resulting in minor personal injury and/or product or property damage.

Graphic, visual, or pictorial devices are strongly encouraged in addition to verbal messages, wherever feasible. Safety label or sign formats shall contain four (4) verbal and/or pictorial messages which communicate:

    8.4 The level of hazard intensity (Danger, Warning, Caution, etc.)

    8.5 The nature of the hazard

    8.6 The consequences that can result if instructions to avoid the hazard are not followed

    8.7 Instructions on how to avoid the hazard (clear, thorough, and efficient)

9.0 Where a conflict arises between an existing code provision and the foregoing, the present "Performance Guidelines for Product Safety" shall be controlling and shall supersede specification or method-of-execution provisions. The present guidelines are intended to supplement, rather than supplant, currently existing specification and/or performance codes, standards, practices, and procedures. They are intended to comprise a frame of reference in the form of a controlling foundation for development, updating, modification, or continued application of specification or method-of-execution-type codes and standards. They should be transmitted with every code and standard.

10.0 Where there is no existing safety code, standard, or code or standard provision pertaining to a particular product, product feature, or safety hazard, the present performance guidelines shall apply, at minimum, across the board, with no exceptions.

11.0 Product design consideration shall include identification, evaluation, and control (i.e., elimination, mitigation, or other highest-order ethically, technically, and economically feasible accommodation) of reasonably foreseeable patent and latent hazards at all stages or phases of the product life cycle. Foregoing Sections 1.0 through 10.0 deal with hazard control, specifically. It shall also include application of technology assessment and technology transfer methodology, in accordance with accepted engineering practice.

12.0 Standard safety engineering methodology shall be applied to the identification and evaluation of product safety hazards prerequisite to their adequate, necessary, and sufficient control through elimination, mitigation, or other satisfactory accommodation based on this methodology. This shall include application of hazard foreseeability techniques, market research and market intelligence procedures and organization (of the overt type), genealogy tracing or life-cycle monitoring, product recall and/or modification, and so on to the extent ethically, technically, and economically feasible. It shall also include application of technology assessment and technology transfer methodology, in accordance with accepted engineering practice.

DEFEATABILITY PRONENESS OF PRODUCT SAFETY AND FUNCTIONAL FEATURES

Provision 6.0 of the foregoing "Performance Standards for Safe Design" uses the term "defeatability proneness." This is a crucially important and practical concept. In fact, it effectively underlies any disciplined attempt to construct and prioritize rules of practice for safe design such as those presented or any other ranking of analogous guidelines.

There are well-known standard design rules for industrial machinery and general product safety. However, up to now this knowledge has not been classified, detailed, or prioritized with respect to one important problem. This has to do with selection of highest-order safeguarding means from the viewpoint of defeatability proneness.

Classical cataloging and grading of machinery and equipment guards, for example, do not facilitate more rigorous ranking and selection of engineering alternatives except within rather broad bands or classes of mechanisms such as (1) fixed versus adjustable enclosure guards, (2) the general class of interlocked guards using assorted energy sources, (3) automatic guards of various types, (4) different varieties of remote control, placing, feeding, and ejecting devices, and so on.

Without taking into account defeatability proneness, less safe, lower-order, inferior, and inadequate guarding modes are more likely to be selected and specified during the product planning, development, and applications engineering process. Only rationalized and disciplined criteria for evaluation are capable of facilitating selection and specification of superior safety hardware and/or software.

The information given in this article restructures and reorganizes relevant guarding information, concepts, and practices in a manner appropriate to solve this problem. It advances, develops, and applies the idea of defeatability proneness in the form of a ranking system utilizing defeatability indexing. The resultant defeatability index will prove valuable to engineering designers and product safety analysts in further assuring that highest-order safeguarding means are selected and specified, consistent with technological and economic feasibility considerations and reasonably foreseeable conditions of service and commerce.

The defeatability index system will also be of assistance in connection with products liability litigation. Both retrospective and contemporary product safety or hazardousness content—depending upon your viewpoint—can be effectively assessed through application of this technique.

The Basic Problem. Designers and manufacturers have long been concerned with preventing or terminating equipment operation in the vicinity of potentially hazardous machine elements in order to preclude or mitigate accidents and injuries due to reasonably foreseeable error, failure, or oversight by human operators.

One aspect of designing for machinery and general product safety relates to an apparent propensity for some equipment operators and other product consumer-user types to purposefully bypass, disable, alter, or otherwise defeat OEM (Original Equipment Manufacturer)-supplied safeguarding means. Such potentially dangerous decision-making and risk-taking behavior, on the surface, might well appear to be a viable defense—depending upon the particular situation—for the equipment designers and manufacturers involved.

However, appropriate, necessary, and sufficient safeguarding means must not themselves present hazards. This particular design rule for machinery and general product safety is a derivative of more fundamental principles, of course.

From this viewpoint, if it is reasonably foreseeable that a particular safety device can or will be bypassed, disabled, altered, or otherwise defeated to permit product function, then it would appear that such a feature, even if it incorporates certain so-called passive or quasi-passive safety elements or aspects, can be said to be unsafe and unfit for its intended purpose. Thereby, it would also be in violation of relevant generic design safety principles and practices. Here we confront issues in safety science relating to industrial engineering and human factors.

The Basic Concept. But now let us be specific and provide the necessary insights to solve our first problem relating to evaluating the likelihood that OEM safeties will be bypassed, disabled, altered, or otherwise defeated. That is, to what extent is it reasonably foreseeable that under reasonably foreseeable conditions of service, a product user or equipment operator will utilize safety features provided by the designer-manufacturer?

One might be tempted to believe (erroneously) that, as a general proposition, the easier it is to use a safety device or, conversely, the harder it is not to use it, the higher should be the prospect of its effectiveness and operation or activation as intended.

These two viewpoints are, however, definitely not identical. For example, owners of certain automobiles are provided with ignition-interlocked occupant restraint systems. In one version the seat belt wraps progressively about the driver as the car door is closed. It is designed to be effective when this standard operating procedure is followed, since the vehicle cannot be started without the interlocked belt in place.

But with this one particular occupant restraint system, the driver (or passenger) can simply leave the belt system interlocked while sitting down on it. The vehicle can be started up even though the driver has purposely defeated and bypassed the device, utilizing a very simple change of recommended standard operating procedure. Nevertheless, this particular interlocked occupant restraint system is advertised and promoted as a bona fide passive safety feature by its designer-manufacturer. In fact, it is even discussed as such in professional, trade, and government publications.

Our point here is that the aforementioned restraint system type (the system actually consists of lap belt and knee pad) is indeed easy to use. The problem is that it is also easy to defeat! This simple example contains the necessary seeds of an elegantly simple, workable, and rather obvious method for evaluating the safety or hazardousness content of interlocked safety features, including those intended to be the passive type.

Once we have constructed the analytical procedure for identifying and evaluating the levels of susceptibility to being defeated, we will possess a critical tool for assessing the extent to which the contemporary ethical imperative has truly been fulfilled by the designer-manufacturer.

In any individual case, on the basis of the methodology following—which is very simple and direct—product safety features under scrutiny may be classified according to their safety-defeatable quality.

We are going to establish a prioritized categorization and ranking of levels or classes of defeatability. The ranked list of defeat modes or conditions is as follows, with the additional property that only Classes 1 through 5 are reversible for purposes of resetting, restoring, or restarting product and/or safety function:

> Class 1—Change in standard operating procedure (SOP) for product and/or safety feature (revert to prescribed procedure to restore)
>
> Class 2—Use of simple tools and/or contrivances
>
> Class 3—Use of special or complex tools or equipment
>
> Class 4—Major product redesign, remanufacture, alteration, rebuilding
>
> Class 5—Product function disruption, impairment, or damage (repair to restore)
>
> Class 6—Product function destruction (irreversible or incapable of having product function reset, restored, or restarted).

Note that a Class 6 defeat mode implies, in essence, a kind of self-destruct feature. That is, as far as the product is concerned, such a defeat mode is a veritable catastrophe or doomsday event, irreversible in nature. At minimum, the economic value of the product per se is irretrievably and totally lost. This is an extreme case that would ordinarily not be encountered in the industrial or household environment. However, its importance in the present context is precisely its extreme nature as a limiting condition.

The Defeatability Index. In any case, armed with our safety defeatability index, we can reasonably determine to what extent any particular product is vulnerable to being operated without the OEM safety feature functioning as intended.

The key to defeatability rating, ranking, or indexing is not solely consideration of ease of use of the safeguarding means, but also parallel concern for the ease with which it may be defeated. Bear in mind, however, that in certain instances both ease of use and comfort level may influence the extent to which an owner or operator may go to attempt to defeat the device. This can be an important companion issue that may have to be addressed and evaluated in any particular instance.

It follows from the above analysis and discussion that the defeatability attribute of a safety device may be, to a greater or lesser degree, passive or automatic in nature. A readily defeatable passive safety device and/or one that is overly cumbersome or uncomfortable to use is a thoughtless, intolerable, reckless, and poor design in violation of standard and fundamental principles and practices of safety engineering, just as much as if there was no safety device present at all. In fact, it is worse. It is a contradiction in terms. Rather than solving the original problem, it effectively creates a new one. Such a device may be said to be defeatability prone. From a human factors standpoint it is a terrible design.

Note that we are including as D.I.=1 (D.I. = Defeatability Index) those devices that can be effectively disconnected or disabled in an unauthorized manner—and restored—by hand without the need for any separate tools or aids.

On the basis of the above analysis and discussion, it would appear that the bottom line in evaluating the legitimacy of safeguarding means represented as passive devices would depend, in large part, upon the technical and economic feasibility of higher-order defeat modes having been provided as original equipment by the designer-manufacturer. Note that we are using the term "higher order" in the sense that the harder to defeat the device, the higher the order.

If, in any particular case, higher-order defeat modes would have been technically and economically feasible for the original equipment manufacturer to have provided, then we must conclude that the product incorporates a design deficiency or defect. Therefore, the product was unfit for its intended use with respect to its accident prevention attributes.

This line of reasoning is entirely consistent with general safety engineering methodology, on a retrospective basis, going back many decades. Therefore, conclusions reached upon application of the above techniques are unassailable from an engineering standpoint.

Let us return for a moment to our example of the interlocked seat belt. The type that becomes ignition-interlocked only after it is latched is actually a combination active-passive device. It must be latched before the car will start. Unfortunately, it is easily defeated if you sit on it while leaving it latched. Therefore, it has a low defeatability index (i.e., it is a Class 1 device by our categorization of defeat modes).

Similarly, although you can literally "sit into" the specific design described earlier in our discussion, as it automatically or passively (without the need for human intervention or human action) wraps around you when the door is closed, one can readily defeat it by sitting down on it while it remains interlocked even though there is no latching action necessary on the part of the driver or occupant.

Thus, this latter style is also a Class 1 device by our indexing system. However, it does not have an active *deactivating* aspect although it incorporates an emergency disconnect latch which, incidentally, will not terminate equipment operation when deployed and which may also be utilized as a bypass.

In short, neither of the above automobile occupant restraint systems is a very effective or positive passive safeguarding means, since a simple procedural maneuver permits it to be bypassed at will. However, the second does represent an advance over the first in that it is reasonably foreseeable that some persons may simply forget to buckle up, for whatever reason, quite aside from any consideration of purposeful defeating of the system. That is, at least the second restraint concept recognizes the more frequently encountered types of human error, failure, and oversight and the fact that not even normally attentive and careful people can be relied upon to remain constantly alert to hazards and to exhibit risk- or hazard-averse behavior over prolonged periods of time. Or, at least, they do not exhibit the intensity of such behavior necessary to generate the action sequence that activates a safety device unless it is the necessary initiating element of a combination active-passive safety.

Utilizing other common terminology, the second restraint concept is more error-accommodating, error-forgiving, or error-tolerant than the first. This is, obviously, a step in the right direction.

But again, if higher-order defeat modes or higher-order safeguarding means are technically and economically feasible, Class 1 devices would appear to be in violation of standard principles and practices of safety engineering, given the ethical mandate or imperative discussed earlier.

In the same way safeguards consisting solely of instructions or training, operating manuals, and/or warning signs and decals—whether verbal, graphic, or a combination—incorporate Class 1 Defeat Modes. Only procedural deviations or nonapplication of safety software is involved in order to bypass or defeat warnings and instructions, whether such reasonably foreseeable misuse is inadvertent or purposeful.

Application of the Defeatability Index. Still another illustration of applying the defeatability index is in relation to enclosure-type guards. A simple barrier guard with quick-detachable fasteners such as wing nuts, spring-loaded lockballs, gravity pins, and so on, can be removed by hand even without the need for simple tools. In fact, these devices are, by design, intended to be removable manually. Therefore, it is procedure alone that is capable of resulting in their deactivation or unfastening as safeguarding elements. Thus, quick-removable, quick-detachable, or quick-releasable guards are intrinsically also Class 1 Defeat Mode-type devices, although acceptable in selected applications where they may be the highest-order safeguard feasible.

In contrast, securing means such as screws, nuts and bolts, cotter pins, keys and keyways, non-quick-release pins, snap rings, and so on would be categorized as Class 2 Defeat Mode-type fasteners. Only simple tools are necessary to remove (and replace) them and, therefore, guards incorporating them.

Guard weldments, glued joints (or fastenings using other adhesive types), riveted connections, and so on would fall into Class 3 Defeat Mode devices. That is, special tools are required to detach, break, or otherwise undo guard connections as well as restore them to intended operational effectiveness. However, product function per se is not impaired.

Since even interlocking and automatic guard types can conceivably be designed and built to incorporate Class 1 or Class 2 Defeat Modes, designers and manufacturers must carefully analyze the vulnerabilities of such safeguard configurations.

As a general rule, it would appear reasonable to conclude that Class 1 and Class 2 Defeat Modes would ordinarily be unacceptable as the basis for designing or constructing product safety features, as usual taking into account technical and economic feasibility.

Exhibit 13-1 is an automatic seat belt advertised as a passive occupant restraint. It is, in fact, a Class 1, D.I.=1 type safety. Exhibit

EXHIBIT 13-1   EXHIBIT 13-2

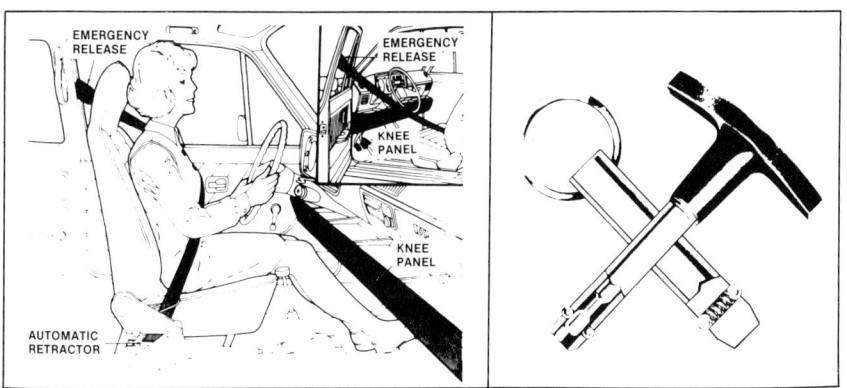

13-2 illustrates another low-order Class 1 Defeat Mode-type safety with D.E.=1, of the so-called quick-release spring-loaded lockball design and construction. In some applications such devices can be acceptably effective despite their defeatability-proneness limitations in others.

Exhibit 13-3 shows some familiar fastener types. The common button may be a Class 1 or 2 quick-release device with D.I.=1 or 2 relative to either intended procedure in buttoning and unbuttoning, or with respect to its disabling, since either manual manipulation or simple tools may be needed to remove it, depending upon how securely it is sewn on. Also, of course, a button can become disabled through ordinary wear and tear—or reasonably foreseeable conditions of service—when the buttonhole becomes frayed or enlarged or the button and thread loosen up and the button falls off.

The paper clip, however, is a Class 1 device since it is, in effect, a quick-release item intended to be removed and replaced at will,

EXHIBIT 13-3

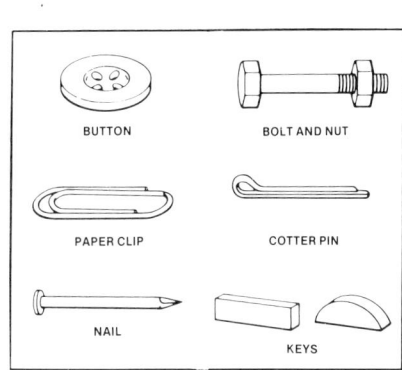

but can also readily fall off or be accidentally dislodged depending upon its condition. A zipper is also a Class 1, D.I.=1 quick-release device that is susceptible to failing or becoming disabled or defeated if it becomes jammed or is not properly sewn or started.

The ordinary paper staple is a D.I.=1 or 2 device, depending upon how heavy-duty it is. But a wood-joining staple would be a D.I.=2 fastener, since simple tools are required to remove it; likewise for the cotter pin.

The nail shown in Exhibit 13-3 would ordinarily be a Class 2, D.I.=2 device, removable and replaceable with simple tools. The key is Class 1.

The defeatability index approach can be utilized to analyze any means of security, fastening or joining system, interlock, and so on where, under reasonably foreseeable Conditions of Service (COS), vulnerability to accidental or purposeful disablement is a potential problem.

In addition to the foregoing examples of items amenable to such analysis, the following serve as illustrations:

1. Chain-link fence ties of different thicknesses and materials
2. Rope and cordage-type rigging knots
3. Wire rope and cable fastenings
4. Valve, pipe, tube, and hose connections, fittings, and adapters
5. Retaining rings and spring clips
6. Mechanical power transmission (MPT) drives and elements
7. Electrical connectors and adapters
8. Mating, interference-fitting elements
9. Assembly and fabrication aids and guides including production jigs, fixtures, tabs, pins and dowels, rails, blades, keys and keyways or keyseats, detents, and so on
10. Adhesives and sealants
11. Torqued fasteners, lock nuts, and prevailing-torque nuts (split)
12. Rivets
13. Clamps
14. Lockwires or lockwired bolts, with and without seals
15. Electrical meter-type wax disk seals
16. All types of mobile vehicle occupant restraint systems, from manually latchable lap belts to air bags
17. Mechanically actuated limit switches and sensors and application configurations

A Crash Course in Safety Engineering That Can Save Your Neck 321

18. Proximity switches, remote sensors, magnetic switches, and application configurations
19. Cams and followers
20. All machinery and equipment operating controls
21. Machinery and equipment process controls and instrumentation, all types

By its nature, the defeatability index system may also be applied to assess malfunction propensity of controls and instrumentation, safety interlocks and activating means, and fastenings or securements under reasonably foreseeable conditions of service. Malfunction is of as great concern as inadvertence and purposeful disabling with respect to defeating of intended function. Exhibit 13-4 shows a wider range of fastener types than Exhibit 13-3.

Approach and Application. Thus, there are three basic uses for the defeatability index system. These include:

1. Analysis of susceptibility to malfunction under reasonably foreseeable conditions of service
2. Analysis of susceptibility to inadvertent or nonpurposeful, reasonably foreseeable misuse or nonuse and reasonably foreseeable conditions of service

EXHIBIT 13-4

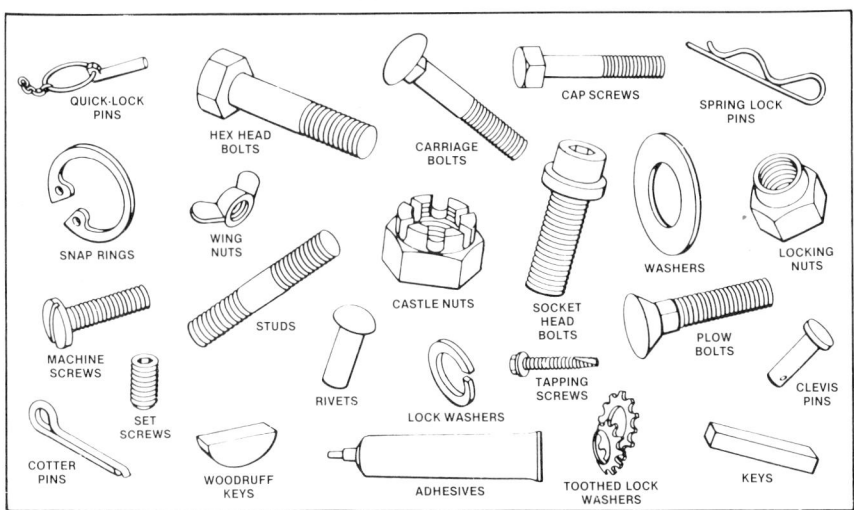

3. Analysis of susceptibility to purposeful misuse, abuse, bypass, disablement, alteration, and other defeat modes under reasonably foreseeable conditions of service

The defeatability index approach should be applied in any design review and product safety audit program. It offers a rational, disciplined framework within which to assess hazardousness content and prioritize design safety alternatives before product safety hardware and software selection and specification are finalized and offered in the marketplace.

Although the defeatability index system really incorporates nothing new, it does represent a reorganizing and reordering of well-known ideas. In this sense the approach is an additional small step in the direction of more scientific product safety management.

In assessing the above—or in fact any control, securement, safeguard, and so on—it is important to bear in mind that:

1. When one is in the midst of having an accident, so-called active safeties may not always be actuated, whether due to lack of presence of mind, bodily instability and lack of leverage, inconvenient position of hands or body relative to safety devices actually provided, or whatever.
2. When an accident is in progress, lower-order safeties will not be as effective as higher-order devices (i.e., safety cable versus automatic body pressure bar or higher-order safety devices) for relevant accident sequences.
3. During an accident, normal or standard operating procedures and warnings may be difficult, if not impossible, to follow (i.e., losing one's balance on top of a tanker truck catwalk where there are not standard handrails, etc.).
4. In any particular case a relatively lower order safety may, nonetheless, be the highest-order safeguarding means that is technically and economically feasible.
5. Depending upon product life-cycle phase or stage, a relatively lower D.I. may be capable of being converted into a higher D.I. value by entities in responsible, effective, or immediate managerial or physical control; that is, what may have been a D.I.=1 safety at the time of design, manufacture, and original sale may be converted into a D.I.=5 at the time of end-of-life scrapping. (A refrigerator or freezer susceptible to a child locking himself inside has a D.I.=1 at the point of original sale, with respect to this particular reasonably foreseeable hazard, whereas this

same product, at the time of scrapping and with the door properly removed, has a D.I.=5 with respect to the same hazard.) This assumes that the OEM D.I.=1 attribute was the highest-order safeguarding means practicable at the outset—an actually incorrect premise in the case of a refrigerator or freezer since interior quick-release-type unlatching features can readily be envisioned that are capable of being actuated by even a panic-stricken youngster who does not necessarily know such a safety device is present as a built-in OEM-supplied feature of design and construction.

6. Reasonably foreseeable conditions of service (COS) may influence the D.I. value that represents highest reasonable condition of safety (COSF) for the product (a) that is technically and economically feasible, and (b) that should be applied as a design condition, criterion, or factor and built into the product as an OEM feature.

7. The competent product safety analyst must be familiar not only with principles and practices of safety engineering, including hazard identification, evaluation, and control, but also the broadest spectrum of actual, real-world accident types experienced in the field on industrial and/or consumer products of all sorts. Product safety analysis or auditing, or design review that concerns itself primarily with experience related to the particular product line being studied can be a terribly misleading indicator of actual product hazardousness content from the technology transfer standpoint. It represents a form of management myopia that can cost a manufacturer dearly. Product safety analysis must be broad-ranging in this sense.

The bottom line is that after product hazards have been identified and evaluated, and when safety features to eliminate or control them are being considered, the designer should ask the question: "How can my safety device be defeated?" Then, by applying the defeatability index procedure, he or she will never again design an inadequate safeguarding means—at least not unintentionally or unknowingly.

DEFEATABILITY INCENTIVES

Let us now be more specific as to why someone might try to defeat a safety device or modify product function. In order to address this aspect of the defeatability-proneness problem, we will talk in terms

TABLE 13-1

DEFEATABILITY INCENTIVES

Class	Motivating pressure, stress, or other incentive	Potential saving, reward, relief, or other objective[a]
1.	Authoritative order, instruction	Subordinated compliance
2.	Illness, handicap, or elderly infirmity and incapability or incapacity, lack of dexterity	Savings in time, comfort, convenience, effort, or energy, or doing the "best one can" (though not good enough)
3.	Aesthetic or ethical repugnance	Aesthetic or ethical acceptability or satisfaction
4.	Custom, habit, pattern, routine	Conformity
5.	Fatigue	Saving or reduction in effort or energy
6.	Discomfort, pain, irritation	Comfort and relief
7.	Anger or other adverse or stressful mood or emotion including overzealousness, overconfidence, and complacence	Outlet for venting emotions, indulging personal hangups, and so on
8.	Being late or being in a hurry	Saving in time, shortcutting
9.	Inconvenience	Convenience or ease
10.	Utility impairment	Utility maintenance or improvement
11.	Profit-detracting or costly	Profitable or inexpensive
12.	Scarce or limited resources	Resource conservation
13.	Improper design of product re: safety feature that makes product unsafe under certain reasonably foreseeable COS or that makes safety feature hazardous itself	Superior safety content or elimination of hazardous aspect of safety feature

Class	Motivating pressure, stress, or other incentive	Potential saving, reward, relief, or other objective[a]
14.	The natural, ornery, cussedness of human nature including recklessness, thrill-seeking propensity of some persons some of the time, and gross assumption of risk	Outlet for exorcising the demon within and pitting "man against himself" (after Menninger[b]), including the "objective" of self-destruction as an abnormal tendency in the most extreme case
15.	Other purposeful, deliberate, conscious, rational, or rationalized incentive, motivation, maintenance of status quo, stress or distress avoidance (fear of being reprimanded), and so on	Other rewards, benefits, savings, relief, freedom from undesired constraint, maintenance of status quo, stress or distress avoidance (physical, physiological, and/or psychological)

[a] As perceived by subject, whether real or imagined on a rationally evaluated net basis.

[b] *Man Against Himself,* Karl A. Menninger (Harcourt, Brace & Company, New York, 1938)

of a companion concept of great importance. It has to do with defeatability incentives or incentives to err and defeat, modify, or bypass product functions or safety features. Again, contrast this with usage incentives and use-inducing or use-compelling attributes.

Table 13-1, Defeatability Incentives, shows motivators to action or efforts that are purposeful or deliberate, whether rationally or emotionally based. The classification is not rigorously ranked in order of how compelling or strong the incentives or inducements are, since this can vary from moment to moment and circumstance to circumstance. This is the same kind of situation we face with respect to accident-proneness of individuals. There is no such thing as an accident-prone individual. It implies that a person may be prone to becoming an accident statistic due to some fixed, perverse characteristic or quirk of the individual. But any of us can be accident-prone or error-prone at certain times and not at others. Or, we can be error-prone in an error-provocative situation, when we are confronted or trapped by an accident-conducive design, and so on. Tolerance threshold is very important also.

However, for comparable or equivalent IMPS (an acronym for some as-yet undiscovered, indefinable unit of quantitative measurement permitting us to measure the amount or level of incentive, motivation, pressure, and stress), they are offered in approximate descending order relating to how irresistible or controllable the incentive, motivation, pressure, stress, and so on is under reasonably foreseeable circumstances or conditions of service for a human being. There may well be violent professional disagreement over this ranking. However, our presentation is merely intended to be illustrative rather than quantitative or rigorous.

Remember, we must also recognize various nonpurposeful reasons for defeatability-proneness of safeguards and product functions, as follows:

1. Preoccupation, diversion, distraction, and so on that interferes with concentration on the task or situation at hand
2. Inadvertence or oversight
3. Unawareness, ignorance, misperception, or misunderstanding
4. Forgetfulness
5. Unfamiliarity with safety features and consequent opportunity for human error, failure, confusion, and disorientation due to newness of the procedure, process, or product
6. Opportunity for error or failure due to inadequacy or absence of warnings, training, instructions, or error-provocative design
7. Other nonpurposeful, nondeliberate, nonrational, or nonrationalized incentive, motivation, pressure, stress, stimulus, and so on
8. Error in judgment, sense of perspective, other perceptual error, and so on

THE FIVE KINDS OF ENGINEERED PROTECTIVE, SAFETY, AND CONTROL FEATURES

Engineers are typically employed in industry to design and redesign products with a view toward improving produceability, productivity, operating economy, flexibility, durability, compactness, maintainability, and profitability. For the present, we are neglecting cosmetic features that improve marketability and features that guarantee or promote lucrative repair and/or spare parts business or future unit sales through planned obsolescence.

Designers and businesspeople have learned to trade off one feature for another in order to achieve product-line success and profitability. Unfortunately, sometimes they also have been pressed to compromise engineering design ideals and practices to achieve maximum product-line investment returns. However, even the most socially irresponsible businesses that employ engineers normally incorporate a substantial variety of engineered protective features into their products. These may be included purely for the purpose of commercial survival, quite aside from those specifically intended to preclude or mitigate personal injuries from product-related accidents and those that are needed to conform to minimum provisions of codes, standards, and governmental regulations.

Protective features can be subdivided into five generic categories. As an example, a typical piece of industrial machinery will ordinarily be designed and built with the following in mind:

Type 1—Functional Performance. A machine must be capable of performing its intended function. It must incorporate the best thinking of the designer-manufacturer to achieve reasonable, functional performance objectives apart from and independent of any other objective. Engineers must be inventive, innovative, even imitative in applying technical concepts, ingenious mechanisms, and systems configurations and features to the production problem at hand. The cumulative technical know-how of virtually the entire engineering and scientific community at large worldwide is accessible to designers and manufacturers.

By their nature, engineering and scientific principles and practices are universally applicable. Soundness of underlying engineering design to achieve the desired product function is quite obviously the first objective that must be addressed. It sets the stage for consideration of other key design objectives—that is, final functional performance levels achieved vis-à-vis original targets or design criteria must be balanced and evaluated compromises or trade-offs among the five generic design objectives. These are:

1. Functional performance
2. Equipment or product internal protection
3. Process control and quality assurance
4. Loss control or asset conservation (i.e., relative to associated property, plant, and equipment exposure)
5. Accident prevention (i.e., personal illness and injury relative to product and workplace safety and health)

Interestingly, attainment of accident prevention objectives seldom necessitates significant compromises with respect to other product-line targets.

The safety aspect lies in the faith of the designer, based upon his or her knowledge and experience, that application of appropriate engineering and scientific principles to original equipment design and development will, in fact, yield desired operational results. A process of design, experimentation, and redesign utilizing the scientific method provides the safest and best-controlled design strategy. The competent product design and development process is a rigorously managed, controlled procedure. It is a protective system or process in its broadest sense, capable of ensuring adequate and appropriate design if properly managed. From this viewpoint, it is a protective technique or a safety management device—that is, with respect to functional performance, the engineer strives for designs and constructions that are function-assuring, function-conserving, or function-preserving.

Although this may be an unfamiliar usage of the term *safety*, from an engineering perspective it is a quite meaningful application. Product function "defeatability-proneness" can give rise to safety and health hazards as well as other problems. If product function cannot or is not assured or controlled to an acceptable degree, accidents can occur to property, plant, equipment, and human beings.

But now, having built a safety or control feature into the product design and development process itself, the designer-manufacturer next must consider manufacturing quality and performance reliability assurance of the machine. Quality control assures that a product's own component parts will perform their intended tasks within the machinery or product system. Only then can we address process and quality control of items made on or in the machine.

Of course, there is overlapping of the two objectives since the end product can only meet specifications if the machine itself is able to function properly. However, for analytical purposes it is more instructive to separate the two situations. We will return to a consideration of process-type quality assurance a little later.

Type 2—Equipment/Product Protection. Industrial, commercial, and residential machinery products must be protected from internal damage due to mechanical and electrical malfunction. Designs must endure and/or prevent reasonably foreseeable overloading and overstressing, vibration, normal wear and tear, corrosion, temperature extremes, feeds and speeds, and operator errors that could result in equipment damage or destruction, and so on. They must be designed against reasonably foreseeable, externally imposed adverse conditions.

Engineering-type factors of safety must be built into practically every designed element. The term *factor of safety* usually refers to the ratio of breaking load on a member, structure, or mechanism and the safe permissible load. This ratio is "allowed" when designing the member, bearing in mind reasonably foreseeable conditions of service. It provides for the possibility of uncertainties of various kinds, including variation of strength possibly resulting from deterioration in service. Another term, the *proof factor of safety,* is the factor of safety based upon the "proof" load. This, in turn, is defined as that load which a structure or mechanism must be able to withstand while remaining serviceable. Alternatively, *proof or test load* can be defined as some load greater than working load to which a structure or mechanism is tested in order to see whether it can be withstood without permanent distortion or damage. The term *safety factor* is also used in nonstructural senses, but the idea is the same.

In short, the product must be designed and built to perform efficiently under all reasonably foreseeable conditions of service, including reasonably foreseeable use and misuse or abuse, and must be properly applications-engineered by its designer-manufacturer-marketer. Numerous types of integral protective features and auxiliary devices are available for internal and external equipment protection purposes. We will further discuss the applications engineering process, trade custom, or practice later.

Type 3—Property, Plant, and Equipment (PP&E) Loss Control.
Machinery and other products must be designed to localize, preclude, or mitigate damage to surrounding property, plant, and equipment to the extent technically and economically feasible. This includes damage due to conditions of normal operation. It also embraces mechanical or electrical malfunction and derivative PP&E damage from altered or damaged process raw materials possessing potentially adverse properties relative to surrounding, exposed PP&E.

PP&E loss prevention is one component of overall "risk management," "loss control," and "asset conservation" as they relate to the prevention or control of property damage in its broader sense.

Type 4—Quality Assurance and Process Control.
A basic industrial machine, and the process or system configuration of which it is a part, must be designed to yield end products of given specifications with defined accuracy and precision. Thus, the production process itself requires engineering-type controls or safe-

guards to assure that desired features of end products are generated on a reliable and sustainable basis with minimal equipment or process disruption and downtime, degradation of product quality over time due to equipment or system inconstancy, and so on. Also, raw materials must not be damaged or destroyed at any stage of production. This category of safety or control feature is wide-ranging and includes in-process and final inspection and testing as elements of the overall quality assurance process.

Built into the manufacturing process must be the capability to identify a product's deviation from specifications for purposes of reworking or rejection. This aspect of industrial production utilizes highly sophisticated engineering, mathematical-statistical-probabilistic, and instrumental techniques. It includes hardware, software, calibration, and testing apparatus and techniques ranging from metrology laboratory to production-line quality.

Type 5—Accident Prevention. Finally, the product, machinery, or process system should prevent and/or mitigate personal injury and illness under reasonably foreseeable conditions of service, including reasonably foreseeable use and misuse. Well-established bodies of knowledge have long been available in the field of safety engineering and from allied technologies to guide designers, manufacturers, and marketers.

Incidentally, personal safety and health-type safeguarding or accident prevention is commonly referred to as *product safety*. Certain recent practices would also consider product safety to be one significant component of risk management and loss control.

In our discussions, we are including health as well as accident hazards. The former are usually considered to be the province of occupational health or industrial hygiene and toxicology. However, there are both industrial and nonindustrial ramifications to the industrial hygiene field. For example, many consumer products give rise to potential health hazards through inhalation, ingestion, and skin absorption routes of entry. These include everything from foods, drugs, and cosmetics to drain cleaners, paints and thinners, lighter fluids, and other toxic substances. Flammability and explosion hazards are also included. Traumatic injury such as amputation, as well as acute and chronic illnesses and burns, are the subject of our interest in the overall product safety field.

Thus, product and/or process control, protection, and safety are broad terms that encompass a wide variety of engineered features, technical objectives, scientific disciplines, and management styles. It is also clear that safety features directed at avoiding personal injury and illness comprise but one of five distinct and specialized classes of controlling, safeguarding, or protective means.

Concern for personal safety and health through responsible and thoughtful design is not unique when compared with other types of product and process control and protection objectives. From technological and economic viewpoints, features assuring personal safety and health are at least as significant to efficient functioning of products and product systems as process control, equipment protection, and overall PP&E loss prevention and control. These have, in fact, become increasingly significant components of total product and system cost.

Arguments questioning the utility or desirability of safeguards as prohibitively expensive or excessively forgiving of human error must be viewed in perspective. For one thing, general technical research and development progress is typically achieved at successively greater incremental cost per unit of improvement. However, it generally has been focused upon increasing machinery and process system capabilities to achieve ever higher levels of productivity, process control, equipment internal protection and overall property, plant and equipment loss prevention.

Significantly, recently legislated aspects of product safety appear to deal more with personal safety, health, and environmental protection issues. It would appear that designers and manufacturers need this degree of official prodding to incorporate at least minimal safeguarding means. This is in contrast to an apparently more natural propensity to focus upon productivity, end-product quality, equipment and process protection, or the more visible bottom-line, shorter-term objectives of commercial profitability and asset conservation. But such management myopia can prove to be dangerous and costly.

The more balanced approach to product design recognizes the full short- and long-term potential of safety engineering. This frequently proves to be an excellent business approach as well. Balanced product design takes into account all types of protective, control, and safety needs as described above. It is the ultimate cost-effective product design and development strategy. In addition, it is a hallmark of social progress consistent with efficient resource management.

Considerable efforts and funds have long been expended by profit-minded managements to assure the efficient and proper functioning of their products in home and factory. The necessary controls, and safeguards to assure efficient and proper functioning and to avoid product damage, production downtime, raw materials loss, unacceptable end product, and so on, have generally been accorded highest priority. But life safety has been short-changed.

In fact, user-operator-consumer safety and health are just as important to man-machine system efficiency, productivity, conservation, and profitability. Indirect costs of accidents are frequently so massive and pervasive that user safety and health should be one of the highest priority objectives of progressive managements, even if only from the viewpoint of enlightened selfishness. Products that promote and assure user safety and health are good business for manufacturers.

From another viewpoint, costs of personal injury-type safeguards are generally comparable and consistent with costs of other product and/or system features dedicated to assuring functional performance, equipment protection, process and quality control, general loss control or asset conservation, and other modes of accident prevention. In addition, product features focused upon environmental protection recently have begun to appear on various types of products and equipment.

To summarize, the five engineered protective, control, and safety objectives can be described as follows:

> Type 1. Functional Performance Assurance. The product must work right in the first place.
>
> Type 2. Equipment Protection and Control. The product must not self-destruct and blow up in your face.
>
> Type 3. Process Control and Quality Assurance. The product must not "eat up" and ruin the raw materials or workpieces.
>
> Type 4. Loss Control. The product must not destroy everything around it or in its path.
>
> Type 5. Accident Prevention. The product must not attack everyone in sight or have an unfettered propensity to use human beings for workpieces and raw materials.

An important additional observation can be made on the basis of the above categorization. The original design, development, and manufacture of many items of machinery, equipment, and consumer products do not reflect some primitive exercise in the irrational, casual application of brute mechanical force. In fact, there may be embodied vast and cumulative know-how in engineering science and/or machinery-making arts. In their functional aspects alone, many industrial and consumer products are veritable masterpieces of ingenuity and technological wizardry. Therefore, a lack of thoughtfulness as to adequate safety in use is all the more disappointing and unacceptable from the safety engineering viewpoint.

Any objection to a safeguard because it interferes with production or decreases productivity must be viewed in perspective. For example, many design trade-offs normally built into a machine or system influence its productivity. Functional design factors, equipment protection features, process and quality controls, and general loss prevention and control features are as important determinants of equipment and overall plant productivity as personal injury-type safeguarding means.

Thus, when considering any apparent productivity decrement, we must view it against the backdrop of decremental trade-offs already built into the product through thoroughly rationalized original equipment design trade-offs with respect to Type 1 through Type 4 engineered protective, control, and safety features. From this viewpoint, apparent decremental productivity occasionally generated by incremental safeties actually may not represent an unacceptable penalty. Comparative analysis can be of assistance in assessing relative productivity characteristics of alternative design philosophies and features.

AN EPIDEMIOLOGICAL APPROACH TO SAFETY ENGINEERING

The concept of defeatability proneness is implicitly built into historically mature standard principles and practices of safety engineering, as observed above, even though it may not be explicitly articulated in the literature.

But there is another significant application for the defeatability-proneness concept. It has to do with a hazard control approach known as Haddon's theory of abnormal energy transfer. This theory of accident causation is directed toward the control of abnormal, unexpected, undesirable energy exchanges and releases capable of causing personal injury or property damage. It is attributed to the late William Haddon, Jr., M.D., who until 1966 was associated with the New York State Department of Health. Dr. Haddon had more recently been president of the Insurance Institute for Highway Safety, Washington, D.C. He found that most, if not all, types of damage to living and inanimate structures fall into a relatively small number of causal groups.

Energy is the capacity to do work or produce an effect; it is consequently central to both function and malfunction. Some of the energy forms that produce or can potentially produce identifiable injury and damage are termed *in vitro energy sources*. That is, they are produced outside of the human organism itself. These sources include electrical, chemical, thermal or heat, mechanical (potential and kinetic) including sound and vibration, ionizing radiation (including electron or X ray, and nuclear such as alpha, beta, gamma, and neutron), nonionizing electromagnetic radiation (including low-frequency, microwave, infrared, visible light, and ultraviolet). However, there is also so-called *in vivo energy of biological processes*, or energy sources within an organism or living body.

The human body has specific tolerance levels or injury thresholds that may be quantified for each energy form. These levels must be known to ascertain the magnitude, frequency, duration, and concentration of exposures for which a controlling means must be provided. Such controlling means can be installed at the source, along the path, or in the carrier mechanism, depending upon the particular problem and the state of the art.

Dr. Haddon's concept of abnormal energy exchanging or release barriers provides a useful checklist for the analyst. These include:

Barrier 1. Limiting the energy or substituting a safer form

Barrier 2. Preventing energy buildup in the first place

Barrier 3. Preventing its release

Barrier 4. Providing for a slower and controlled or safer energy release

Barrier 5. Diverting or rechanneling the energy with respect to either time or space

Barrier 6. Placing a barrier or safeguard on the energy source

Barrier 7. Placing such a barrier between the source and the receiver

Barrier 8. Placing a barrier about the receiver through blockage or attenuation

Barrier 9. Increasing the injury, illness, or damage threshold of the receiver or host

Barrier 10. Treating or repairing the injury or damage

Barrier 11. Rehabilitating the damaged receiver

Barrier 12. Some combination of the above to provide at least the minimum level of control required to ensure and preserve the integrity of the receiver, as defined by predetermined criteria as to its desirable, sustainable, or required function and condition. Such function and condition may be defined in either or both quantitative or qualitative terms (i.e., injury to one's front teeth may not be critical to life or health but presents a cosmetic and potentially critical social problem).

In the Haddon approach, adverse energy transfer effects are classified into five levels, as follows:

Effect I. None or safe
Effect II. Negligible or no injury or damage except possible irritation, "near miss," or other intangible problem
Effect III. Slightly or marginally injurious or damaging
Effect IV. Seriously or critically injurious or damaging
Effect V. Fatal or catastrophic event

The procedure is well suited to both scientific accident reconstruction and accident prevention work. It provides an important checklist approach for the designer in hazard foreseeability analysis. The approach is also incorporated into more comprehensive schemes such as MORT (Management Oversight and Risk Tree). It is also applicable to epidemiological and statistical studies. The defeatability-proneness or defeatability-resistance properties of Haddon's twelve barriers are evident.

CHAPTER 14

SPECIALIZED APPROACHES IN SAFETY ENGINEERING

HAZARD AND RISK FORESEEABILITY

A number of essentially epidemiological tools and procedures relating to product safety are available for both accident reconstruction and accident prevention applications. These can be broadly categorized as hazard and risk foreseeability techniques.

Several were specifically developed to aid in accident reconstruction, while others have found widest application in connection with accident prevention. Since a comprehensive scientific accident reconstruction (SCAR) approach attempts to converge on actual accident causation, both types supplement each other, approaching the problem from opposing viewpoints. When they are used in conjunction, there may be added benefits in that judgments are possible with respect to whether an accident was reasonably foreseeable or practically preventable, and what trade-offs were intentionally or inadvertently made by management that tended to increase or decrease product hazardousness content, user imperilment, and accident frequency, severity, and criticality. The most common trade-off encountered is in the matter of priorities assigned to user safety versus near-term profitability of the firm.

There is considerable overlap among some of the techniques and their application; each has a number of variations; and one or more may be combined in a single SCAR analysis. However, each has retained its identity and conceptual integrity in the literature and in practice. Complete descriptions and examples can be found in a number of excellent texts and reference books on safety engineering, product safety engineering, and risk assessment and management. Here we will merely list them. These techniques are as follows:

1. Fault tree or logic diagram analysis (FTA)
2. Management oversight and risk tree (MORT)
3. Process charting
4. Failure modes and effects analysis (FMEA)
5. Technique for human error rate prediction (THERP)
6. Job safety analysis (JSA)
7. Preliminary or gross hazard analysis (PHA or GHA)
8. Burke procedural and classification chart for arson investigators
9. Haddon's theory of abnormal energy exchange (HEX)
10. Human factors analysis (HFA)
11. System safety analysis (S^2)
12. Critical incident technique (CRIT)
13. Technology transfer and assessment (TETRA)
14. Failure analysis (FANNY)
15. "What if . . . ?" trains (WIT)
16. Hazards and operability review (HAZOP).

In effect, each features a checklist approach. But the checklist may be presented as a diagram, flowchart, table, or some other visual format that has more "dimensions" than a mere one-dimensional list.

Exhibit 14-1 repeats Exhibit 12-2. It is a checklist for facilitating the discovery and determination of reasonably foreseeable product safety hazards. Whether you engage in product design and develop-

EXHIBIT 14-1 (EXHIBIT 12-2 REPEATED)

PRACTICAL TECHNIQUES AND SOURCES FOR DISCOVERY AND DETERMINATION OF REASONABLY FORESEEABLE PRODUCT SAFETY HAZARDS*

1. Existing published safety codes, standards, and regulations.
2. Company, competitor, and general industry field experience including misuse, abuse, and error patterns; product failure and malfunction patterns; aftermarket and downgraded service patterns; and so on.
3. Products liability case law and legal publications (periodicals and legal association special services for trial attorneys, etc.)
4. Technical reference manuals, textbooks, data sheets, manufacturer trade literature, professional and trade association proceedings and periodicals, and so on.
5. Product safety checklists, cookbooks, scientific accident reconstructions and malfunction and failure analyses of reported incidents.
6. Consumer usage testing under simulated, reasonably foreseeable conditions of service and commerce by independent testing/polling organizations, industry trade associations, manufacturers, and so on.
7. Human factors analysis and application of the rules of practice for safe design (performance standards).
8. Hazard foreseeability, technological forecasting, and technology transfer techniques and procedures (including the Delphi method, brainstorming, comparative analysis [interfirm/interindustry/interproduct], etc.).
9. Other market research and commercial intelligence, including overt sources as the press, dealers and distributors, customer complaints, U.S. government and industry trade association accident statistics, commercial clipping services, other media coverage.
10. Applications engineering, defeatability proneness, safety serendipity, and form versus function analysis.
11. Laboratory experimentation, analysis, testing, mockup and general modeling (including dimensional analysis and simulations), quality assurance and reliability analysis.
12. Covert intelligence (industrial espionage methodology).

(*) Also facilitates discovery and determination of reasonably foreseeable conditions of service and commerce.

ment for an employer or on your own as a consultant, you should also attempt to collect field and competitive intelligence of an epidemiological nature relative to safety and/or hazardousness content.

More than ever before, today's business environment is information oriented. Whoever possesses the best information—sometimes the most, as well—is potentially in the most competitive position. Commercial intelligence facilitates efficient internal and external management planning, analysis, and control.

Product safety-related applications of field and competitive intelligence qualify as market research. The following information, as an extension of market intelligence in product safety management engineering, would be particularly relevant, if not critical:

1. Accident frequency
2. Accident severity
3. Accident descriptions
4. Records of "unusual" accident occurrences involving the product line
5. Scientific accident reconstructions
6. U.S. government studies, publications, and analysis by OSHA, CPSC, DOT, FDA, NBS, DEP, and so on.
7. Product defect or involved hazard descriptions
8. Regional and local accident incident analysis
9. Use or misuse and abuse patterns
10. Average age and age range of standing stock (i.e., product units already in the field)
11. Number of product units in the field
12. Number of product units in the field in each product age category
13. Product accident experience of your client versus its competitors
14. Types of modifications made by customers
15. Types of product failures/malfunctions that occur
16. User/operator errors typically encountered in the field
17. Comparative analysis of competitive safety features
18. Types of downgraded service as product gets older
19. Number of lawsuits and outcomes with damage award data, cases won versus cases lost, out-of-court settlements, and so on.
20. Customer complaint history, analysis, and processing effectiveness

Specialized Approaches in Safety Engineering 341

21. Geographical location of standing stock by estimated product unit population
22. Traceability of standing stock
23. Average condition and condition range of standing stock
24. After-market characteristics
25. Field servicing effectiveness
26. Product recall history and retrofit patterns
27. Media coverage of accident events and public opinion indications
28. Clipping service indications
29. Other accident epidemiology data

A WORD ABOUT WARNINGS AND THE UNREASONABLY DANGEROUS PRODUCT

As we saw in our discussion of performance standards for safe design, warnings and other safety "software" are bona fide methods for safeguarding against accident hazards. However, warnings are unacceptable as the primary or sole safety feature if higher-order safeguards are technically and economically feasible. Under such circumstances, use of warnings alone renders a product, workplace, or facility *unreasonably dangerous.*

That is, a product, workplace, and so on is unreasonably dangerous if it contains a reasonably foreseeable hazard that it would have been technically and economically feasible for its designer-manufacturer-marketer to have eliminated or mitigated. These are very pragmatic design guidelines. They can save consulting engineers and manufacturers a great deal of money and grief.

Only knowledge, resolve, and commitment are necessary to lift a design from the ranks of the unreasonably dangerous to the reasonably safe (or reasonably dangerous, since we are not suggesting that all hazards are necessarily technically or economically feasible to eliminate; our choice may not always be between risk and no risk but, rather, risk A versus risk B).

To drive home the inherent problem with warnings as the sole safeguarding strategy, look at Exhibit 14-2. The unavoidable potential for human error, failure, and oversight (typically 1 to 2 percent under the best of conditions) due to the high defeatability-proneness of warnings is evident, though our example is extreme. Not all warnings are as vague and noninformative.

The use of warnings alone, instead of more effective, higher-

EXHIBIT 14-2

THE WARNING

Now remember
and don't forget . . . BE CAREFUL. DON'T MAKE ANY MISTAKES.
NONE. NOT EVEN ONE. EVER.

EXHIBIT 14-3

THE MESSAGE

DANGER! WARNING! CAUTION! NOTICE!

BE CAREFUL! DANGEROUS HAZARD PRESENT UNDER REASONABLY FORESEEABLE CONDITIONS OF SERVICE AND COMMERCE!

THERE ARE NO SAFETY FEATURES INCORPORATED INTO THIS PRODUCT AS ORIGINAL EQUIPMENT. READ ALL INSTRUCTIONS CAREFULLY AND THOROUGHLY. SAFETY WARNINGS AND INSTRUCTIONS ARE ALL THAT STAND BETWEEN YOU AND CERTAIN, TOTAL, PERMANENT INJURY AND/OR DEATH.

WE ARE RELYING UPON YOUR GOOD JUDGMENT TO PROTECT YOURSELF UNDER ALL CIRCUMSTANCES! WE ARE DELEGATING PERSONAL SAFETY TO YOU AND YOU ALONE. YOUR SAFETY IS IN YOUR OWN HANDS AND ENTIRELY UNDER YOUR OWN CONTROL.

THAT'S HOW MUCH WE TRUST YOU AND YOUR ABILITY TO BE CAREFUL AND NEVER MAKE A MISTAKE! NONE! NOT EVEN ONE! EVER!

(It also shows how little we really care about you and how little we have concerned ourselves with safe design and reasonably foreseeable human error, failure, and oversight. By the way, this dangerous product complies with all published safety codes, standards, and government regulations. So don't give us a hard time about hazards not covered by the code, or hazards that the codes don't treat properly or adequately.) **THE MANAGEMENT**

order accident prevention strategies (when feasible) is also well illustrated in another example. Exhibit 14-3 articulates what a warning really says to a product, workplace, or facility user who is not provided with feasible higher-order safeguarding means.

MAINTAINABILITY SAFETY AND SAFETY THROUGH MAINTENANCE

Product design features facilitating safe maintainability are no less important than those that assure safe functional operation or production efficiency. But, as a practical matter, manufacturers do not typically do as good a job of protecting human beings from maintenance-oriented work compared with production-oriented operations.

Yet maintenance personnel, and consumers who attempt to maintain products they buy and use, are probably vulnerable to a wider variety of exceptionally severe hazards than any other single category of person. There is a high correlation between safety and maintenance; however, because of its wide variety, maintenance work rarely follows a set pattern. This is one of the factors contributing to its peculiarly great potential hazardousness.

The need for proper maintenance applies to all kinds of facilities and products. Maintenance and servicing work ranges from housekeeping and equipment setup to in-process adjustment, parts replacement, repair, and refurbishing. Ordinarily, what is included under this category stops short of what might be considered partial or total rebuilding, reconstruction, and the like. However, the same principles apply with respect to identification, evaluation, and control of safety hazards relating to life and/or property.

The concept of the "machine tender" is also applicable to many nonoperational work sequences for both industrial and nonindustrial products. The terms *machine operator, machine tender,* and *maintenance employee* cover almost all tasks of interest. Maintainability safety includes "tending" safety. However, there can be overlap between equipment maintenance and tending, on the one hand, and equipment tending and operation on the other.

For example, setup and make-ready tasks are sometimes routinely performed by machine operators, rather than by separate personnel. Machine tenders or tending personnel typically perform setup, make-ready, cleanup, controls monitoring, start-up, shutdown, parts transfer, parts or materials feeding, and other tasks.

Maintainability safety thus includes safety during preventive maintenance, corrective or restorative maintenance, and machine tending, adjustment, and so on. Note that "adequate" maintenance includes both necessary and sufficient care and servicing attention, and the application of necessary parts, material, and labor to ensure proper and safe functional operation as well as maintenance or serviceability safety itself.

We must also include housekeeping-type chores, which are performed for hygiene, safety, functional readiness/soundness, and even aesthetic purposes.

Since maintenance is a reasonably foreseeable necessity, so is maintainability safety. In the real world, there is simply no such animal as a maintenance-free or no-maintenance product.

The foregoing analysis points to four fundamental concepts

and activity profiles when we think of maintenance-related safety, as follows:

Maintainability Safety through Product Safety Content. This deals with safety of the maintenance worker while he or she is engaged in the maintenance task. It derives from specific "hard" design approaches to, and features of, the machine or other product, workplace, or facility being maintained. It is safety hardware-oriented and is product, workplace, or facility designer-builder-manufacturer-focused. Lack of safety features to assure maintainability safety can result in improper, deferred, or even no maintenance. Products that are made inconvenient or expensive to maintain due to designer error, failure, or oversight will be user error-provocative and maintenance-resistant. From the defeatability-proneness standpoint, maintenance objectives, requirements, and procedures are too defeatability-prone since there are excessively high maintenance process defeatability incentives. The product, workplace, or facility is accident-conducive and user-hostile.

Maintenance Safety through Job Safety Content. This, again, deals with safety of the maintenance worker while he or she is engaged in servicing tasks. But here we are not concerned with physical or "hard" product, workplace, or facility design and construction features. Rather, our focus is on "soft" or nonphysical factors such as work methods and procedures, maintenance and servicing programs, managerial and supervisory control, instructions and training, warnings and labels, personal protective gear, and administrative practices and controls. This maintenance safety-related perspective is both designer-builder-manufacturer- and owner-operator- or owner-user-oriented. On-product warnings and instructions are the responsibility of the OEM. In-plant maintenance safety, such as housekeeping around the maintenance task workplace, are owner-user-oriented unless made difficult by some feature of product or machine design and/or construction.

Safety through Maintenance. This aspect of maintenance-related safety relates to assuring that the product, workplace, or facility will be safe to utilize due to its continued functional soundness, housekeeping attentiveness, and so on, under reasonably foreseeable conditions of service and commerce. It is achieved through application of both "hard" and "soft" safety assurance techniques. This concept embraces product workplace and facility safety, occupational health (also termed *industrial hygiene*), fire safety, and so on, and is owner-employer-operator-user-oriented.

Maintenance Failure Accommodation. Here we confront possibly the most important problem in the maintenance field. This is the reasonably foreseeable neglect of product, machinery, or facility maintenance, for whatever reason. This would include forgetfulness, inattention, deferral, inconvenience due to poor design and/or construction, and production pressures. For example, it is reasonably foreseeable that deferred maintenance may be the rule during a business recovery period when capital is scarce and the sales pace is starting to quicken. Or it may be reasonably foreseeable that equipment will not be maintained in the strictest accordance with manufacturer recommendations and that the product is somewhat sensitive to servicing effectiveness and/or timeliness. That is, reasonably foreseeable maintenance management error, failure, and oversight may well result in an inadequately maintained product, machine, or facility that, thereby, may become unreasonably dangerous under reasonably foreseeable conditions of service and commerce. This adverse human factors aspect should be viewed by designers, manufacturers, and marketers as a design condition or design parameter. It should be designed against so the product or machine becomes maintenance error-forgiving and user-tolerant rather than user-hostile and error-provocative. To the extent feasible, therefore, reasonably foreseeable hazards deriving from reasonably foreseeable maintenance neglect should be precluded, mitigated, or otherwise controlled or accommodated through fail-safe features, for example. Generally speaking, protection against maintenance failure-induced hazards through this and other means is merely a reasonable and ordinary extension of the standard or generic rules of practice for safe design. This maintenance-related safety consideration is OEM-oriented.

APPLICATIONS ENGINEERING

Although the need to engineer the product or system for the application, long known as applications engineering, is old hat to most engineers, less well known are the safety ramifications of this basic concept. Proper applications engineering, by definition, should result in hazard identification, evaluation, and control. However, as a practical matter, applications engineering too often is limited to product or system function assurance. When properly utilized, however, applications engineering methodology is capable of defining all man-machine-milieu (human factors) interfaces and designing against reasonably foreseeable interface failures under reasonably foreseeable conditions of service and commerce. In fact, the methodology of applications engineering intrinsically embraces and anticipates human factors considerations and the man-machine-milieu interface.

In fact, applications engineering, or engineering the application, is really at least five things. It is, at the same time:

1. An engineering concept and technical process
2. A management technique, practice, process, and procedure
3. A tradition, custom, or practice in the "engineering" or capital goods industries
4. An engineering function and/or organizational cell
5. A partnership among customer, vendor, distributor, and other entities involved in the successful, efficient, economical, safe, and so on use of capital goods and many other kinds of products, subsystems, and systems vis-à-vis reasonably foreseeable conditions of service and commerce, as well as a product design and development imperative (as suggested above, items 1 and 2)

The marketing of many kinds of products, not only heavy engineering equipment and other capital goods, where the concept first arose, carries with it unique problems and opportunities with respect to product safety. Products with high technical content must be "applications engineered" to ensure proper selection and specification compatible with reasonably foreseeable conditions of service (COS). Both pre- and post-sale collaboration among customer, vendors, and subcontractors can be vital to assure that optimum results are realized from such purchased equipment, whether new or rebuilt. In effect, applications engineering means engineering the application.

The machinery builder, rebuilder, or dealer cannot unilaterally design, build, and market complex machinery and then delegate proper and safe installation, operation, and servicing to the customer. Nor can it be assumed that a customer has the necessary and sufficient expertise to select and specify this equipment properly and incorporate it into a process or other system.

From start to finish the specification, selection, design, construction, installation, operation, and servicing of engineering equipment must be a cooperative endeavor. While certain durable industrial goods are more akin to familiar varieties of retail merchandise, heavier engineering equipment must, of necessity, be custom built or custom tailored to the specific application. Although the machinery maker is a specialist, heavier equipment cannot usually be premanufactured or prestocked like foods, cosmetics, clothing, appliances, automobiles, power tools, lawn mowers, residential air conditioning equipment, heating and ventilating products, and so on.

The capital goods or engineering industries run the gamut from laboratory models, prototypes, and jobbed or custom apparatus—which may even be sold to the customer on a "trial and test" basis—to pre-engineered, premanufactured, and prestocked machines. It is only this last category that is comparable to the mass production or mass assembly consumer industry concepts.

Premanufacturing permits full manufacturing information, patterns and tooling, parts, and subassemblies to be stocked. Even complete machines may be built and stocked where demand permits, whereas pre-engineering makes available only standard drawings, patterns, and other manufacturing data that can be used if and when the order comes in. There is no economic justification for stocking parts or component assemblies.

However, even many types of premanufactured and prestocked items must be applications engineered to ensure optimum materials of construction, feeds, speeds, strength, efficiency, economy, dependability, and safety in use consistent or compatible with design or service conditions. Certain types of heavy equipment are more subject to ordering in larger quantities than others, although mass production in the technical sense is not involved. For example, a customer may order a fleet of 1,000 tractor-trailer rigs or container chassis units; 100 centrifugal gas compressors, 100 forklift trucks, or 100 portable highway compressors; or 50 sewage comminutors, 10 vertical boring mills, or 5 punch presses. In these instances, both customer and vendor work hand in hand to determine product feature requirements vis-à-vis reasonably foreseeable COS. But the OEM usually has final effective control over functional and safety content of the delivered product.

By the same token, the builder, dealer, and even M&E (machinery and equipment) appraiser have analogous responsibilities to at least communicate to the purchaser the presence of problems relating to functional or safety performance. Of course, such responsibility exists to the degree that such entities are critically situated in the equipment life cycle and the customer can reasonably be said to depend upon their services in optimizing performance results.

One very old, large, and highly regarded international manufacturer of a wide variety of capital goods equipment is organized into applications engineering, sales engineering, contract engineering, and field service and erection departments. The sales engineer represents the firm in the field or marketplace, making the initial contact leading to the ultimate order. The sales engineer interfaces with manufacturing division applications engineers, who provide necessary heavy engineering support to assure that customer requirements are properly detailed and analyzed. Customer engineers and other personnel are also consulted for in-depth assessment of conditions of service.

The contract engineering department enters the picture where large systems of equipment are involved and other machinery, instrumentation, and controls producers supply other process components or subsystems, or where the firm is not itself the prime contractor.

After the sale is made and the applications engineering phase is completed, the field service and erection department shifts into high gear. Installation, operating, and maintenance manuals are prepared and departmental personnel go into the field to assist in erecting the system on-site once delivery is made, to train customer personnel, test and "shakedown" the equipment, and stand by to provide start-up assistance until the equipment or system is in a state of production readiness.

Where multiple vendors are involved, all have representatives on-site to assist the customer and each other in getting the system going and in resolving start-up problems relating to their own products.

Obviously, all aspects of the preceding description will not apply to M&E rebuilders, dealers, appraisers, independent consultants, and contractors. However, these entities should be in an excellent position to provide important functional, safety, and other specialized information as part of their normal professional service or commercial package.

The comprehensive technical marketing process detailed above is typically intense and compressed within a demanding time frame. This is due to the fact that capital goods and heavy engineering equipment and systems are purchased on a highly rational and disciplined basis, unlike typical retail customer merchandise. Every day of equipment or process start-up delay or downtime can cost thousands or even millions of dollars in lost production. "Turnkey" projects represent an ultimate type of responsibility where a large plant or mill is built from scratch. Schedules are critical if the completion date is to be met. Collaboration among all vendors and subcontractors is on the most disciplined and rigorous basis. The term *turnkey* refers to total "design and build" responsibility. Project planning procedures are frequently computerized and most sophisticated. A shortage of, or missed delivery schedule for, an apparently minor and low-cost element can set back a project for days or weeks, since many phases may be synchronized and interrelated with one another.

There is another major issue relating to safety-in-service of capital goods. It is reasonably foreseeable that many pieces of general and even special-purpose equipment may be removed from one system and placed into another to which they have to be adapted. Such modifications are not at all uncommon, but, with regard to safety, they may be defective.

Further, it is reasonable to presume that machinery makers possess both specialized and diverse experience and expertise in the application and safety ramifications of both their own equipment and/or related or other equipment typically used in conjunction with their products.

Therefore, appropriate warnings, instructions, and guidelines related to product and/or system safety should be provided. They should be affixed directly to the equipment or to "captive" documents (utilizing a lanyard or chain, etc.) contained in metal cabinets or pockets built onto or within the equipment (with appropriate referencing labels outside, in the latter case). Applications safety manuals should be supplied as well.

Thus, safety integrity must be maintained throughout the product's economic applications life cycle. This includes general hazards or dangers commonly encountered in the field or otherwise associated with the equipment as a self-standing product or a systems component.

The applications engineering process is both a technological and a managerial control tool. It requires in-depth cooperation and collaboration by all engineering design, equipment manufacturing, and end-user entities.

Only through thoughtful anticipation of service conditions in the form of applications engineering data and guidelines, warnings, and instructions can product and/or system safety be assured.

Highest-priority safeguards must be used, and they must be technically and economically feasible to bring about. But without either safety hardware or software provided by the builder, rebuilder, or dealer, the machinery and equipment customer is at a distinct disadvantage. He has not even been given the opportunity to make rational and disciplined product safety analyses or decisions with respect to a specific application or conditions of service. In fact, he has not even been advised to contact the original equipment manufacturer to obtain product safety information, recommendations, or guidelines relating to his proposed equipment and/or system use or reuse configuration.

Industrial accidents comprise a very large area of interest in the safety engineering field. Therefore, industrial machinery and equipment are of great concern in relation to hazardousness content.

Here is a summary of the five levels of technical service or support characteristic of capital goods and other engineering products, based upon the inherent nature of typical customer applications and requirements.

> *Level 1.* Supply of technical information only, relating to performance specifications and/or applications data to facilitate proper product selection and specification by the customer.
>
> *Level 2.* Presale survey of customer needs by technical sales representatives or sales engineers with respect to applications requirements facilitating proper product selection and specification.
>
> *Level 3.* Presale applications engineering in collaboration with customer engineering and/or production personnel to develop and/or modify equipment and/or processes to satisfy customer requirements, plus contract engineering interfacing with other equipment, controls, and instrumentation suppliers as required.
>
> *Level 4.* Presale applications and contract engineering plus on-site field inspection, erection, testing, startup, and initial standby services to ensure attainment of performance specifications and commitments in relation to required conditions of service.
>
> *Level 5.* Presale applications and contract engineering, installation, testing, startup and standby, plus postsale demonstration to and training of customer operating and supervisory personnel, ready availability of trained field service and repair or troubleshooting technicians and engineering personnel over the life of the equipment, plus expeditious availability of spare or replacement parts.

Typically, for Levels 3 through 5, the original equipment manufacturer (OEM) will also retain archival records including original sales orders, specifications, bills of material, engineering drawings and original tracings, engineering design change or revision orders, manuals, and so on to facilitate field servicing and parts replacement for machinery and systems already in place in the field. It must be remembered that, by their very nature, capital goods products are long-lasting. In many cases, they may be completely serviceable for many decades with attentive and proper maintenance care. In the capital goods field, as much machinery is taken apart as falls apart. Some items of equipment are known to have been in useful service for well over fifty years.

Technical and commercial records retention also facilitates ongoing marketing and sales efforts such as implementation of marketing strategies. The cumulative effect of such records retention is to:

1. Define and document cumulative technological know-how
2. Provide in-depth support for applications engineering of new orders
3. Provide the capability to duplicate or modify product or system performance characteristics previously applications-engineered
4. Expedite competent resolution of field problems
5. Permit sound and efficient evaluation of field intelligence and competitive positions
6. Generally facilitate commercial intelligence activities
7. Provide the necessary data base for product life cycle monitoring, traceability, or "genealogy tracing" with respect to retrofitting, updating or modernization, and recall of equipment already in the field

This technological know-how is also typically used in conjunction with product development activities.

These same archival records retention and order engineering capabilities can be effectively utilized for purposes of upgrading safety in service of products in the field as well as for the purposes enumerated above. Obtaining a competitive edge in selling, however, has been and remains the primary purpose for the vast majority of manufacturers.

The applications engineering process and reapplication of cumulative technological know-how for order engineering purposes can be interpreted as that component of machinery-making activity that places the capital goods industry within the so-called "knowledge industry."

The typical M&E producer is as much of a knowledge company, or a marketer of information and "intelligence," as any book or newspaper publisher. In the latter two instances, knowledge is imparted in the form of software-type products (i.e., books or newspapers). In the case of the manufacturer of engineering equipment, this technological know-how is built into the applications engineering process that ultimately shapes the steel sinews and electronic controls hardware of heavy machinery and other capital goods. But the knowledge component of the product and the business is identifiable nonetheless, if you know what you're looking for. And the knowledge component of the product must include adequate, necessary, and sufficient consideration of safety.

HUMAN FACTORS ENGINEERING

Human factors engineering in its various forms is also called human performance engineering, man-machine-systems engineering, human engineering, biomechanics, ergonomics, engineering psychology, and so on, depending upon the background or orientation of the practitioner and the segment of this broad-ranging field that is involved. However, all variations on the theme deal with human physical, physiological, and psychological strengths and weaknesses. Typically, bodies of knowledge in the allied fields relating to human factors focus upon the man-machine-milieu (or environment) system and the multitudinous opportunities for breakdown and failure at the interfaces.

The human factors field takes the stand that products, machines, workplaces, and so on should be designed to "fit" or be compatible with humans, rather than forcing people to adapt to designs that are poorly suited to them. Human frailties (and strengths) thus become design conditions. There is important emphasis upon the propensity of human beings to err. Reasonably foreseeable human error, failure, and over sight are critical considerations to be accommodated through acceptable, thoughtful, and superior design.

Exhibit 14-4 portrays a set of operational definitions for human factors engineering. It is a broad subject because it is so multidisciplinary. The reader is urged to purchase several good books on the subject for reference purposes.

A partial sampling of human performance parameters to be considered in engineering design includes the following:

1. Anthropometric or dimensional range of movement, skeletal weight, musculature, and nervous system aspects
2. Human strength limitations
3. Equilibrium requirements
4. Reaction times

EXHIBIT 14-4

OPERATIONAL DEFINITIONS OF HUMAN FACTORS ENGINEERING

1. HF recognizes the physical/physiological/psychological capabilities and limitations of human beings as product/equipment/facility/systems operators/users.
2. HF recognizes human error/failure/oversight as ordinary/normal/reasonably foreseeable/inevitable/predictable human attributes/deficits.
3. HF recognizes that ordinary care and ordinary carelessness are two sides of the same coin; that even a normally attentive and careful person cannot be relied upon all the time.
4. HF recognizes that an error-provocative or accident-conducive design can result in an error- or accident-prone human being, and that all people can be "trapped" and rendered temporarily error- or accident-prone by such user-hostile, "unforgiving" designs even though there is no such thing as an innate, permanent personal predisposition to having a high accident rate (contrary to old-fashioned, obsolete ideas on the subject).
5. HF recognizes that M^3 (i.e., man-machine-milieu or man-machine-medium) interface failures are inevitable in the absence of adequate, necessary, and sufficient product design based upon HF principles and practices.
6. HF recognizes that human nature, habits, and practices are difficult to change, modify, and maintain; but physical, mechanical built-in product gains are permanent.
7. HF recognizes that product safety can best be promoted by designing against human frailties, limitations, and deficits as design criteria/conditions of service (COS).
8. HF recognizes Murphy's Law as an immutable law of nature, or at least one that applies, with the greatest certainty, whenever you are most confident of favorable results.
9. HF recognizes that, when things don't work out, human beings will generally be found to be among the weak links in the failure chain.
10. HF recognizes the systems ramifications of the M^3 interface.
11. HF recognizes that adequate engineering is holistic engineering.
12. HF recognizes that human engineering is humane engineering.

Specialized Approaches in Safety Engineering

5. Control response quantity and quality and human information-processing capabilities
6. Effects of heat, acceleration, vibration, shock or impact, impulse, altitude, and gravity, and so on
7. Work, fatigue, and physical energy expenditure patterns
8. Visual (illumination, color), auditory (noise), tactile (touch, texture), and other sensory (smell, taste) abilities and limitations
9. Electric current effects
10. Effects of chemicals through absorption, inhalation, and ingestion
11. Effects of atmospheric conditions and various radiation forms
12. The intransigent human propensity to err
13. Effects on human performance of inherent biological cycles or biorhythms (i.e., intellectual, physical, and emotional) and daily, circadian or diurnal rhythms corresponding to the twenty four-hour day-night cycle (such as metabolism, heart rate, body temperature, alertness, concentration, physical work, and output potential)
14. Ramifications of the foregoing at different ages or life-cycle stages and under different circumstances in work and nonwork environments (i.e., children are extremely curious and unsteady on their feet; adults don't always read directions; shift work interferes with circadian rhythms, etc.)
15. "Normal" habit patterns and behavioral expectancies and population stereotypical expectancies and responses
16. Effects of cultural and economic bias on warnings and instructions effectiveness and efficiency
17. Physical, physiological, and psychological deficits of the handicapped (deficits of the young and the aged are included under item 14 above)
18. Human performance under overload conditions in one or a combination of the foregoing categories

Product life-cycle stages that must be considered in engineering design, including their human factors ramifications, are as shown in Exhibit 14-5.

There are at least seventeen potential sources of product failure that you should take into account:

1. Design deficiency
2. Incorrect materials section and specification

EXHIBIT 14-5

PRODUCT LIFE-CYCLE STAGES

Group	Stage	Description
I	1	RDE&T (Research/Development/Engineering/Testing)
II	2	Manufacturing/Production/Fabrication/Quality Assurance or Quality Control/Rework for Product Hardware Elements
	3	Packaging/Labeling/Manuals/Customer Training Programs
	4	Storage/Warehousing/Materials Handling/Shipping/Transportation
	5	Marketing/Sales/Leasing/Sales or Applications Engineering
III	6	Product Life-Cycle Management/Life-Cycle Monitoring/Genealogy Tracing/Archival Records Retention/Product Recalls/Product Advisories
IV	7	Installation/Erection/Incorporation into System
	8	Running-In/Burning-In/Shaking Down/Acceptance Testing/Field Trials/Start-up/Trial and Test
V	9	Intended Use: Operations within Design COS
	10	Intended Use: Maintenance/Repair/Servicing/Adjustment/Tending/Setup/Cleanup within Design COS
VI	11	Reasonably Foreseeable Modification/Alteration/Retrofitting
	12	Rebuilding/Reconstruction/Reconditioning/Resale and Reuse as Rebuilt/Reconditioned/Resold
	13	Reasonably Foreseeable Alternative Application/COS
	14	Reasonably Foreseeable Downgraded/Degraded Use/Service/Application Including Periods of Nonuse
VII	15	Other Reasonably Foreseeable and/or Unintended COS/Use/Misuse/Abuse/Misapplication Outside Intended Stream of Commerce/Uses and Abuses *not* Reasonably Foreseeable
VIII	16	Salvage of Selected Parts/Materials
	17	Scrap
	18	Recycling of Selected Materials

Specialized Approaches in Safety Engineering

3. Poor workmanship
4. Material imperfection
5. Assembly or fabrication error
6. Defective processing or reworking
7. Inadequate inspection, testing, and quality control
8. Improper materials handling or warehousing
9. Neglectful or improper maintenance or field servicing
10. Incorrect applications engineering
11. Harmful environmental exposure
12. Improper conditions of service
13. Improper operation
14. Improper rebuilding or reconditioning
15. Improper disposal
16. Improper recycling
17. Attempted utilization of products beyond the outer, end limits of their reasonable economic life

Carefully take note of the inherent human performance error that is suggested by each of these foregoing failure or defect sources, including both design error and management or supervisory error.

SOME ADDITIONAL ASSISTANCE FOR THE SCIENTIST AND DESIGN ENGINEER-TURNED-CONSULTANT

Checklists are handy to have. They are not always that easy to use, as they all have their limitations. However, Exhibit 14-6 should be of interest to design engineers. It categorizes and lists various types of safeguarding means.

Notice that there is no attempt in Exhibit 14-6 to classify the safety devices or techniques from a "defeatability proneness" standpoint. Again, you must array your alternatives and assign defeatability-proneness—or defeatability resistance—priorities to them for purposes of decision making.

Exhibit 14-7 shows pictorially a number of common mechanical hazards. Almost every book on safety will be profusely illustrated with such diagrams and many others.

EXHIBIT 14-6

PRODUCT SAFETY-RELATED CHECKLIST OF SELECTED SAFEGUARDING MEANS

1. Safety "Hardware" (Built-in Physical Design Features)
    A. Controls and Sensors
        (1) Location (re: inadvertent actuation by operator body, hands, palms, fingers, elbows, feet, etc.)
        (2) Recessed buttons
        (3) Shrouded buttons
        (4) Arrangement/actuation mode/shape/size/texture/and so on
        (5) Spacing (including foot pedals/controls, and interpedal height/level)
        (6) Local jogging/inching
        (7) Reset
        (8) Two-hand/concurrent
        (9) Sequential
        (10) Multistation/multi-operator
        (11) Time delay to start-up
        (12) "Sniffers" (built-in analytical instrumentation)
        (13) Overhead/suspended/pendant
        (14) Rigid conduit (versus flexible power cord)
        (15) Attitude/rotation/movement sensors/arrestors/inhibitors/controls
        (16) Pattern recognition/spectral sensing
        (17) Palm buttons/pressure pads/dorsal fins
        (18)

    B. Guards/Restraints
        (1) Handrails/guardrails/guide rails (including recessed and padded)
        (2) Barriers/enclosures/cages
        (3) Gates/accesses covers
        (4) Interlocks (including "deadman" safeties)
        (5) Lockouts/tagouts
        (6) Pullouts/pushaways
        (7) Awareness barriers
        (8) Presence-sensing devices
        (9) Distance/location/position guarding (also see G23–24)
        (10) ROPS/FOPS (including retractable for between decks/low headroom service)
        (11) Stabilizers/outriggers (including interlocked)

Specialized Approaches in Safety Engineering

(12) Pressure-sensitive body bars/contact barriers (stationary equipment)
(13) Pressure-sensitive body bumpers/skirts/sleeves/kickplates (mobile equipment, escalators, etc.)
(14) Pulls/safety cables/pushrods/chains
(15) Occupant restraints (lap belts/shoulder harnesses/air bags/active versus passive)
(16) Heat shields
(17) Noise suppressors/attenuators
(18) Vibration suppressors/attenuators
(19) Toxic vapor containment
(20) Environmental control features
(21) "Quick-stop"/"instantaneous" equipment shutdown/rotation termination
(22) Product jackets (to contain broken glass fragments, cushion impact)
(23)
(24)
(25)
(26)

C. Access/Ingress/Egress/Climbing/Fall Arresting/Walking and Working Surfaces
  (1) Three-point access/climbing systems
  (2) Handholds (including recessed and padded)
  (3) Safety cages (climbing)
  (4) Skid-resistant treads/nosings/abrasive strips
  (5) Safety lines/lifelines/safety harnesses/lanyards
  (6) Open edge and side barriers/handrails/guardrails/toeboards
  (7) Wall/floor opening covers/guards/railings
  (8) Portable chain/tape and stanchions
  (9) Obstructions removed
  (10) Raised edges/corners removed
  (11) Stringers (inclined versus vertical for equipment access)
  (12) Protrusions removed
  (13) Depressions removed
  (14) Other preventatives re trips/slips/pitchouts/pitchovers/falls
  (15) Self-draining/self-cleaning surfaces/gratings
  (16) Abrasive strips/surfaces
  (17) Ramps versus steps/steep ramps/steep steps

(continued)

EXHIBIT 14-6 (Continued)

    (18) Shallow risers/small elevation differences removed
    (19) Catwalk protection
    (20) Mats/runners/carpets (remove raised edges & corners/use recesses/watch seams & joints)
    (21) Abraded steel plates (to reduce slipperiness)
    (22) Avoid bare steel-surfaced diamond or cleated tread/walking plates
  (23)
  (24)
  (25)
  (26)

D. Materials Handling
    (1) Work manipulators/positioners/feeding tables/holding tables/powered lifts (including built-in/integral)
    (2) Load monitors
    (3) Slack cable sensors
    (4) Catching devices/wedge-clamp safeties/centrifugal brakes/retarders (for cable-mounted vertical motion equipment, and lifts)
    (5)
    (6)
    (7)
    (8)
    (9)

E. Fasteners
    (1) Quick-release
    (2) Captive
    (3) One-way/nonreversible
    (4) Reusable
    (5)
    (6)
    (7)
    (8)

F. Form Follows Function
    (1) Rounded edges/corners
    (2) No points
    (3) No protrusions
    (4) No snagging features/catch points

Specialized Approaches in Safety Engineering

(5) No pinch points/nip points (generally under section 1B above)
(6)
(7)
(8)
(9)
(10)

G. Other Safeguards
  (1) ZMS ("Zero Mechanical State") safeties
  (2) Redundancy
  (3) Error-resistant features (including "idiot-proofing" and "fool-proofing")*
  (4) User-friendly features
  (5) Environmental control features n.e.c.
  (6) Controlled failure
  (7) Vandal-resistant/pilfer-proof features
  (8) Alternative materials of construction (i.e., stainless steel versus glass)
  (9) Rupture/frangible discs/safety plugs/safety fuses/fusible plugs
  (10) Safety and relief valves/pressure regulators/control valves
  (11) Electric fuses and circuit breakers
  (12) Check valves/other flow controls/overflow devices
  (13) Other overpressure/overtemperature safeguards
  (14) Overrunning/slip clutches
  (15) Brakes (including quick-stop capabilities)
  (16) Locks
  (17) Liquid/solid level controls
  (18) Indicating/recording instruments and controls
  (19) Scientific instrumentation/sensors/controllers in-process
  (20) Easy-opening lid/cover/hatch/cap (but beware defeatability problems)
  (21) Child-resistant closures (but beware elderly-hostile designs)
  (22) Elderly-friendly designs and constructions
  (23) Deflectors/baffles/other path modifiers except lengtheners
  (24) Extensions/spacers/other path lengtheners
  (25)

(*) Misnomers that tend to obscure and confuse the real issues relating to HF.

(continued)

EXHIBIT 14-6 (Continued)

2. Safety "Software"
    A. Unclassified Catalog
        (1) Backup/reverse alarms/beepers/flashers/beacons/lights/sirens/passive*
        (2) Yellow warning tapes/strips/stripes/markings/pavement markings
        (3) Pressure-sensitive warnings/strips/abrasive strips
        (4) Manuals/instruction/operating/maintenance/servicing/parts/installation/applications/safety/handling and transportation
        (5) On-product warnings/labels/tags/signs/flags
        (6) Packaging/containers/closures/burst containment shells/jackets
        (7) Tamper-evident/fail-evident devices
        (8) Advertising (all media)
        (9) Warranties
        (10) Trade literature/catalogs
        (11) Press releases
        (12) Sales contracts
        (13) Captive software (chain/lanyards/cable/plastic cord)
        (14) "Idiot" lights (misnomer)
        (15) Peril point/hazard point software location
        (16) Directional signals/hazard lights/blinkers (mobile equipment, passageways)
        (17) Temporary sign/signal devices/structures/barricades/cordons/ribbons/tapes
        (18) Talking/speaking/automatic voice response devices
        (19) Voice-actuated/voice recognition equipment
        (20) Pattern recognition equipment
        (21) Interactive safeties
        (22) Operator-actuated/manual horns/bells/whistles/lights/sirens/active*
        (23) Annunciators
        (24)
        (25)
        (26)

(*) Passive = Automatic; Active = Manual

EXHIBIT 14-7

PICTORIAL "CHECKLIST" OF MECHANICAL HAZARDS

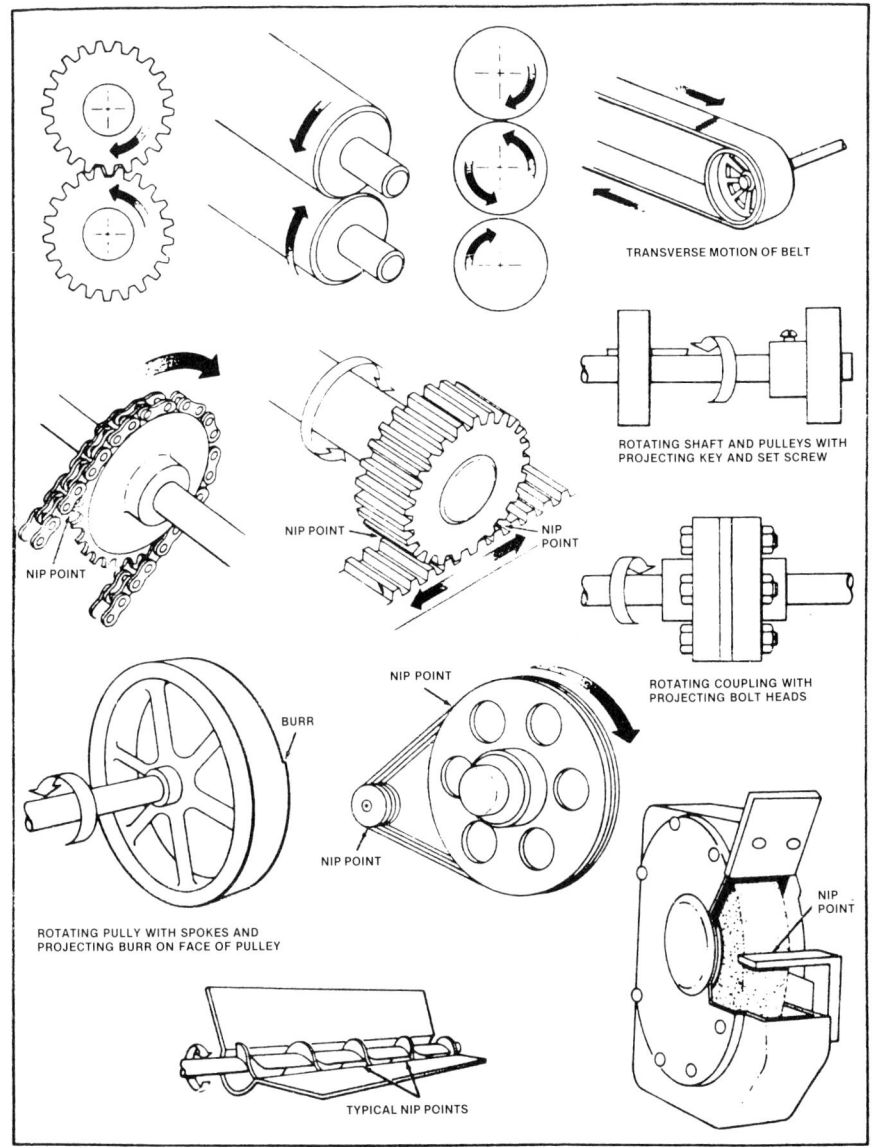

EXHIBIT 14-7

PICTORIAL HAZARD "CHECKLIST" (CONTINUED)

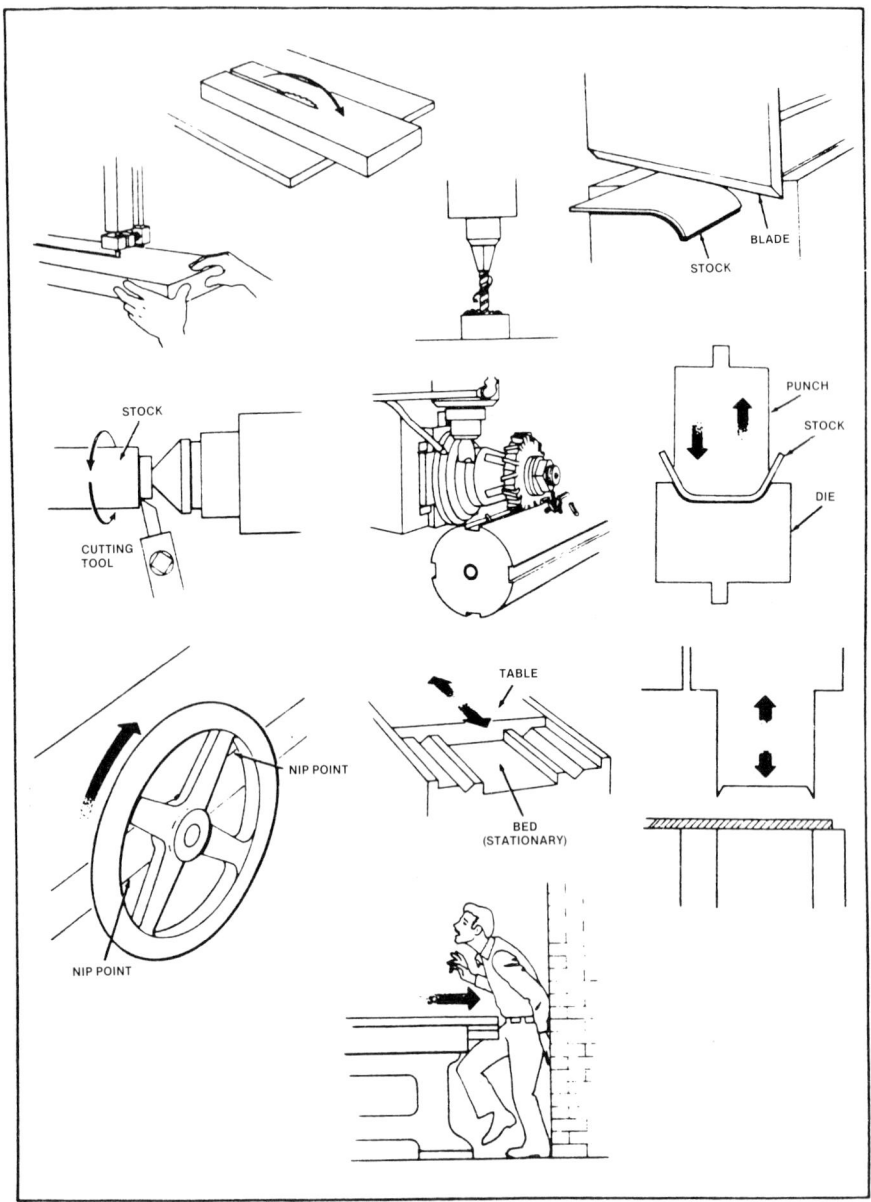

Specialized Approaches in Safety Engineering 365

THE REST DEPENDS ON YOU

With the foregoing information, you are in an excellent position to survive the jungle of litigation that can otherwise envelop and overwhelm even the most talented and diligent engineering designer-consultant. Naturally, there is a lot more you should know about the fields of safety engineering, safety management, and general accident prevention than it is possible to include here. You may elect to make a more intensive and extensive study of the field. Excellent textbooks and handbooks abound, in addition to various types of manuals, monographs, and special publications of professional, technical, and trade organizations in the allied safety and health fields.

This primer in safety engineering provides you with a most important starting point. You now have an operational understanding of the foundation philosophy of safe design, and a framework to fill in the detail, using the diverse array of engineering problem-solving tools that are at your disposal in your own specialty.

As noted at the outset of the previous chapter, one of the fundamental and critical rules is not to rely upon published safety codes and standards. They are almost all deficient in any one or more of over twenty-five ways. Memorize the twelve rules of practice for safe design. They are controlling in all situations; to compromise them is to invite accidents and lawsuits. Also, understand the concept of defeatability proneness.

In recent years there has been a great national controversy and debate relating to the subject of tort reform in our civil justice system. Industry has effectively pitted itself against the consuming public. The legal profession and the insurance industry have hardened their positions and are at loggerheads. The engineering and medical professions have become more susceptible to malpractice actions. There have been numerous proposals for uniform federal legislation relating to products liability lawsuits. Most of the legislation proposed is, itself, defective from several standpoints, not the least of which is failure to recognize the "human factors" elements in product designs that induce injuries and fatalities.

However, with all of the activity and controversy in the field, there has been more public speaking than private thinking. And with it all, the subject of tort reform remains, for all practical purposes, irrelevant for purposes of engineering design.

Products liability prevention must begin with product safety assurance. Sound product safety assurance consists of effective and efficient product safety engineering and product safety management.

This is the bottom line for the engineering profession. The application of the principles and practices presented in these last two chapters ultimately is crucial for purposes of protecting the public interest, precluding undue exposure of your client to products liability litigation, and protecting yourself against malpractice-type negligence lawsuits.

CHAPTER 15

HOW TO EXPAND YOUR PROFIT OPPORTUNITIES IN CONSULTING

SELF-ACTUALIZATION THROUGH EXPANSION

In this chapter we present a number of strategic and tactical measures for enlarging your professional and financial potential. There are also several strategic issues you must consider very carefully due to their potentially adverse affects on the economics of your consulting practice. For example, consulting for consultants offers one route to establishing a consulting practice. But a consulting partnership can be a curse rather than a blessing.

You want your consulting efforts to be profitable in a significant way. Perhaps you want to break away and do consulting full-time. You also want to be able to express yourself through your work and the practice. It provides satisfaction and a sense of accomplishment. Despite the fact that we engineers are highly trained technically, we remain, to a greater or lesser degree, a little bit the artist. In fact, engineering, like all of the sciences, is almost as much art as science.

The best way to lay the groundwork for an expanded practice is to keep up your activity. Become more visible. Publish. Lecture. Teach. Give talks. Attend technical society meetings. Attend conventions at which you know that you will meet and interact with prospective clients.

IT'S A TELEPHONE BUSINESS

Consulting is a "telephone business." You will be on the phone for endless hours, it will sometimes seem. Clients will call. You will call them. A lot of work gets done on the phone. A lot of expertise passes through those wires. You don't have to be a telecommunications wizard or have a $15,000 telephone system with all kinds of fancy features to be successful. But you had better develop a good telephone personality and know how to speak, listen, sell, and follow-up effectively on the telephone and use this indispensable tool efficiently. When you are talking with clients, no matter what the subject or the reason, remember that you are being assessed, even if subconsciously. Therefore, you are really always selling. There's no body language. There are no facial expressions. There are no props. Your voice is your only resource, so back it up with all it takes. And your "marketplace" is only as far away as the nearest telephone. With a telephone your market is anywhere, anytime, and anyone. You can travel all through your territory, even if it is coast to coast, via a telephone. It can get expensive, but it's worth every penny. It is an investment.

The elements of successful telephone selling, even in a professional practice if you are calling prospects, are finding the decision maker, getting through that third-party "screener" or "barrier," speaking with authority without being abrasive, leaving an effective message, returning calls in a timely manner, qualifying your prospects in some relevant way, following up, delivering a planned but flexible presentation that also is time-conserving for your prospect, sounding positive and being prepared to handle both real and false objections, being diplomatic, offering alternative service approaches, and being reasonably accessible if your client or prospect calls you. In addition, your sense of humor should come through without being offensive. That person on the other end should clearly "see" your "telephone smile."

Believe it or not, the person at the other end will "see" it.

Perhaps most important, again, learn to listen. Be a responsive listener on the telephone. Don't be totally silent, but don't interrupt flows of thought. Use negative responses to discover prospect or client needs and real, as opposed to merely expressed, objections. If you listen, you will be surprised how much useful and actionable information clients or prospects will give you gratuitously—information you can use to provide superior needs-oriented service to mutual advantage. And since you are an expert in your specialty, be careful not to bully or intimidate a prospective client with an excessive display of knowledge. Your self-image will come through loud and clear on the telephone, as well as your attitude.

EXPANSION AND DIVERSIFICATION

Develop more activities to service a greater number of needs of current clients and also to attract new client categories. In your own small way, become more integrated and diversified, both vertically and horizontally. You can develop a multidisciplinary capability to handle more specialized aspects of problems with which you typically deal, or you can develop it in order to expand the types of problems you can solve. Vertical integration or diversification is *intensive,* horizontal is *extensive.* A large automobile company was vertically integrated at one time. It made steel, glass, paint, and spark plugs as well as the cars in which these products were used. But it was also horizontally diversified. It made radios; TVs; automobiles; trucks; defense, aerospace, and electronics products and equipment.

From another standpoint, you can grow by giving yourself greater depth through a group-type, multidisciplinary practice. Or you can diversify into allied engineering services. You can outfit a small laboratory and offer testing services. Or you can start to sell engineering equipment or other products as a manufacturer's representative. You can even go into a little bit of customer proprietary manufacturing of your own. You can outfit a shop. But you may spread yourself too thin.

Or you might, by design, develop your consulting practice into a broader range of services or products until the business outgrows you. You might wind up with something to sell beyond having solely a successful consulting practice—which is, by the way, very difficult to sell if you decide to retire from it. After all, who's going to buy *you?* What are your files really worth? Will your clients be willing to work with someone else, merely because that someone is a new owner of your practice? In this way a practice differs somewhat from a business: A professional practice is a highly personal thing; a "regular" business is capable of surviving you. But, of course, that's the whole idea of a "professional practice"—it's personalized.

A Ph.D. organic chemist left his employer some years ago to strike out on his own. It took three or four years before it began to work for him, but he kept at it. He now provides consulting services, and also sells pollution control chemicals. His competitive edge is providing high-powered environmental chemistry consultative services along with the chemicals he distributes under his own private label. He continues to sell pure consulting services, however. This is how he frequently is able to open the door in the first place. Then he follows up through selling the chemicals. He has bid on, and won, a number of nice contracts over the years. A growing percentage of his customers have continued to utilize both his services and his products. He is beginning to develop a good customer base. No doubt repeat business will eventually make his business grow beyond his wildest expectations.

OTHER AVENUES FOR GROWTH

Writing newsletters to be sent to clients and/or prospects can be another way of staying close to your market. You might even be lucky enough to sell subscriptions to these newsletters. A lot of people make over $25,000 a year just writing and publishing newsletters of all types in practically every industry. There are even books written about the opportunities.

Or you can attempt to find a profitable way to provide services that your competitors shun. Just be sure you know exactly what you are doing. Generally speaking, you have to do this on the volume rather than on the margin. Usually, competitors shun the work because it is either too unprofitable or else too complicated. Be sure you know which it is and whether you can successfully fill the gap. Sometimes a large company won't find it profitable to service a low-volume-per-customer corner of the market because there's just not enough business there, or it is not profitable enough, taking both volume and margin into account, to mount a full-blown sales offensive. However, maybe you can do it chiefly because you are also small. For you, there may be all the business in that little corner of the marketplace that you can possibly handle for the next five years. It could make you a millionaire.

Develop a unique specialty that can become your hallmark. Or give a discount on peripheral services that you offer in order to attract the main body of business you are really after. Such loss leaders can be very effective. Can you invent something? Watch out, this can be a quagmire. Patent protection, with its pitfalls, may be necessary.

Periodically assess your practice to determine what business you are really in. Is the business changing? Is the client profile changing? Is the competition changing? Is the fee structure or mode changing? Who's new? Who's visible? Who's retiring? Who died? Did everybody move to Florida? Who's been acquired? Who's been fired or laid off? Who's expanding? Who's failing? New laws? Fewer laws? Who's getting sued? Why did you experience such a tremendous upsurge in business over the last two months? Where have all your customers gone? Assess. Reassess. Regroup. Identify, evaluate, and eliminate or mitigate soft spots.

DON'T FORGET HOW YOU STARTED

In consulting you will enjoy a certain amount of natural growth through word-of-mouth referrals. But you must get out there and push. Knock on doors, pound pavements, and blow your own horn by advertising. A good ad is your best emissary, no matter how small or inexpensive it may be. But there is absolutely nothing like your own physical presence, wherever possible.

DECONTROL THE TRIVIA

Remember that being efficient is not the same as being effective. You can be as efficient as it is humanly possible to be in meeting your objectives or attaining your goals. But if you have set the wrong objectives or goals, you have not been effective. Be sure you analyze business needs the same way you would your engineering work. Use the scientific method: Plan, acquire data, and then plan some more. Try not to leave important things to chance.

RAISING YOUR FEES

Eventually you will reexamine your fee structure. You will want to seriously consider raising fees, if only selectively. Some of the signals that should prompt you to review your fee structure, generally speaking, are as follows:

1. If you have not raised them in several years
2. If you started with a low fee structure to buy into the business
3. If the economy has recently gone through a severe inflationary or deflationary period
4. If you have been gaining too many clients for you to handle on a timely basis no matter how you try to stretch your time and resources
5. If you are losing clients whom you feel you should have kept
6. If you have developed a new level of expertise or are now at the top of your field professionally
7. If you have indications that your fees are lower than what the traffic will bear
8. If you have gotten into new types of work, or the character of your work has changed such that your old fee structure is inappropriate or even inapplicable
9. If you decide to specialize, and practitioners in your new specialty almost invariably utilize a specific fee type and/or level
10. If you have solid evidence that your most able competitors are getting a significantly different fee, using a significantly different fee arrangement, or are getting some of your own clients or clients you feel you should have gotten or wanted, and so on

Whenever you raise your fees you run the risk of losing some clients. But if you feel your new fee levels are justifiable, stick to your guns. Never be afraid to raise them, either. A certain percentage of your clients will moan and groan and threaten to abandon you for cheaper competition. They will try to hit you over the head and intimidate you. And, of course, there will always be clients who try to "nickel and dime" you.

However, remember that you cannot manufacture available hours. You only have so many hours in a day, a week, a month, or a year. The only way you can increase your income is to raise your fees. If you can learn to practice more efficiently, so that it takes you fewer hours to do a particular job, then naturally you can fit more work into the available time. But if all you do is bill out your old hourly fee rate for the fewer hours per job, then your total income will remain the same. It may get you more clients, but your total income is stagnant. At some point, then, you simply have to raise your fees, such as by charging the same overall amount for your increasing skill, which in turn permits you to do the work in less time. As a matter of fact, you can even lower your fees somewhat, although not so much that you're making no more then your old income. Using this latter strategy, everybody gains. Your clients pay somewhat less and you can make more money.

Or you may tend to leave your rates pretty much the same for rather long periods of time, even years. Eventually you simply raise them significantly. In the interim, clients will have been treated more than fairly. Maybe you come out ahead or behind, depending upon the engagement, but overall your income should continue to increase.

MULTIPLE FEE RATES

You can have different fee rates and structures for different kinds of work. For example, in forensic practice you might have one type of charge for reports and inspections, and another for more demanding trial and deposition testimony. There is nothing as demanding as testimony. Depending upon the class of work, you may charge by the hour, by the job or engagement, by the day or half day, and so on.

For seminars you might charge by the length, by the number of attendees, or by the person, depending upon how the program is set up. On a book project, of course, you will work on an advance and royalty basis, generally using the publisher's standard contract.

GETTING PUBLISHED

While we're on the subject of writing books, there are basically two ways to go. You can publish your own book by paying one of the so-called "vanity" houses. It could cost you $7,000 to $10,000 or more for a rather slim volume. Such publishers will promise you the moon, but be wary. Frequently, they will not deliver. Your book will appear in some catalog that has a limited distribution. More often than not, you will wind up with thousands of books in cartons that take up all of your garage and part of your living room.

If you have something you want to say in a book, continue to try to interest a publisher. It is not as easy to find publishable authors as you might think, so don't give up. Write articles, write anything, but write. Get into the habit of writing and submitting your work. There are no guarantees, but maybe you'll hit it lucky. Contact only leading publishers with proven direct-mail capabilities. That's the name of the game in the technical, business, and professional publishing business.

THE CONSULTANT'S CONSULTANT

At the outset you may find consulting opportunities on a subcontracting basis working for other solo consultants or even for larger consulting firms. Such assignments will pay well, as long as you steer away from any suggestion that you go on the payroll as a salaried professional, even if only for several months.

If you are approached by a pure "body shop" that asks you what you charge and then adds a flat percentage, watch out. You will get no support services, and this method of consulting can cause you grief. Such firms are merely brokers rather than true consulting practices. You may also have to wait a good long time to get paid.

More and more industrial companies are also using "outsourcing," so-called contingency workers and temporaries. This is a virgin field for the engineer wanting to strike out on his or her own.

TESTING LABORATORIES

Testing laboratories also use outside consultants. A very able metallurgist I know derives significant income by subcontracting for various testing companies. This works to their mutual advantage. For him, consulting income from testing laboratories is a routine part of his activity. For the labs, it affords them greater flexibility because they do not have to keep a lot of full-time professional personnel, at professional-level salaries, on their permanent staff. They can very effectively work with a skeleton professional crew.

To these laboratories this metallurgist is a "variable cost." If he was on permanent staff he would be an item of fixed expense, overhead, or "burden."

Both this metallurgist and the labs who engage him on an "if, as, and when needed" basis benefit in still another way. He gains diversified experience and they gain a better-rounded professional. Since the test programs are mostly routine, there is no real conflict of interest in his working for several labs. But even if there were, he would just have to be very careful not to disclose proprietary information of one client to another. However, by the nature of consulting practice, there may arise potential conflicts of interest. Always be on guard in relation to ethical considerations of this sort.

Here again, you may be asked to sign a nondisclosure agreement. Scrutinize it carefully, making changes if warranted. These agreements are not chiseled in stone. If you truly have the best interests of the client at heart but simply wish to protect yourself also, you can usually come to some meeting of the minds with respect to nondisclosure of specific proprietary content of engagements or processes, and so on.

YOUR CONSULTING ASSOCIATES

Consulting associates should be true "associates." They should be your colleagues. You should all work together very closely on engagements. Although you must never alter the purely technical content of an associate report if the specialty differs from yours, you will invariably confer about engagement dimensions, nature of the problem from multiple standpoints, methodologies and instrumentation, equipment or other resources to be applied, and so on. You should require an oral report right after any inspection, and at successive milestones. Also, require a rough-draft report for your personal review and ongoing conferences with the consulting associate, even if by telephone, to be certain you all remain on the same wavelength. Additional client contacts may also be necessary.

In other words, although you should studiously avoid putting words into the mouth of an associate, since he or she is the real expert in the particular field, control the quality of the engagement very carefully. Due to this sometimes torturous, but always highly disciplined process, you will be able to preclude, or at least minimize, situations where your joint efforts are less responsive than the client expected or desired. In these cases, there should be no problem at all in satisfactorily resolving problems before any damage is done. This is the essence of control. Never let an engagement get away from you.

You are particularly vulnerable when you bring subcontracted consulting associates into the picture. Be certain you keep track of casework at every step along the way that constitutes a bona fide milestone. The more multidisciplinary the engagement is, the more vulnerable you are. But, with this control, if a client then complains to you and you know you have actually outdone yourself, you will know how to react to him next time. He'll be back, all right, but you may just decline his case.

In some fire and explosion cases, four or even five consulting associates may be necessary. Such engagements may have to be covered from safety engineering, product safety, and management standpoints, in addition to the reconstruction problem. You may need a chemist or chemical engineer, an architect, an electrical engineer, or even a metallurgist, depending upon the problem. In this type of casework, it is the need to reconstruct accident events that demands multiple expertise. Of course, ordinary design problems can also have very cross-disciplinary aspects if attacked properly.

PRACTICE WITHIN YOUR SPECIALTY

The reason for multidisciplinary casework structure is obvious, of course. Nobody can possibly be really expert in all phases of engineering or science. You may think that you are pretty good, and maybe you really are. But be ultrasensitive to the problem that can arise if you are practicing out of your specialty or specialties, even to a minor degree. You can be asking for big trouble, especially in today's litigious environment. Unfortunately, a lawsuit is sometimes justified. Learn to recognize when you are actually practicing out of your specialty, to what extent, and with what limitations.

Being sensitive to practicing only within your specialty has a very bright side: It can lead to an expanded engagement. If you assemble some really good people around you with diverse specialties, you may be able to do a real job for the client. Usually, the incremental casework hours will not be greatly increased, but the inputs will be totally competent. Attempt to avoid duplication of effort. This is where your client may become unreceptive, if there is any hint that he will have to pay for duplicated hours.

But if you make this approach to problem solving a way of life, you will also find that you broaden your own knowledge and potential. You will be able to speak more intelligently and authoritatively to clients about their problems. Best of all, however, you will learn to work with your consulting associates as a team. This is really the bottom line. Your technical problem-solving capabilities will be top drawer—and they will be expanded. Your entire practice will benefit.

THE MULTIDISCIPLINARY PRACTICE

The multidisciplinary, multiprofessional style practice will give you the greatest professional and competitive leverage over the years. It works; it's real—it's also demanding, and not easy to control. But it will be profitable for your clients, your consultants, and you.

If you consult for consultants, look for an organization that operates in a professional manner, pays professional-level fees, and is not merely a broker or clearing house. The numerous services that will happily list your name, address, telephone, and specialty for a fee, and have it all tied into a computer, are really making their money on the book or computer printout that they then peddle to subscribers. For you to pay a fee for such a listing is just so much money down the drain.

Instead, spend the same few hundred dollars a year printing up your own qualifications and activities profile, even if it is relatively plain at first. Use the money telephoning prospects and following up with a mailing piece. Make an investment in yourself rather than in somebody else's profitable subscription directory or computer service business.

MERGERS AND ACQUISITIONS

At a later point in your consulting career, you may want to explore the possibility of merging with one of your competitors, or acquiring another consulting or related business, and so on.

In one instance a consultant approached a competitor and suggested that they rent a common office, with secretaries and copiers and other equipment to be used jointly. It was a relatively casual conversation. An extensive library would go along with the deal. However, the competitor opted to remain independent in this case, even though he knew some economies would come from some kind of consolidation, even if only of physical facilities. Some diseconomies of solo practice have to be viewed as beneficial from strategic standpoints. They really are not *dis*economies, but may actually protect your best interests.

In another instance a large laboratory division of a major technologically based conglomerate approached a small consulting firm. The big company executive who called said that they wanted to consider the possibility of expanding and commercializing their expertise into the field of the small firm. Would the consulting firm be interested in talking with them, with the thought of ultimately being acquired by them and helping them set up branch operations around the country?

Well, in the mergers and acquisitions game there are certain ways of making contact. The trick, obviously, is to find out as much about your competition as you can without revealing anything about yourself. If you are the acquirer or potential parent company, you obviously are going to be in a position to request complete details on the operations of prospective acquisition candidates. After all, who wouldn't be flattered to have one of the country's leading technical conglomerates call and tell you they're interested in buying you out?

The fact of the matter is, however, that corporate types don't understand the entrepreneur. They don't understand the fiercely independent and competitive small businessperson or professional. They really don't understand the way the entrepreneurial mind works. And they don't really care. The entrepreneur is a maverick in the corporate scheme. He or she doesn't think like a regular employee.

The small businessperson must learn to be "street-wise." Don't be taken in by intimidating approaches, whether it involves potential acquisition or other matters of importance to you in protecting your best interests.

DO YOUR OWN THING . . . AND DREAM

If you concentrate on doing your own thing and retain an intense interest in what goes on around you in engineering, and in the way engineering interacts with the rest of the world, you will identify commercial and professional opportunities that you cannot even dream of now. The important thing is to keep your eyes and ears open. Don't let anything pass you by. Even if you never leave that high-paying job you have now, with all of the fantastic ways it affords you to express yourself and to be rewarded for outstanding loyalty to your company, make believe that you want to go into consulting practice. Doing all of the things that would make an independent practice possible will keep you on your toes, if nothing else. You will become of even greater value to your employer and, therefore, to yourself. Best of all, you will have come out of your shell. Whether you go into consulting practice or not, there can never be a return to the cloistered environment or myopic vision that typifies so many professional employees, even executives. Even if you stay in a good job, progressing within your company to an extent that is palatable for you, the experience will also benefit your employer.

PARTNERS

Another question that you should consider both at the very beginning of your consulting adventure as well as when you are considering growth opportunities after you have achieved some success is whether to take in partners. On the surface it sounds great. Aren't two heads better than one? If there is enough business out there for one, why not for two—or even more? Doctors, lawyers, and accountants do it. Why not engineers? Doesn't taking in partners give you a true multidisciplinary capability on the most professional level imaginable? Just think of it . . . a partnership of peers? It does sound great.

Incidentally, when we use the term "partnership" here we are not referring to the legal business form. You should always incorporate yourself. By "partnership" we merely mean that you are splitting up ownership of the business among more than one person. In a corporation, of course, each partner or stockholder would have shares. We're using the term in a more operational than legal sense.

You cannot prejudge whether the kind of consulting you have in mind will be more effective if you and one or more partners decide to set up shop together from the very start. There are no hard and fast rules. But having a partner will most probably cramp your style. Eventually, as your practice grows, you may even find that stresses and strains develop. How do you measure contributions to the business? Split everything equally? Does that mean that both of you are contributing equally on a continuing basis? Who brings in the clients or other business? Is he or she worth the same as the "inside man"? Whose spouse gets to be secretary? Whose son runs errands and makes a little bit for himself on the side? What wages do you pay family members? For what? When does who go on vacation? For how long? Who gets reimbursed for what in connection with business expenses and entertainment? Is there a twilight zone? Who gets to collect for convention expenses? How much? Who gets to keep the books that are bought out of the business? Who spends how much time at the business? And so on.

Get a contract drawn up with buy-out arrangements. Hire a good lawyer. Suppose you don't always see eye to eye. How do you resolve disagreements and arguments? Suppose you have to dissolve the partnership. You get an appraisal and then you go to arbitration. That means you have to hire a business appraiser. These kinds of professionals are also high-powered consultants, just like you and me. How much is your stock worth? Who gets the books? Who gets the typewriter? Who gets the business station wagon? What is it worth? Who gets the copier? Or maybe you can't or won't arbitrate and just decide to sue each other.

If you take in a partner, where are you going to work? Out of your respective houses? What kind of partnership is that? Where's the synergy going to come from if you are physically apart? And where are the economies going to come from if you can't have a single office? And if you do have a single office, is your partner going to work in your house? Are you going to work in his? Is your partner male or female? Oh, the opposite sex. How does your spouse feel about that? And where are the economies going to come from if you have to rent an office when you don't really need it? Well, maybe you can share a secretary, a typewriter, a computer, your library, the rent of course, a copier, office supplies, telephones, and all that. It could sound good on paper.

But remember that one of the important reasons for considering a consulting engineering practice is that you don't want to answer to a boss. A partner is very much like a boss. He or she will always be looking over your shoulder, and vice versa. You really don't have complete freedom to do your own thing, especially if its evenly split up among your partners.

Just think it through. On balance, will the disadvantages and increased costs in some sectors be more than offset by benefits and economies in others?

STAY SOLO

On balance, forget it. Stay solo. In fact, remain an independent solo practitioner for as long as possible, and even a little longer than that. Engineers, lawyers, and doctors who have gone into partnership-type arrangements—incorporated, of course, or else professional associations—sometimes return to solo practice when things get out of hand due to problems that develop in their own relationships or among members of their families, and so on. Be wary and take in a partner only if it is the only possible way that you can accomplish some special objective, or if it is a last resort.

UNLESS . . .

On a more optimistic note, of course, if your practice really grows explosively, you may almost be forced to surround yourself with top-notch professional colleagues whom you must really depend upon. In such a case you still have two alternatives. You can invite some of them to become partners or else give them some meaningful recognition or executive status within the firm that you still own, or at least of which you retain significant control.

You could create several classes of stock, including some nonvoting categories. You could offer profit sharing on some basis. Or you may actually elect to bring in some partners. Architectural and engineering firms have long had several classes of partners, for example, each carrying a different status, compensation mode, and so on. One can progress all the way up to "general partner."

You have to play it by ear. One reason for considering a partnership arrangement or equivalent early on might be the need to have several colleagues in branch-type locations in a number of major cities where you anticipate significant market potentials. This is legitimate. Under such circumstances there is a much more reasonable and actionable basis to apportion control, identify authority and responsibility, account for revenues, expenses, and so on. This could work neatly under the right circumstances, if you have chosen the right kinds of partners.

However, the business has to lend itself to this kind of an arrangement. If you have built up a highly successful practice all by yourself, and then set about establishing branch offices, the problem becomes a little stickier. Of course, a larger firm is easier to sell.

FRANCHISING

Suppose you consider the possibility of franchising. Exactly how do you franchise a professional practice? And how do you control professional service output? It is one thing to sell a product through many locations, where the branch office is merely a marketing and sales operation. But how do you control the quality of a franchisee? *You* are the "parent company." How do you "parent" in a professional practice? There are some unique problems. Do you know of a single law firm, accounting firm, management consulting firm, architectural engineering firm, and so on that "franchises"? Why not? Have you ever heard of a medical practice that has either branches, franchisees, or other remote and far-flung representatives?

There are good reasons why professional practices are organized in certain ways and not in others. This doesn't mean that you should not consider being innovative, but you should tread very carefully when you start considering expansion opportunities.

Maybe you will become the General Motors or Gulf & Western of the engineering consulting business. The sky's the limit; just be certain that you keep your feet planted well on the ground at all times.

CHAPTER 16

ETHICS AND ISSUES

THE LIMITS OF NEGOTIATION

The content of engineering problem solving, decision making, risk taking, objective setting, resource allocating, and so on involves multiple elements. There is pure technical or technological content, managerial content, financial-economic content, and even political content since large-scale or "macro" engineering projects are frequently in the public domain, and since private industry plans and operations can impinge upon the communal and institutional infrastructure of society.

But there is also ethical content. In this chapter we will touch upon a few of the more important and interesting aspects of the ethical content of engineering practice that you will encounter sooner or later. Upholding the engineering Code of Ethics or Ethical

Canons is one obvious ingredient of ethical-professional conduct, even though violations may not always be clear-cut. And whistle blowing is an obvious extreme case of ethical conduct in the face of substantial pressures to compromise the public safety, health, or welfare.

However, less obvious are the numerous occasions when interaction with clients for purposes of objective setting, problem solving, and so on may involve potential conflicts of interest with serious ethical overtones. Negotiations with clients as to fees may involve greater or lesser attention to project and personnel safety and health. It is easy to negotiate away safety and health measures in order to come in with a lower bid. The risk of an accidental occurrence may seem remote, and the pressure to cut corners can be great.

In forensic engineering work, it is easy to tell a client what he wants to hear. It is easy to level a vicious attack on an ethical engineer whose opinions are actually correct, responsible, humanitarian, and competent. But the jury doesn't know one engineer is ethical and the other isn't. All a jury knows is that two apparently qualified engineers have provided conflicting opinions. There must be two schools of thought on the subject. Confusion reigns. The wrong party may prevail due to unethical behavior.

In fact, since businesspeople, including professionals, spend so much of their time negotiating in one form or another, let us devote our initial efforts to this subject. We will then move on to some of the more visible ethical concerns and issues.

As an engineer you must understand the negotiation process, its ploys and convolutions. This is necessary for you to be able to maintain the integrity of your professional opinions, your personal commitment to the ethics of the engineering profession, and even the financial equilibrium, vitality, and viability of your professional practice.

Always remember that there is only one day left until tomorrow. So make it a good one; make it count. Don't negotiate it away unwittingly or unreasonably. Let today stand as tall as possible even in minor matters, let alone major ones.

Negotiating is viewed as a game—albeit perhaps a war game—by politicians, lawyers, and businesspeople. But the pragmatic object of negotiation, like oratory, is not necessarily truth, but persuasion to facilitate a desirable outcome. Lawyers, for example, are expert negotiators. Of course, even the trained or experienced negotiator can outsmart himself and sometimes does. On occasion, this can be observed in the courtroom during cross-examination. You shouldn't negotiate away what you consider to be your minimum position . . . your bottom line. There are certain things that you don't want to negotiate at all. So don't be intimidated into giving up what you rationally believe must be your minimum ultimate position.

Understand all aspects of any situation in which you are negotiating. If you don't have facts, you can't negotiate from strength. A negotiator can negotiate even with somebody who intends not to by getting behind a firm position and undermining or softening it.

Be careful about the issues that you permit to be negotiable and the issues on which you will not negotiate, or will negotiate only minimally. Stick to your guns. Understand tactics and strategies of deception used by your adversary. One type of negotiating entails the recognition that "Half a loaf is better than none." Other types "win" through deception, intimidation, and bluffing. Pride can help or hurt; false pride always hurts. Another strategy meshes one position with another. The outcome need not necessarily be "I win, you lose" or vice versa. Both sides can win if the objectives can be restructured and reframed. And always try to deescalate potentially escalating conflict through objective, empathetic conciliation.

Is the issue fundamentally negotiable? Are negotiating parties able to exchange value for value in a give-and-take? Can compromise or bargaining work if parties are dealing in good faith? What are the real, imagined, and perceived needs of all parties? Who is prepared and who is unprepared? Is there mistrust and, if so, why? Is it theoretically possible for everyone to come out a winner if the negotiation is truly successful in the broader sense? Can models be constructed of alternative outcome scenarios? Can they become working tools at the bargaining table? What are the risks of using such models for actual bargaining?

A good feel for, or knowledge of, human behavior is useful if not critical in negotiating. Can you accurately size up your opponent and determine if he or she is prepared, being rational or merely rationalizing, playing some role or part and acting instead of truly dealing? What is being hidden? Who is bluffing and to what extent? Are there multiple goals and interests? What is really at stake? What is the real issue being negotiated? What are the costs of a "no-deal" situation or a stalemate to each party? Who are the real parties to the negotiation as opposed to those sitting at the bargaining table? What are the tacit or hidden assumptions? How do they differ from those stated and perceived?

Satisfaction of human needs is always an element of the negotiation environment. A useful hierarchy of human needs and motivating factors has been developed by psychologists. It places basic biological-physiological needs first, including food, clean air, comfortable temperature range, and so on. Second is the need for stimulation and arousal, including sex, activity, experimentation and exploration, even information and intellectualizing. The third level includes the various types of personal safety and security of others within one's personal social framework. The fourth stage is the need and quest for self-esteem. This entails a series of stepping-stones such as education, achievement, competence, independence, freedom, status, recognition, and control. The final personal goal or need is self-actualization or self-fulfillment. At this state we attain our potential. We are aesthetically and intellectually content, although we will still reach out for new ways to develop and express ourselves. With this in mind, for example, can you convey an attitude of active caring about your opponent's needs and interests, even though you have to protect your own best interests as well? Can your best interests accommodate those of the parties at the bargaining table?

As a practical matter, all of the above motivating needs coexist to a lesser or greater degree. They are not usually or entirely mutually exclusive. We are constantly balancing one against another and making trade-offs. The boundaries between different levels are not perfectly clear-cut. Nor is their ranking or priority, except under dire circumstances where survival itself is at stake, such as where a person loses a job or in a life-threatening situation. Of course, the more competitive we are, the greater will be our perception of survival threats, even at relatively higher levels within this hierarchy of needs.

A negotiating situation, even if not formal, such as at an arbitration or mediation, typically arises when you are face to face with your adversary. You must be alert to nonverbal communication forms, such as body language. Other forms of nonverbal or indirect communications may include knowledge about your opponent gleaned from intelligence that you either have actively sought to develop or that has fortuitously fallen your way. But be careful of planted or leaked information purposefully intended to influence you. You must know how to interpret and apply it in the negotiating situations you are confronting. Attempt to comprehend strategies and tactics of your adversary.

Also attempt to discern how much leverage you have in the situation. Where are the limits? At what point does your leverage disappear? At what price to the adversary? At what price to you? At what point does your opponent begin to have leverage? When does it become overwhelming? How can you neutralize his leverage? Can you soften it through compromise? Can you make concessions that are only apparent rather than real? Can you utilize a ploy or bluffing to advantage? What are the real risks to you of bluffing? Are there broader risks to the entire negotiating situation? If you bluff and lose, are your best interests retrievable or is the damage irreversible?

Just as you should be able to separate love from respect for a person, or respect for a person from respect for his or her knowledge or achievements, and so on, you should also separate problems and premises from people and personalities, and these from postures, positions, and promises. Focus on solutions of substance that truly reconcile underlying interests and goals. Beware of merely stated or even tacit posturing or positioning that sounds hardened and specific but which, in reality, is noninterests and nongoals-oriented. What is the probability that your perception of the real needs of your opponent is accurate? Be certain you have thought through your options.

Attempt to attain agreement based upon some objective standard or criterion. Invite analysis and critique. Avoid direct attack. Be both reasonably open and open to reason. Try to get behind disagreement, rather than confronting it in a hardened and uncompromising manner. Always seek mutual benefits. Threats are counterproductive and simply evoke counterthreats or withdrawal. Be emphathetic. This does not mean that you have to appear easygoing. Just attack the problem sincerely, in a spirit of concern and collaboration, and firmly. Unfortunately, many people interpret being nice or easy going as a sign of weakness. Be on guard in this connection.

Your goal should be an attitude of "enlightened selfishness." Agreement results from the efforts of negotiating parties to work toward both their own respective needs and also toward the needs of their opponents. The basic problem is to identify possible settlements in which common interests outweigh conflicts of interest. But negotiations only work to the extent that, in fact, convergent interests actually exist. The trick is to identify this common ground and its boundaries.

Be aware of the fact that face-to-face confrontation, even in a reasonably friendly adversarial climate, has a disadvantage. It may be hard to turn down or negotiate a position or need when your opponent is looking you squarely in the eye. You have to be tough inside and soft outside, firm but empathetic, while remaining conciliatory and mutual-needs-oriented. Telephone conversations can also be difficult.

You also have to be prepared to recognize and soften the frequently encountered negotiating situation where your opponent will immediately attempt to set out his demands and specify their outer limits, expecting you to do the same. This would ordinarily be followed by a sort of "soft-spot probing," where opponents seek weaknesses and crumbling in adversary positions. Ultimately, a crisis stage is reached when negotiation is either going to work or else the whole thing will be called off. This type of process is more of a brute-force exercise than true negotiating, however. It does not create a true, mutual-needs-oriented environment. It is a primitive approach that promotes resolution by conflict, rather than through empathy.

In resolution by conflict, no one is really satisfied because of the way the negotiation is approached and viewed in the first place. It's like laying your guns on the table prior to a shootout. But where empathy is the basis of resolution, there may be resultant synergy if the negotiating parties attempt to be statesmanlike rather than petty and tricky. If both sides can gain each other's confidence, negotiations can be much more constructive. In the real world, however, true statesmanlike negotiations—even between two individuals—are rarely empathetic and usually not synergistic. But attempt empathetic, synergistic negotiating as the preferred and initial strategy. Both you and your opponent have much more to gain than through conflict-based negotiating, because longer-term relationships may be at stake in addition to fees or other currently pressing issues.

AVOID CONFLICTS OF INTEREST

Be alert to potential conflicts of interest between prospective engagements and the best interests of your present employer. Only remember that if there is no real conflict of interest, and if you have not signed any agreement that limits your outside activity—if such an agreement is legal at all—your employer has no hold on you; your employer does not *own* you. It was not a condition of employment. In fact, if your employer decides to fire you for engaging in such nonconflicting independent consultant activities, you may even have a bona fide basis for legal action against the company. Remember that you are a professional. Nobody but you "owns" your skills, potential, and time, even though you may "rent" your skills and time to some employer for a limited number of hours per week, month, or year. If a company thinks it owns you or is renting your skill for hours beyond a given workday or work-week length, then you ought to be getting paid for such after-work hours that you are not being permitted to utilize to your own professional and financial advantage, such as for independent consulting purposes.

Make it your business to learn where the line is drawn between company trade secrets or other proprietary properties, on the one hand, and mere employment experience. Obviously, preserving the secrecy of proprietary information, from a security standpoint, and preserving the confidentiality of information acquired as an employee are important aspects of the problem. There can be a fine line between the rights and obligations of employers versus employees. Therefore, you have to watch your step.

On the other hand, as a practical matter it is usually not hard for an engineer or scientist employed full-time to gravitate toward consulting activities that are sufficiently different from the full-time employment focus as to avoid the problem altogether. If you have a question, consult your attorney. If your consulting is sufficiently different from your full-time employment responsibilities, the chances are excellent that your employer will not even mind if the consulting will give you broader exposure and greater depth. In fact, it can work to his advantage.

Of course, the confidentiality, proprietary information, and trade secrets problem can arise even during the course of a full-time consulting practice. For example, a highly specialized and focused practitioner may have several clients in similar businesses. The consultant must be particularly careful to avoid conflicts of interest in order to protect the best interests of each and every client. In fact, for the highly experienced, well-traveled consultant the confidentiality and trade secrets situation may even be more of a potential problem than for the typical engineering or scientific employee who has one full-time job but decides to do some part-time consulting to augment his or her income and make more meaningful and creative use of professional expertise. Also keep in mind that mere refinements and modifications in already established technologies are frequently considered to be part and parcel of the ordinary knowledge and experience of the scientist or engineer, rather than the property of an employer upon whose premises such minor refinements and modifications took place. Enhancement of general employee skill and experience can also be used freely, without any restriction, even though the technical employee gains expertise.

On the other hand, divulging confidential personnel information, nonpublic employer capabilities, identities and/or capabilities of suppliers, advertising, marketing and sales promotion information, operational details, closely held financial information, competitive bidding information and bidding policies or strategies, contract data, credit information, customer lists and information relating to customer relationships are clearly proprietary and confidential. Most of the time the potential for conflicts of interest would be quite obvious. But again, where it is not, you should either contact your attorney or avoid consulting subjects or clients potentially posing conflict-of-interest problems.

If you have signed a patent and invention assignment agreement, check it again. It may have a built-in confidentiality and disclosure clause.

The bottom line is that every engineer and scientist must recognize the extent to which his or her own optimum professional functioning is based upon information both given and received during the course of ordinary employment. Signing away one's right to what one learns and/or contributes may well severely impair potential for independent consulting work in the future.

With this in mind, every engineer and scientist should consider carefully, at the very beginning of his or her professional career, about what the present and future may hold in terms of both full-time employment, on the one hand, and independent consulting, on the other hand. One career strategy would be to enter industry on a full-time basis only to the extent that broad and diverse background knowledge and experience will be acquired, with the thought of identifying some specialty that will permit independent consulting later. On the other hand, if one becomes a highly focused specialist during the course of full-time employment, then the individual should probably seek out—indeed, plan for—alternative specialty involvement on one's own time to avoid conflict at the point where independent consulting begins, whether part- or full-time.

PROFESSIONAL ETHICS VERSUS PROFESSIONAL FLEXIBILITY

Whatever you do you must do on the highest professional plane. Don't settle for less. Consulting is not going to be satisfying, or ultimately even financially rewarding, if you cannot practice ethically, as well as competently and responsibly. Believe it or not, it's possible, even practical, to set this goal yourself and still be highly successful. Also, you'll be able to live with yourself, sleep nights, and become respected for it in your field.

If you have found your way into a consulting specialty where ethical conduct is not the rule, however, beware. If you cannot make it by being ethical, look someplace else. Don't be intimidated into playing the game.

Be flexible, but don't compromise your professional integrity. Not even once. Being flexible does not imply a willingness to look the other way in situations where engineering ethics are involved. To suggest that you should be flexible in order to be successful as a consultant is merely to remind yourself to adjust your thinking and retain an open mind all the way up to, but not including, the point where ethical judgments are involved. You must maintain your professional integrity throughout your career, at all costs. The record will show that more engineers lose their licenses—and their jobs—for being unethical than for making technical blunders.

Remember, however, that unethical practice can result in technical blunders. If you practice your profession the way it should be practiced, fulfilling your mandate to protect and promote the public safety, health, and welfare, then adequate, necessary, and sufficient technical judgments will flow from your efforts. Otherwise, adverse technical compromises have a nasty way of resulting from ethical ones.

The New York State Education Department had this to say about professional conduct:

> A MESSAGE TO LICENSED PROFESSIONALS PRACTICING IN NEW YORK STATE
>
> State regulation in 31 fields of professional practice, including issuance of licenses and enforcement of rules of professional conduct, is vested in the Board of Regents in New York.
>
> Practice of a profession is a public trust, earned through educational preparation, experience, passing rigorous examinations, and commitment on the part of the practitioner to public service. The professional carries out that trust in accordance with ethical standards developed through years—often decades and even centuries—of the best professional traditions, and with state law and regulations. To you who have earned a professional license in New York, the Board of Regents offers congratulations. We urge that professional practitioners be conscious always of the very special obligations of public service and of ethical conduct that the privilege of licensure creates.

Engineers, unlike lawyers and physicians, are not required to take an oath of service at the time of graduation. However, most engineering associations have their own codes, rules, or canons of ethical and professional conduct which members are expected to faithfully follow. There is also an ethics section on the Professional Engineering examination.

Unfortunately, however, the engineering profession is not totally independent of economic, political, and managerial pressures, prerogatives, or priorities. Thus, the safety, health, and environmental ramifications of product, workplace, and facility design and construction are not always adequately under the control of the engineer. Consequently, the mandate to the engineering profession from society is not always properly fulfilled.

But in private consulting practice, where you have better control over the safety and health content of your own designs and problem solutions, you must be ever mindful of your professional obligation to promote, first and foremost, the public safety, health, and welfare. This can even include, on those rare occasions where conscience so dictates, the solution of last resort: whistle blowing.

Legislation protecting employees and others from retaliation for disclosure of illegal or improper activities is a growing trend and force that has particularly meaningful ramifications for professionals who have a legal and/or ethical mandate to promote and hold paramount the public safety, health, and welfare.

INDEX

Accident prevention, 330–33
Acquisition, of another consulting business, 377
Advertising, 61–63
Age, as factor for consulting engineer, 9
Analysis
 "bumble bee can't fly," 276–77
 comparative, 132–34
 inspection, field or product, 215
 limited scope, 277–78
 sound and acceptable, 275
 unsound, 276–77
Answering service, 41
Applications engineering, 346–53
Attorney grievance form, 250–51
Avant-garde assault, in expert reports, 278–79

Bad habits, 31–33, 36–37
Benefits, summary of, 179–80
"Bumble bee can't fly" analysis, 276–77
Business-building reports
 appearance of, 189–90
 evaluation of, 192–93
 hints on writing, 185–86
 rules of writing, 186–88
 technical reports, 188–89
 visual aids, 189–90
 writing, 183–93
Business cards, 50
Business library, of consulting engineer, 20–21

Capital, outside, 147
Capitalgraph, working, 124–26
Case analysis, in forensic engagements, 216
Casework files, 44–45
Cash flow, 111–14, 147–49
Cash Flow Profit Planner, 153, 154, 155, 157
Clients
 attracting, 49–52
 client lists, 61
 client retention factors, 67–68
 contact with, 68–69
 educating, 69–70
 follow-up, 69
 limited engagements, 71–72
 needs, responsiveness to, 193
 prospecting, 53–60
  part-time, 60–61
 qualifying, 70–71
"Clipping," 41
Code compliance fallacy, in expert reports, 275–76
Codes and standards, safety, limitations of, 299–307
Common body of knowledge, 18–19
Comparative analysis, 132–34
Competition
 containment of, 199–200, 200
 identification of, 195–96
 surpassing, 202–204
Computer, as office equipment, 41
Conflicts
 conflicts of interest, 388–90
 in management, 98

393

Consulting business
  comparison of consulting types, 17–18
  multidisciplinary practice, 376–77
  opportunities for, 38–39
  part-time versus full-time, 198
  practicing with your specialty, 376
  productivity in, 38
  technical content of, 15–17
  types of, 75–76
Consulting engineer
  in private practice
    acquisitions, 377
    age factor, 9
    and associates, 375–76
    business library, 20–21
    common body of knowledge, 18–19
    comparison of consulting types, 17–18
    dabbling as, 66
    effectiveness as, 47
    as entrepreneur, 10–11
    ethics, 383–92
    franchising, 381
    income, 4–8
    and "intrapreneuring," 12
    keeping busy, 34–36
    launching your business, 25–26
    mergers, 377–78
    need for, 1–3
    partnerships, 378–80
    planning process, 102–103
    profession, 3
    profit opportunity for, 367–71
    requirements, 11
  setting income goals, 12–13
  as subcontractor, 374
  as systems manager, 26–27
  technical library of, 19–20
  telephone use, 368–69
  ten steps to begin practice, 21–25
  for testing laboratories, 374–75
  testing the market, 8–9
Contingency fees, 86–87
Contracts, 84–86
Control
  job safety, 345
  loss, 329
  managerial, 96–97
  process, 329–30
  quality, 89–90
Copier machine, 41
Cop-outs, in expert reports, 281
"Court-wise" expert, 262–66
Credentials, required for forensic consulting, 207–208
Credibility, 67
Cross-examination, during expert testimony, 266–70

Defeatability proneness
  basic problem, 313–14
  concept of, 314–16
  defeatability incentives, 323–26
  defeatability index
    application of, 318–21
    approach and application, 321–23
    definition of, 316–18
  of product safety, 312–23
Defects
  code and standard, 300–307
  irrelevant, 275
Defense report review, in forensic engagements, 218
Deficits, 111–14, 149–50
Depositions, in forensic engagements, 217–18
Designing for confusion, in expert reports, 277
Differential relationships (see Leverage relationships)
Diversification, 369–70
Documentation review, in forensic engagements, 213–15

Engagement quality control, 89–90
Engagements
  forensic, 216–221
  limited, 71–72
Engagement viability, 90–91

Engineering
  applications, 346–53
  forensic, 211–53
  human factors, 353–57
Enterprise structural profiles, 114–22
Entrepreneuring, 10–11
Equipment protection, 328–29
Ethics
  conflicts of interest, 388–90
  versus flexibility, 390–92
Expansion, profit opportunities, 367–81
Expertise, identification of, 208–209
Expert Opinion Rating and Reconciliation Work Chart, 261
Expert report, types of, 275–86
Expert testimony
  battleground of, 257–62
  cross-examination, 266–70
  crossfire in, 291–94
  deadly sins of, 270–74
  image versus substance, 262–66

Fees
  arrangements for, 76–77
  competition and, 201–02
  components, 77–79
  contingency fees, 86–87
  in forensic engagements, 220–21, 246–53
  invoicing, 86
  levels, 79–81
  multiple fee rates, 373
  raising, 371–73
  retainers, 83–84
  working cheap, 82–83
Field or product inspection, in forensic engagements, 215
Files, 41
  casework, 44–45
  cross-referencing of, 43–44
Financial engineering
  cash flow, 147–49
  deficits, 149–50
  leverage, 155–73
  outside capital, 147
  pickup characteristics, 150–51
  profitography system, 151–54
  scattergraphs, 150
  summary of benefits, 179–80
  surpluses, 149–50
Financial fact sheet, 130–31
Financial Makeup Analyzer, 156
Financial planning
  cash flow, 111–14
  deficits, 111–14
  risk financing, 109–111
Firepower, 198–99
Flexibility, versus ethics, 390–92
Follow-up, 69
Forensic consulting
  credentials required, 207–208
  engagements, 210–20
  expert testimony, 257–62
  identifying your expertise, 208–209
  marketing your specialty, 209–10
Forensic engagements
  case analysis, 216
  case research, 216
  defense report review, 218
  depositions, 217–18
  documentation review, 213–15
  fees, 220–21, 246–53
  field or product inspection, 215
  initial consultation, 211–13
  interviews, 216
  laboratory testing, 215–16
  lead times, 224–35, 246
  postreport consultation, 217
  rebuttal reports, 218
  rebuttal testimony, 219–20
  report preparation, 216–17
  trial appearance, 218–19
Forensic engineering
  index to investigated subjects, 236–46
  specialties, 206
Franchising, 381
Full-time consulting, versus part-time, 60–61

"GIGO effect," in expert reports, 281

Index

"Hassle-free" management, 99–100
Hazard and risk foreseeability, 337–41
Home office, 39–40
Hourly rate, 81
Human factors engineering, operational definitions of, 354

ICBM reports, in expert reports, 278
Inadequate foundations, in expert reports, 277
Incremental relationships (see Leverage relationships)
Initial consultation, in forensic engagements, 211–13
Interviews, in forensic engagements, 216
Intrapreneuring, 12
Inventory, determination of, 46
Investments, 63–66
Invoicing, 86

Job safety control, 345

Laboratory testing, in forensic engagements, 215–16
Lead times, in forensic engagements, 224–35, 246
Leverage
  concepts of, 155–65
  relationships, examples of, 165–73
Limited engagements, 71–72
Limited scope analysis, in expert reports, 277–78
Log book system, 41
Long-term offers, 72–73
Loss control, 329

Maintainability safety
  maintenance failure accommodation, 346
  safety through maintenance, 345–46
  through job safety control, 345
  through product safety content, 345
Maintenance failure accommodation, 346
Management
  by objectives, 96
  conflicts in, 98
  function of, 96
  "hassle-free," 99–100
  "intensity" factors, 102
  managerial control, 96–97
  milestone, 88–89
  moral of, 99
  "mountain goat," 98–99
  overtaken by events, 104
  performance measurement, 103
  practical planning, 105
  resources, 42
  rules of business survival, 95–96
  scientific method, 95
  styles of, 98
  systems, 26–27
  "voodoo," 100–101
Managerial control, 96–97
Market share, increasing, 199–200
Market testing, 8–9
Measuring equipment, 41
Mechanical hazards, pictorial checklist of, 363–64
Mergers, 377–78
"Merry measurement" syndrome, in expert reports, 276
Milestone management, 88–89
"MLC" report, in expert reports, 279
"Mountain goat" management, 98–99
Multidisciplinary practice, 376–77
Multiple fee rates, 373

Net opinions, in expert reports, 277
Noncomplying formats and contents, in expert reports, 277
Nondisclosures, 84–86

Office equipment, 41
Opportunities, for consulting, 38–39
Organizations, need to join, 40
Outside capital, need for, 141–47
Overtaken by events, 104

Partnerships, 378–80
Part-time consulting, versus full-
 time, 60–61
Performance, efficiency of, 126–32
Performance measurement, 103
Performance standards
 functional, 327–28
 for safe design, 307–12
Personal computer, 41
Photographic equipment, 41
Pickup characteristics, 150–51
Planning, 105
Postreport consultation, in forensic
 engagements, 217
Process control, 329–30
Product failure, potential sources of,
 355–57
Productivity, 38
Product life-cycle stages, 356
Product protection, 328–29
Product safety content, 345
Product warnings, 341–43
Profitgraph, working, 122–23
Profit levels, 79–81
Profitography system, of financial
 analysis, 151–54
Profit opportunities
 developing a specialty, 370–71
 diversification, 369–70
 expansion of, 367–81
 newsletters, 370
 telephone use, 368–69
Profit planning, strategic versus
 tactical, 151
Project indexing, 44–45
Proposals, 84–86
Prospecting, 53–60
 advertising, 61–63
 clients and, 53–60

client lists, 61
part-time, 60–61
Protective features
 of products
  accident prevention, 330–33
  equipment/product protection,
   328–29
  functional performance, 327–28
  loss control, 329
  process control, 329–30
  quality assurance, 329–30

Qualification, of clients, 70–71
Quality assurance, 329–30
Quality control, of engagements, 89–
 90

Ratio analysis, basic, 173
Rebuttal reports, in forensic engagements, 218
Rebuttal testimony, in forensic engagements, 219–20
Relationships, 87–88
Report preparation, in forensic engagements, 216–17
Reports
 business-building, 183–93
 defense, 218, 285
 expert, 274–84
 preparation of, 216–17
 rebuttal, 218
 technical, 188–89
 writing of, 183–88
Research, in forensic engagements,
 216
Resources, management of, 42
Resume, 50
Retainers, 83–84
Return on capital, 134–39
 graph, 135, 137
Risk financing, 109–11

Safe design, rules of practice, 308

Safeguarding means, checklist of, 358–62
Safety, through maintenance, 345–46
Safety engineering
  code and standard defects, 300–307
  epidemiological approach to, 333–35
  performance standards for safe design, 307–12
  safety standards as design guidelines, 299–300
  specialized approaches, 337–66
Safety standards, as design guidelines, 299–300
Salary levels, 79–81
  hourly rate, 81
Scope, of work, 88
Service, and competition, 201
Small business
  characteristics of, 27–28
  details of running, 33–34
Specializing, 376
  developing a specialty, 370–71
  marketing your specialty, 209–10
Stationery, 50
Staying power, 198–99
Subcontracting, 374
Success, 30–31
"Superman" requirement, in expert reports, 275
Support systems, 39
Surpluses, 149–50
Switched defect trick, in expert reports, 280–81
Systems management, 26–27

Technical library, of consulting engineer, 19–20
Technical report writing, 188–89
Telephone use, 368–69
Testing laboratories, consulting engineer for, 374–75
Trial appearance, in forensic engagements, 218–19
Trial deposit log, 249

"UFO" report, in expert reports, 279–80
Unethical practice, in expert report, 277
Unreasonably dangerous product, 341–43
Unsafe modification, in expert reports, 282

Visual aids, for reports, 183–93
"Voodoo" management, 100–101

Weekend projects, 101
Writing
  evaluation of, 192–93
  hints, 185–86
  newsletters, 370
  rules of, 186–88
  technical reports, 188–89